A Level Physics for OCR Revision Guide

Gurinder Chadha

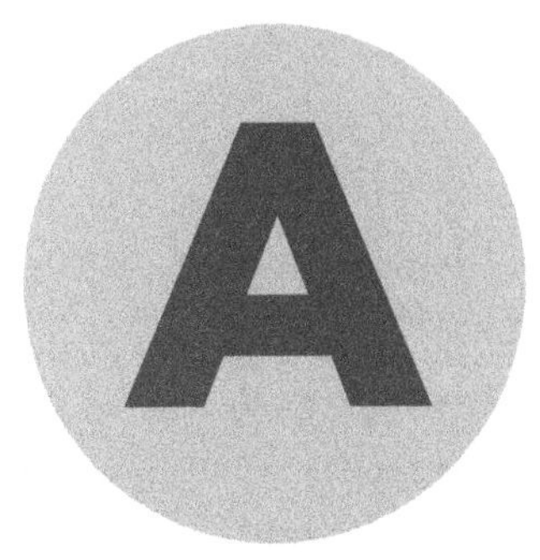

Great Clarendon Street, Oxford, OX2 6DP, United Kingdom

Oxford University Press is a department of the University of Oxford. It furthers the University's objective of excellence in research, scholarship, and education by publishing worldwide. Oxford is a registered trade mark of Oxford University Press in the UK and in certain other countries

First published in 2016

British Library Cataloguing in Publication Data
Data available

978 0 19 835220 4

10

Paper used in the production of this book is a natural, recyclable product made from wood grown in sustainable forests. The manufacturing process conforms to the environmental regulations of the country of origin.

Printed and bound by CPI Group (UK) Ltd, Croydon, CR0 4YY

Artwork by Q2A Media

This book has been written to support students studying for OCR A Level Physics A. It covers all A Level modules from the specification. These are shown in the contents list, which also shows you the page numbers for the main topics within each module. There is also an index at the back to help you find what you are looking for. If you are studying for OCR AS Physics A, you will only need to know the content in the blue box.

AS exam

Year 1 content

1 Development of practical skills in physics
2 Foundations in physics
3 Forces and motion
4 Electrons, waves, and photons

Year 2 content

5 Newtonian world and astrophysics
6 Particles and medical physics

A level exam

A Level exams will cover content from Year 1 and Year 2 and will be at a higher demand. You will also carry out practical activities throughout your course.

How to use this book v

Module 2 Foundations of physics 1
2 Foundations of physics 1
2.1 Quantities and units 2
2.2 Derived units 2
2.3 Scalar and vector quantities 4
2.4 Adding vectors 4
2.5 Resolving vectors 6
2.6 More on vectors 6
Practice questions 8

Module 3 Forces and motion 9
3 Motion 9
3.1 Distance and speed 10
3.2 Displacement and velocity 10
3.3 Acceleration 12
3.4 More on velocity–time graphs 12
3.5 Equations of motion 14
3.6 Car stopping distances 14
3.7 Free fall and *g* 16
3.8 Projectile motion 18
Practice questions 20

4 Forces in action 21
4.1 Force, mass, and weight 22
4.2 Centre of mass 22
4.3 Free-body diagrams 22
4.4 Drag and terminal velocity 24
4.5 Moments and equilibrium 26
4.6 Couples and torques 26
4.7 Triangles of forces 28
4.8 Density and pressure 30
4.9 $p = h\rho g$ and Archimedes' principle 30
Practice questions 32

5 Work, energy, and power 33
5.1 Work done and energy 34
5.2 Conservation of energy 34
5.3 Kinetic energy and gravitational potential energy 36
5.4 Power and efficiency 38
Practice questions 40

6 Materials 41
6.1 Springs and Hooke's law 42
6.2 Elastic potential energy 42
6.3 Deforming materials 44
6.4 Stress, strain, and the Young modulus 44
Practice questions 46

7 Laws of motion and momentum 47
7.1 Newton's first and third laws of motion 48
7.2 Linear momentum 48
7.3 Newton's second law of motion 50
7.4 Impulse 50
7.5 Collisions in two dimensions 50
Practice questions 52

Module 4 Electrons, waves, and photons 53
8 Charge and current 53
8.1 Current and charge 54
8.2 Moving charges 54
8.3 Kirchhoff's first law 56
8.4 Mean drift velocity 56
Practice questions 58

9 Energy, power, and resistance 59
9.1 Circuit symbols 60
9.2 Potential difference and electromotive force 60
9.3 The electron gun 60
9.4 Resistance 62
9.5 I-V characteristics 62
9.6 Diodes 62
9.7 Resistance and resistivity 64
9.8 The thermistor 64
9.9 The LDR 64
9.10 Electrical energy and power 66
9.11 Paying for electricity 66
Practice questions 68

10 Electrical circuits 69
10.1 Kirchhoff's laws and circuits 70
10.2 Combining resistors 70
10.3 Analysing circuits 70
10.4 Internal resistance 72
10.5 Potential divider circuits 72
10.6 Sensing circuits 72
Practice questions 74

11 Waves 1 75
11.1 Progressive waves 76
11.2 Wave properties 76
11.3 Reflection and refraction 78
11.4 Diffraction and polarisation 78
11.5 Intensity 80
11.6 Electromagnetic waves 80
11.7 Polarisation of electromagnetic waves 80
11.8 Refractive index 82
11.9 Total internal reflection 82
Practice questions 84

12 Waves 2 85
12.1 Superposition of waves 86
12.2 Interference 86
12.3 The Young double-slit experiment 86
12.4 Stationary waves 88
12.5 Harmonics 88
12.6 Stationary waves in air columns 88
Practice questions 90

13 Quantum physics 91
13.1 The photon model 92
13.2 The photoelectric effect 92
13.3 Einstein's photoelectric effect equation 94
13.4 Wave-particle duality 94
Practice questions 96

Module 5 Newtonian world and astrophysics 97
14 Thermal physics 97
14.1 Temperature 98
14.2 Solids, liquids, and gases 98
14.3 Internal energy 100
14.4 Specific heat capacity 100
14.5 Specific latent heat 100
Practice questions 102

15 Ideal gases 103
15.1 The kinetic theory of gases 104
15.2 Gas laws 104
15.3 Root mean square speed 106
15.4 The Boltzmann constant 106
Practice questions 108

16 Circular motion 109
16.1 Angular velocity and the radian 110
16.2 Centripetal acceleration 110
16.3 Exploring centripetal forces 112
Practice questions 114

17 Oscillations 115
17.1 Oscillations and simple harmonic motion 116
17.2 Analysing simple harmonic motion 118
17.3 Simple harmonic motion and energy 118
17.4 Damping and driving 120
17.5 Resonance 120
Practice questions 122

18 Gravitational fields 123
18.1 Gravitational fields 124
18.2 Newton's law of gravitation 124
18.3 Gravitational field strength for a point mass 124
18.4 Kepler's laws 126
18.5 Satellites 126
18.6 Gravitational potential 128
18.7 Gravitational potential energy 128
Practice questions 130

19 Stars 131
19.1 Objects in the Universe 132
19.2 The life cycle of stars 132
19.3 The Hertzsprung-Russell diagram 132
19.4 Energy levels in atoms 134
19.5 Spectra 134
19.6 Analysing starlight 136
19.7 Stellar luminosity 136
Practice questions 138

20 Cosmology (the Big Bang) 139
20.1 Astronomical distances 140
20.2 The Doppler effect 140
20.3 Hubble's law 142
20.4 The Big Bang theory 142
20.5 Evolution of the Universe 142
Practice questions 144

Module 6 Particles and medical physics 145

21 Capacitance 145
21.1 Capacitors 146
21.2 Capacitors in circuits 146
21.3 Energy stored by capacitors 146
21.4 Discharging capacitors 148
21.5 Charging capacitors 148
21.6 Uses of capacitors 148
Practice questions 150

22 Electric fields 151
22.1 Electric fields 152
22.2 Coulomb's law 152
22.3 Uniform electric fields and capacitance 154
22.4 Charged particles in uniform electric fields 154
22.5 Electric potential and energy 156
Practice questions 158

23 Magnetic fields 159
23.1 Magnetic fields 160
23.2 Understanding magnetic fields 160
23.3 Charged particles in magnetic fields 162
23.4 Electromagnetic induction 164
23.5 Faraday's law and Lenz's law 164
23.6 Transformers 164
Practice questions 166

24 Particle physics 167
24.1 Alpha-particle scattering experiment 168
24.2 The nucleus 168
24.3 Antiparticles, hadrons, and leptons 170
24.4 Quarks 170
24.5 Beta decay 170
Practice questions 172

25 Radioactivity 173
25.1 Radioactivity 174
25.2 Nuclear decay equations 174
25.3 Half-life and activity 176
25.4 Radioactive decay calculations 176
25.5 Modelling radioactive decay 178
25.6 Radioactive dating 178
Practice questions 180

26 Nuclear physics 181
26.1 Einstein's mass-energy equation 182
26.2 Binding energy 182
26.3 Nuclear fission 184
26.4 Nuclear fusion 184
Practice questions 186

27 Medical imaging 187
27.1 X-rays 188
27.2 Interaction of X-rays with matter 188
27.3 CAT scans 190
27.4 The gamma camera 190
27.5 PET scans 190
27.6 Ultrasound 192
27.7 Acoustic impedance 192
27.8 Doppler imaging 192
Practice questions 194

A1 Physics quantities and units 195
A2 Recording results and straight lines 195
A3 Measurements and uncertainties 196
Physics A data sheet 198
Answers to practice questions 200
Answers to summary questions 209

How to use this book

Specification references

→ At the beginning of each topic, there are specification references to allow you to monitor your progress.

This book contains many different features. Each feature is designed to support and develop the skills you will need for your examinations, as well as foster and stimulate your interest in physics.

Worked example

Step-by-step worked solutions.

Revision tips

Prompts to help you with your understanding and revision.

Common misconception

Common misunderstandings clarified.

Synoptic link

These highlight the key areas where topics relate to each other. As you go through your course, knowing how to link different areas of physics together becomes increasingly important. Many exam questions, particularly at A Level, will require you to bring together your knowledge from different areas.

Maths skills

A focus on maths skills.

Model answers

Sample answers to exam-style questions.

Summary Questions

1 These are short questions at the end of each topic.

2 They test your understanding of the topic and allow you to apply the knowledge and skills you have acquired.

3 The questions are ramped in order of difficulty. Lower-demand questions have a paler background, with the higher-demand questions having a darker background. Try to attempt every question you can, to help you achieve your best in the exams.

Chapter 2 Practice questions

Practice questions at the end of each chapter including questions that cover practical and math skills.

1 Which is the correct definition for a scalar quantity?
A scalar quantity ......

A has no units.

B has direction only.

C has magnitude only.

D has magnitude and direction. *(1 mark)*

2 What is force measured in base units?

A N

B kg

C $kg\,s^2\,m^{-1}$

D $kg\,m\,s^{-2}$ *(1 mark)*

3 An object experiences two forces of magnitudes 3.2 N and 5.0 N. The forces are at right angles to each other.
What is the magnitude of the resultant force on the object?

A 3.8 N

B 5.9 N

C 15 N

D 35 N *(1 mark)*

Module 2 Foundations of physics
Chapter 2 Foundations of physics

In this chapter you will learn about ...

- [ ] SI units
- [ ] Base units
- [ ] Derived units
- [ ] Prefixes
- [ ] Scalars and vectors
- [ ] Vector triangles
- [ ] Resultant vectors
- [ ] Resolving a vector
- [ ] Components of a vector

2 FOUNDATIONS OF PHYSICS
2.1 Quantities and units
2.2 Derived units

Specification reference: 2.1.1, 2.1.2

2.1 Quantities and units

A **physical quantity** in physics means a measurement of something and its unit.

A physical quantity has a number followed by a unit. For example, the length of a mobile phone is 0.125 m. There are a few exceptions where a physical quantity does not have a unit. Examples of this include tensile strain and efficiency.

Quantities and SI base units

There are many different units for length – inch, foot, yard, mile, and so on. In physics we agree to work with the metre (m). The metre is an example from a list of seven internationally agreed base units, or SI units (*Système International d'Unités*). Table 1 summarises the six physical quantities and their SI base units you need to know.

▼ **Table 1** *Physical quantities and their SI base units*

Physical quantity	SI base unit
length	metre (m)
mass	kilogram (kg)
time	second (s)
electric current	ampere (A)
temperature	kelvin (K)
amount of substance	mole (mol)

Prefixes

To cope with the very large and very small numbers that are used in the study of physics, it is often useful to write numbers as powers of ten. Prefixes are used as abbreviations for some powers of ten – you need to learn the prefixes listed in Table 2.

The sub-multiples (for smaller measurements) are femto, pico, nano, micro, milli, centi, and deci, and the multiples (for larger measurements) are kilo, mega, giga, tera, and peta.

▼ **Table 2** *Important prefixes*

Prefix name	Prefix symbol	Factor
peta	P	10^{15}
tera	T	10^{12}
giga	G	10^{9}
mega	M	10^{6}
kilo	k	10^{3}
deci	d	10^{-1}
centi	c	10^{-2}
milli	m	10^{-3}
micro	μ	10^{-6}
nano	n	10^{-9}
pico	p	10^{-12}
femto	f	10^{-15}

Common misconception

The prefix mega is *capital* M and the milli is *lower case* m. If you have poor writing, then these two prefixes can look the same. Just be vigilant.

Maths: Standard form or scientific notation

A number in standard notation is written as $n \times 10^m$, where n is a number greater than 1 but less than 10 and m is a positive or a negative integer. A distance of 320 km in metres may be written as 320×10^3 m but in standard form it must be written as 3.2×10^5 m.

Worked example: Speed

Speed is defined as the rate of change of distance. Calculate the speed of a particle that travels a distance of 9.1 Mm in a time of 50 ms.

Step 1: Write the magnitude of each quantity by substituting for the prefix.

distance = 9.1×10^6 m   time = 50×10^{-3} s

Step 2: Calculate the speed and write your answer in standard notation.

$$\text{speed} = \frac{\text{distance}}{\text{time}} = \frac{9.1 \times 10^6}{50 \times 10^{-3}} = 1.82 \times 10^8\ \text{m s}^{-1} = 1.8 \times 10^8\ \text{m s}^{-1}\ (2\ \text{s.f.})$$

Note: The data is given to 2 significant figures (s.f.), so the final answer must also be written to 2 s.f.

2.2 Derived units

You have already met an example of a derived unit: m s^{-1}. This unit for speed can be determined from the equation for speed and the base units for length and time. Since speed = $\frac{\text{distance}}{\text{time}}$, the derived unit for speed must be metre (m) divided by second (s). This can be written as m/s, but at advanced level this is written as m s^{-1}.

Revision tip: Negative powers

In mathematics, $\frac{1}{x^n}$ may be written as x^{-n}. This convention is adopted when writing units. Therefore $\frac{\text{kg}}{\text{m}^3}$ is equivalent to kg m^{-3}.

More on derived units

Table 3 shows some examples of derived units. In each case, the equation used to determine the final units is shown.

▼ **Table 3** *Derived units*

Physical quantity	Equation	Derived unit	
area	area = length × width	m × m	$\rightarrow \text{m}^2$
density	density = $\frac{\text{mass}}{\text{volume}}$	$\frac{\text{kg}}{\text{m}^3}$	$\rightarrow \text{kg m}^{-3}$
acceleration	acceleration = $\frac{\text{change in velocity}}{\text{time}}$	$\frac{\text{ms}^{-1}}{\text{s}}$	$\rightarrow \text{ms}^{-2}$
force	force = mass × acceleration	$\text{kg} \times \text{m s}^{-2}$	$\rightarrow \text{kg m s}^{-2}$

Summary questions

1. The distance between two towns is 210 km. Write this distance in metres. *(1 mark)*
2. The speed of an electron within a wire is $1.2\ \text{mm s}^{-1}$. Write this speed in m s^{-1}. *(1 mark)*
3. It takes 12 ns for light to travel across a room. Write this time in standard form and in seconds. *(1 mark)*
4. The diameter of an atom is 2.3×10^{-10} m. Write this diameter in pm and in nm. *(2 marks)*
5. Use Table 3 to determine the unit for work done by a force, where work done = force × distance travelled in the direction of the force. *(2 marks)*
6. The distance between the Sun and the Earth is 150 Gm and the speed of light is $300\ \text{Mm s}^{-1}$. Calculate the time taken for light to travel from the Sun to the Earth. *(3 marks)*
7. The area of a sheet of paper is $620\ \text{cm}^3$. Calculate this area in m^2. *(2 marks)*

2.3 Scalar and vector quantities
2.4 Adding vectors

Specification reference: 2.3.1

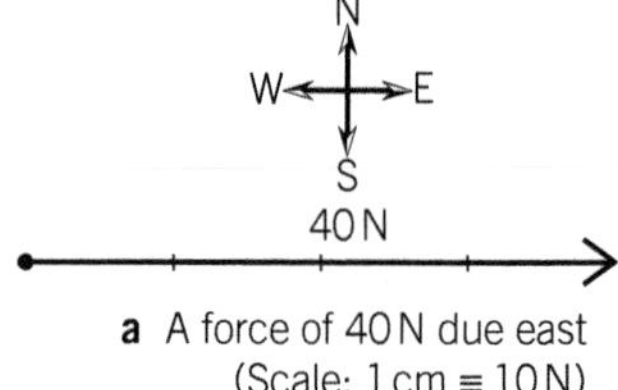

a A force of 40 N due east (Scale: 1 cm ≡ 10 N)

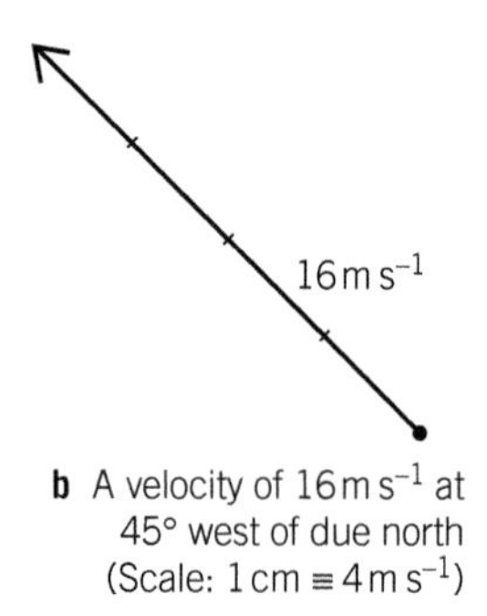

b A velocity of $16\,m\,s^{-1}$ at 45° west of due north (Scale: 1 cm ≡ $4\,m\,s^{-1}$)

▲ **Figure 1** *Representing vectors*

Revision tip

Distance is a scalar and displacement is a vector.

Speed is a scalar and velocity is a vector.

2.3 Scalar and vector quantities

Many of the physical quantities in physics can be divided into two categories – scalars and vectors. The way we add, or subtract, these quantities is very different, so you must be able to identify whether a quantity is a scalar or a vector.

Scalars

A **scalar quantity** has only magnitude (size).

Distance, speed, time, area, volume, mass, energy, temperature, electrical charge, and frequency are all scalars. Two similar scalar quantities can be added or subtracted in the normal way. For example, if you have a beaker of water on a digital balance, reading 250 g, and you add 50 g of water to the beaker, the total mass of the beaker and water will now be 300 g.

Vectors

A **vector quantity** has both magnitude and direction.

Displacement, velocity, acceleration, momentum, and force are all vectors. A vector can be represented as a line of suitable length drawn in the direction of the vector, see Figure 1.

2.4 Adding vectors

You have to take the directions of the two vectors into account when adding, or subtracting them.

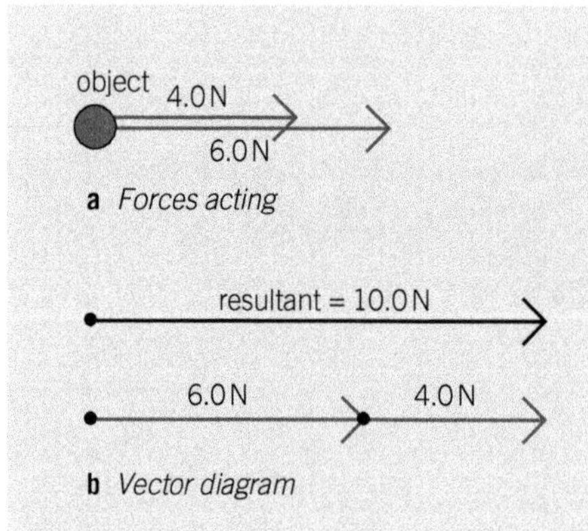

▲ **Figure 2** *Adding forces in the same direction*

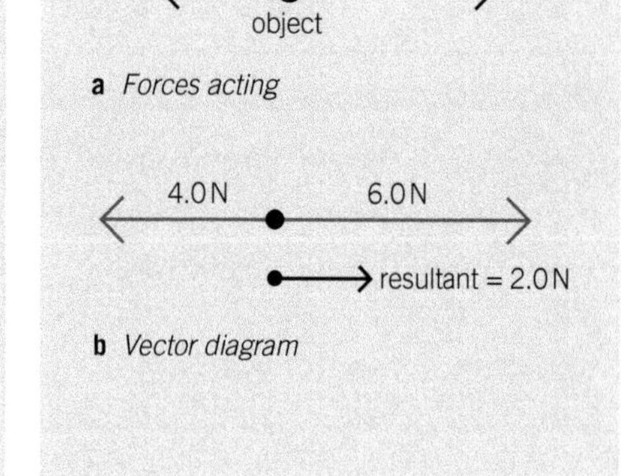

▲ **Figure 3** *Adding forces in opposite directions*

Figure 2 shows two forces acting on an object in the same direction and Figure 3 shows two forces acting in opposite directions. The resultant is just a matter of adding or subtracting the magnitudes.

What happens when the forces are at an angle to each other? This is where you have to carefully construct a vector triangle force – the resultant can either be determined from a scale drawing or by calculation.

Vector triangle

Figure 4 shows two forces, at right angles to each other, acting on an object and a vector triangle used to determine the resultant force. The vector triangle can be drawn to scale using the following rules:

- Choose a suitable scale.
- Draw a line to represent the first vector.
- Draw a line to represent the second vector from the end of the first vector.
- The magnitude and direction of the resultant vector can be found by drawing a line from the start to the end.

You can use these rules for any other vector and even when the angle between the vectors is not 90°.

$F_1 = 10.0$ N

object O

$F_2 = 7.0$ N

▲ **Figure 4a** *Two forces acting on an object*

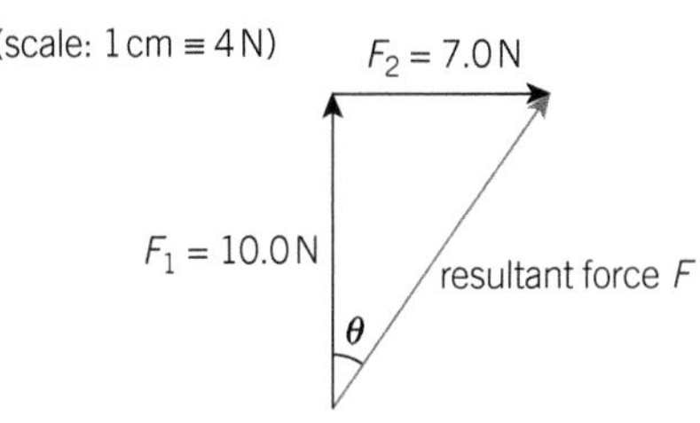

▲ **Figure 4b** *The vector triangle to determine the resultant force*

Worked example: Resultant force

Use Figure 4b to determine the magnitude and direction of the resultant force.

Step 1: Draw a clearly labelled vector triangle.

This has already been done for you in Figure 4b.

Step 2: Use Pythagoras' theorem to determine the magnitude of the resultant force F.

$$F^2 = F_1^{\ 2} + F_2^{\ 2} = 10.0^2 + 7.0^2$$

$$F = \sqrt{149} = 12 \text{ N}$$

Note: The resultant force must be the largest force, and it is.

Step 3: Calculate the angle using trigonometry.

$$\tan\theta = \frac{7.0}{10.0} = 0.700$$

$$\theta = \tan^{-1}(0.700) = 35° \text{ (2 s.f.)}$$

Revision tip: Get the mode right

Make sure that your calculator is in the correct mode when determining the angle. An angle in radians is not the same as the angle in degrees.

Maths: Pythagoras' theorem

This theorem only works when one of the internal angles of the triangle is 90°. You can use the sine and the cosine rules when this is not the case. All this information is provided for you in the Data, formula, and relationships booklet.

Summary questions

1 Name two scalar quantities and two vector quantities. *(2 marks)*

2 Suggest what is wrong with the calculation 13 kg + 100 N. *(1 mark)*

3 A car is travelling at 20 m s^{-1} on a circular track. The wind speed is 5.0 m s^{-1}.
Calculate the maximum and minimum magnitudes of the resultant velocity of the car. *(2 marks)*

4 Two forces of the same magnitude of 4.0 N and at right angles to each other are added together.
Draw a vector triangle. *(2 marks)*

5 Calculate the resultant force for the two forces in **Q4**. *(2 marks)*

6 A swimmer can swim at 1.8 m s^{-1} in still water.
He aims to swim *directly* across a river flowing at 1.0 m s^{-1}.

a Draw a labelled vector triangle. *(2 marks)*

b Calculate the magnitude of the resultant velocity of the swimmer. *(2 marks)*

2.5 Resolving vectors
2.6 More on vectors

Specification reference: 2.3.1

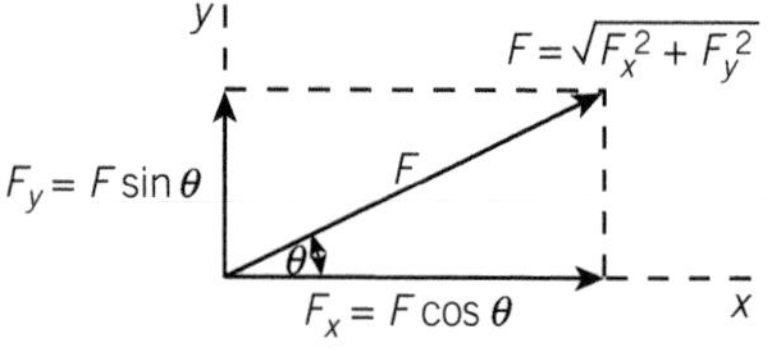

▲ **Figure 1** *Resolving a vector into two mutually perpendicular components*

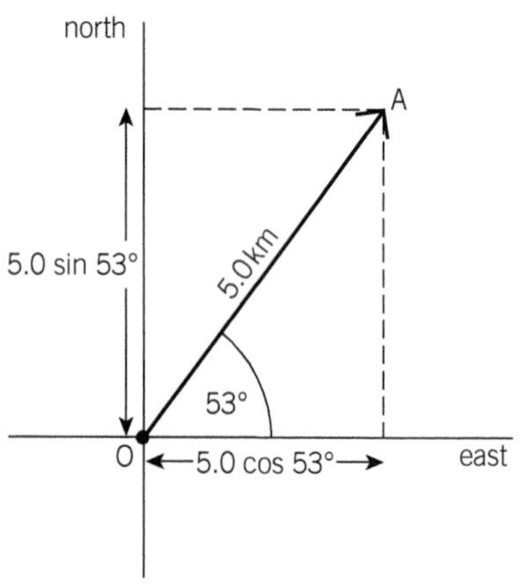

▲ **Figure 2** *The components of a displacement*

2.5 Resolving vectors

It is often useful to resolve, or split up, a vector into two components at right angles to each other. This process is particularly useful when analysing the motion of a projectile on the surface of the Earth – the horizontal and vertical velocities are independent of each other.

Components of a vector

Figure 1 shows a vector, in this case a force of magnitude F, being resolved in the x- and y-directions. You can use the rules below for any vector (velocity, acceleration, etc).

- The components of the vector are:
  - $F_x = F\cos\theta$ in the x-direction and $F_y = F\sin\theta$ in the y-direction.
- The magnitude F of the vector can be determined using $F^2 = F_x^2 + F_y^2$.
- The angle θ made by the vector to the horizontal can be determined using $\tan\theta = \frac{F_y}{F_x}$.

Figure 2 shows a displacement of 5.0 km resolved into its two components. The displacement due east is 30 km and 40 km in the north direction.

Maths: Cos and sin

It is useful to know that $\cos(90 - \theta) = \sin\theta$. Try this on your calculator.

When the x-component is $F\cos\theta$, then the y-component is equivalent to $F\cos(90 - \theta)$.

Revision tip: Just remember the cosine

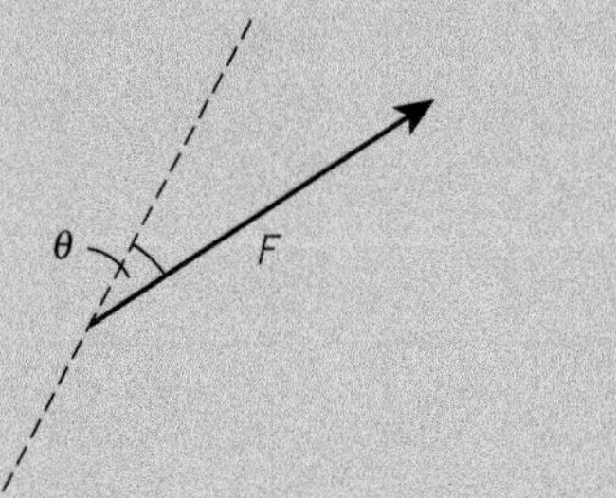

In general, the component of a force (or any other vector) at an angle θ is given by $F\cos\theta$.

2.6 More on vectors

Adding non-perpendicular vectors

Vectors are often not at right angles to each other. You can still determine the resultant vector using the four bulleted rules in Topic 2.4, Adding vectors, by drawing a vector triangle. You will need a very sharp pencil and be able to measure length and angle size accurately, see Figure 3.

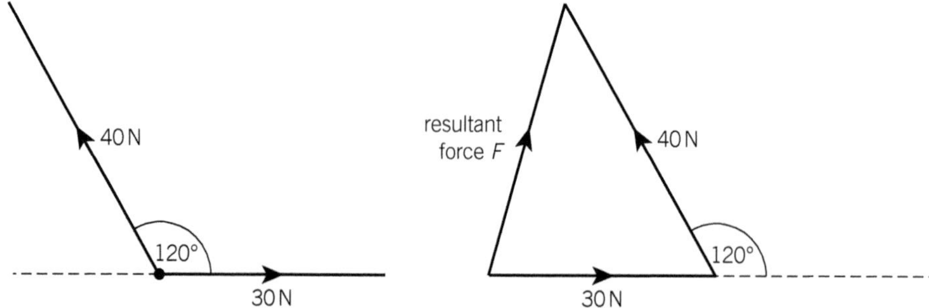

▲ **Figure 3** *You can draw a scaled vector triangle to determine the resultant vector*

You can replicate the process illustrated in Figure 3 to determine any resultant vector, including velocity and momentum.

Calculations

You can also use the sine rule and the cosine rule to determine the magnitude and direction of the resultant vector. A decent sketch of a vector triangle is necessary before you can do any calculations. See the worked example below.

Model answers: Using the cosine rule

Use the cosine rule to show that the magnitude of the resultant force F in Figure 3 is 36 N.

Answer

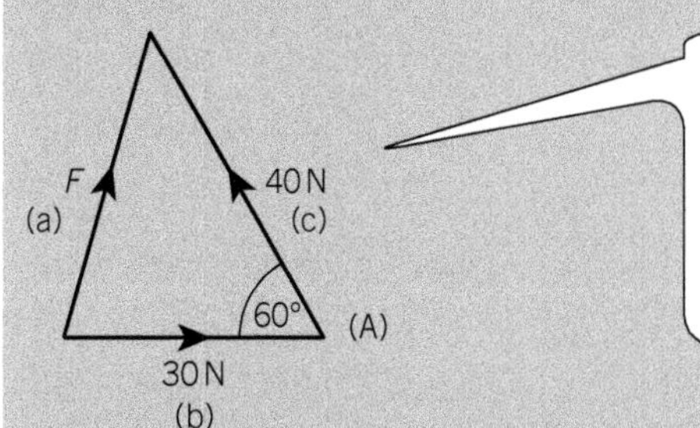

A clearly labelled vector triangle is essential and you need to know the angle 'opposite' to the 'side' F of this triangle for the cosine rule.

cosine rule: $a^2 = b^2 + c^2 - 2bc\cos A$

$a^2 = b^2 + c^2 - 2bc\cos A$

Therefore

$F^2 = 30^2 + 40^2 - 2 \times 30 \times 40 \times \cos 60°$

It is important to show all your working.

$F = \sqrt{1300}$

$F = 36\text{ N}$

The data is given to 2 s.f. It is good practice to write the final answer also to 2 s.f.

Note: In order to calculate any of the other internal angles of the triangle, you can use the sine rule

$$\frac{a}{\sin A} = \frac{b}{\sin B} = \frac{c}{\sin C}$$

Subtracting two vectors

Subtracting two vectors is very much like adding two vectors, except you reverse the direction of one of the vectors. This is illustrated in Figure 4. Where you want the resultant vector of **A** – **B**, you can apply the following rules:

- Draw a line to represent the vector **A**.
- Draw a line in the opposite direction to represent the vector **–B** from the end of the vector **A**.
- The magnitude and direction of the resultant vector can be found by drawing a line from the start to the end.

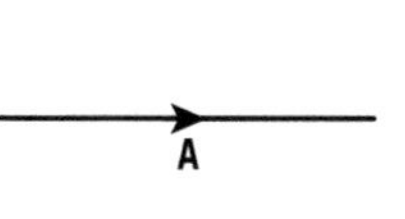

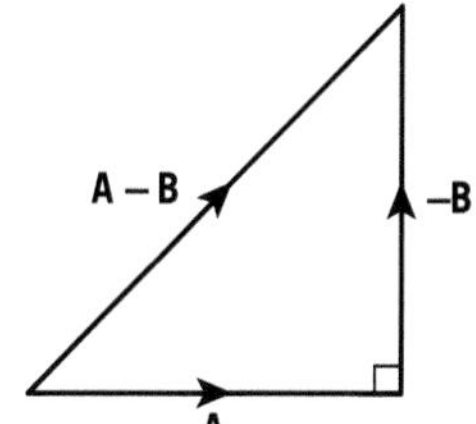

▲ **Figure 4** *Subtracting vectors*

Summary questions

1 A rocket has a velocity of 300 m s^{-1} at an angle of 65° to the horizontal. Calculate the horizontal and vertical components of this velocity. (*2 marks*)

2 A ball rolls down a frictionless ramp. The acceleration of free fall is 9.81 m s^{-2}. The ramp makes an angle of 30° to the vertical. Calculate the acceleration of the ball down the ramp. (*1 mark*)

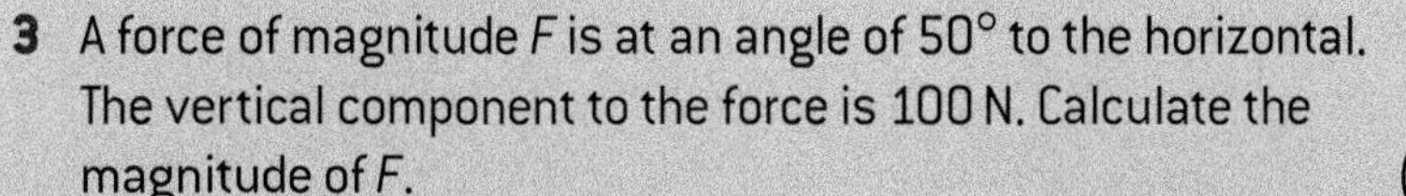

3 A force of magnitude F is at an angle of 50° to the horizontal. The vertical component to the force is 100 N. Calculate the magnitude of F. (*2 marks*)

4 Calculate the resultant force in each case shown in Figure 5. (*4 marks*)

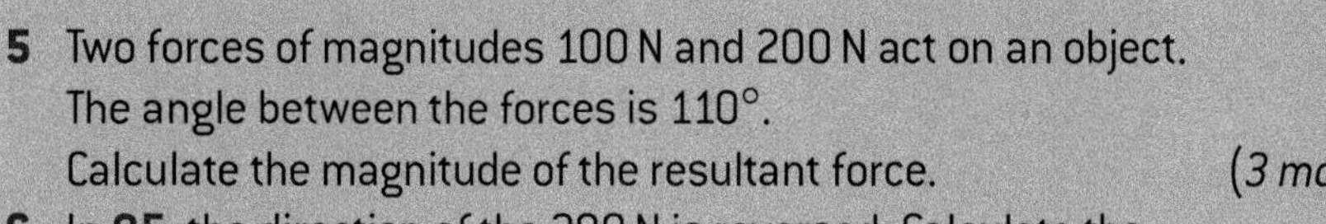

5 Two forces of magnitudes 100 N and 200 N act on an object. The angle between the forces is 110°. Calculate the magnitude of the resultant force. (*3 marks*)

6 In **Q5**, the direction of the 200 N is reversed. Calculate the magnitude of the resultant force now. (*3 marks*)

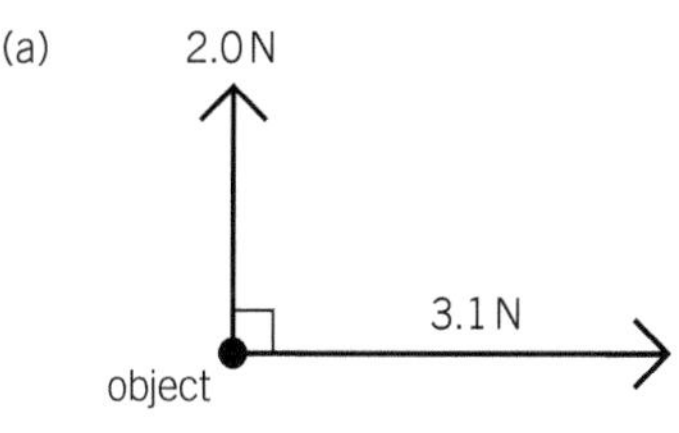

(b) 2.0 N object 3.0 N 1.0 N

▲ **Figure 5**

Chapter 2 Practice questions

1 Which is the correct definition for a scalar quantity?
A scalar quantity ......

A has no units.

B has direction only.

C has magnitude only.

D has magnitude and direction. *(1 mark)*

2 What is force measured in base units?

A N

B kg

C $kg\,s^2\,m^{-1}$

D $kg\,m\,s^{-2}$ *(1 mark)*

3 An object experiences two forces of magnitudes 3.2 N and 5.0 N. The forces are at right angles to each other.

What is the magnitude of the resultant force on the object?

A 3.8 N

B 5.9 N

C 15 N

D 35 N *(1 mark)*

4 The horizontal component of a 10.0 N force is 8.0 N.

What is the angle between the 10.0 N force and the horizontal?

A 0°

B 37°

C 39°

D 53° *(1 mark)*

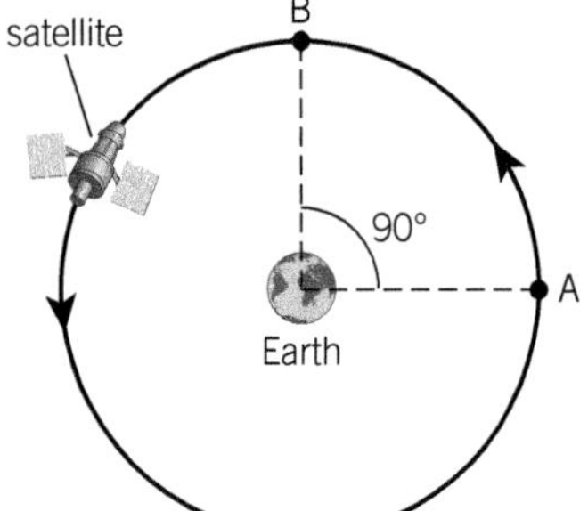

▲ **Figure 1**

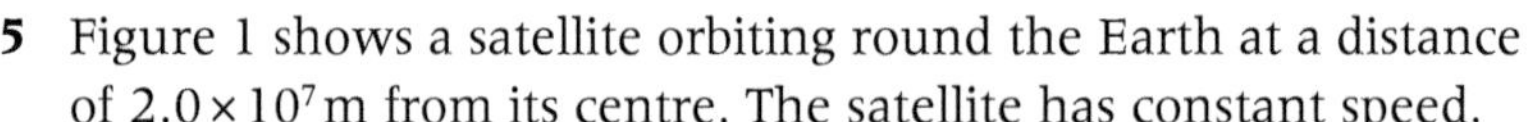

5 Figure 1 shows a satellite orbiting round the Earth at a distance of 2.0×10^7 m from its centre. The satellite has constant speed.

a Explain why the velocity of the satellite is not constant in its orbit. *(1 mark)*

b The satellite travels from point A to point B. Calculate:

i the distance travelled by the satellite; *(2 marks)*

ii the displacement of the satellite from point A. *(3 marks)*

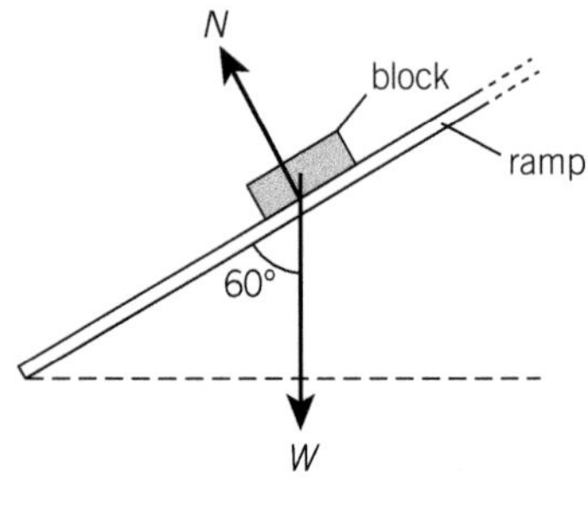

▲ **Figure 2**

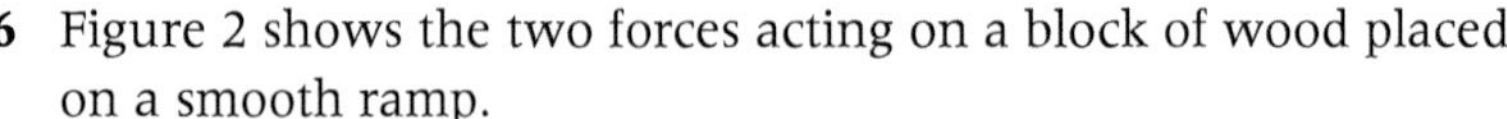

6 Figure 2 shows the two forces acting on a block of wood placed on a smooth ramp.

The weight *W* of the block is 8.0 N and the normal contact force *N* is 6.9 N.

a Explain why *N* has no affect on the motion of the block down the ramp. *(1 mark)*

b Calculate the component of the weight down the ramp. *(1 mark)*

c Draw a labelled vector diagram and determine the resultant force acting on the block. *(4 marks)*

Module 3 Forces and Motion

Chapter 3 Motion

In this chapter you will learn about …

- [ ] Speed
- [ ] Distance–time graphs
- [ ] Velocity
- [ ] Displacement-time graphs
- [ ] Acceleration
- [ ] Velocity-time graphs
- [ ] Equations of motion (*suvat* equations)
- [ ] Thinking, braking, and stopping distances
- [ ] Free fall and *g*
- [ ] Projectiles

3 MOTION
3.1 Distance and speed
3.2 Displacement and velocity

Specification reference: 3.1.1

3.1 Distance and speed

You can measure the distance travelled by an object using a metre rule or a measuring tape. The speed of an object is related to distance and time.

Speed

Speed is defined as the rate of change of distance. You can use the following word equation to determine the average speed v of an object:

$$\text{average speed} = \frac{\text{total distance travelled}}{\text{time taken}}$$

or

$$v = \frac{\Delta x}{\Delta t}$$

where Δx is the total distance travelled in a time Δt. The SI unit for speed is $\mathrm{m\,s^{-1}}$. You can also use this equation to determine the **constant** speed of an object and it can be modified to determine the instantaneous speed of an object, as shown below.

$$v = \frac{\delta x}{\delta t}$$

where δx is the very small distance travelled by the object in a very small time interval δt.

Maths: Change in

The Greek letter capital delta Δ is shorthand for 'change in...'. So Δt means 'change in t'.

In physics lower case δ is used to signify 'very small change in...'.

Worked example: Speed

A space probe orbits round a planet in a circular path of radius 5.6×10^5 m at a constant speed. It takes 96 minutes to orbit once round the planet. Calculate the speed of the space probe.

Step 1: Write down all the quantities given in SI units.

$r = 5.6 \times 10^5\,\mathrm{m} \qquad t = 96 \times 60 = 5.76 \times 10^3\,\mathrm{s}$

Step 2: Calculate the speed, remembering that the distance is the circumference.

$$v = \frac{\Delta x}{\Delta t} = \frac{2\pi r}{t} = \frac{2\pi \times 5.6 \times 10^5}{5.76 \times 10^3}$$

$v = 611\,\mathrm{m\,s^{-1}} \approx 610\,\mathrm{m\,s^{-1}}$ (2 s.f.)

Distance–time graph

The motion of an object can be displayed on a distance–time (x–t) graph.

- The speed of the object at time t can be determined from the gradient of the graph at this time t.
- For a curved distance–time graph, the speed can be calculated by determining the gradient of a tangent drawn to the graph, see Figure 1.

Maths: Gradient

On a y against x graph, the gradient $= \frac{\text{change in } y}{\text{change in } x}$

$$\text{speed at Y} = \frac{PQ}{QR} = \frac{192 - 52}{20} = 7.0\,\mathrm{m\,s^{-1}}$$

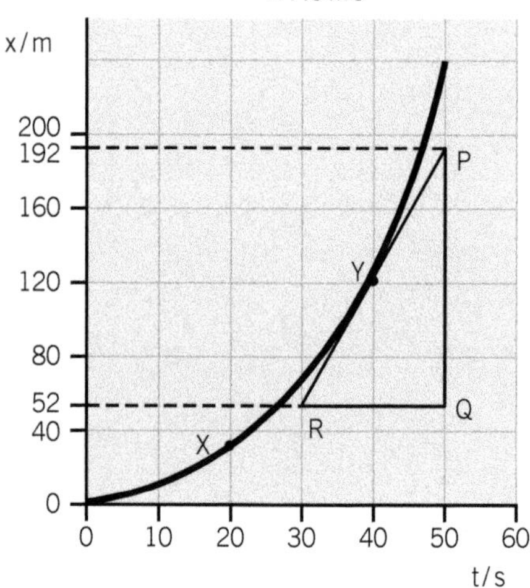

▲ **Figure 1** *You can determine the speed of an object from the gradient of the graph*

3.2 Displacement and velocity

Displacement is defined as the distance in a given direction.

It is a vector with both magnitude and direction. The magnitude of a displacement is the same as the distance travelled when an object travels in a straight line in a given direction, for example, a falling apple.

Velocity

Velocity is defined as the rate of change of displacement. The SI unit for velocity is the same as that for speed. The word equation for velocity is

$$\text{velocity} = \frac{\text{change in displacement}}{\text{time taken}}$$

The equation for velocity v is

$$v = \frac{\Delta s}{\Delta t}$$

where Δs is the change in displacement and Δt is the time taken.

Displacement–time graph

The motion of an object can be displayed on a displacement–time (s–t) graph.

- The gradient of an s–t graph is equal to the velocity v of the object.
- A negative gradient represents motion in the opposite direction.
- For a curved s–t graph, the instantaneous velocity is found by drawing a tangent to the graph and then determining the gradient of this tangent.

Summary questions

1 Calculate the average speed in m s^{-1} of a trolley travelling a distance of 120 cm in a time of 3.4 s. *(1 mark)*

2 A car is travelling at a constant speed of 20 m s^{-1}. Calculate the distance travelled in 0.40 hours. *(2 marks)*

3 Figure 2 shows the displacement–time graph for three objects **A**, **B**, and **C**. Describe the motion of each object. *(3 marks)*

4 A car travels 2.0 km due north for 30 minutes at a constant velocity. Then it travels 1.0 km due south for another 30 minutes at a different constant velocity.
Sketch a displacement–time (s–t) graph for the car. *(2 marks)*

5 An electron travels in a circular path of radius 2.0 cm at a constant speed of 5.0×10^5 m s^{-1}.
Calculate the time taken by the electron to complete one orbit. *(2 marks)*

6 Figure 3 shows the displacement–time graph for a ball thrown vertically upwards on the Earth's surface.

a Describe the motion of the ball at time $t = 0.5$ s and time $t = 1.5$ s. *(2 marks)*

b What is the magnitude of the velocity of the ball at $t = 1.0$ s? Explain your answer. *(2 marks)*

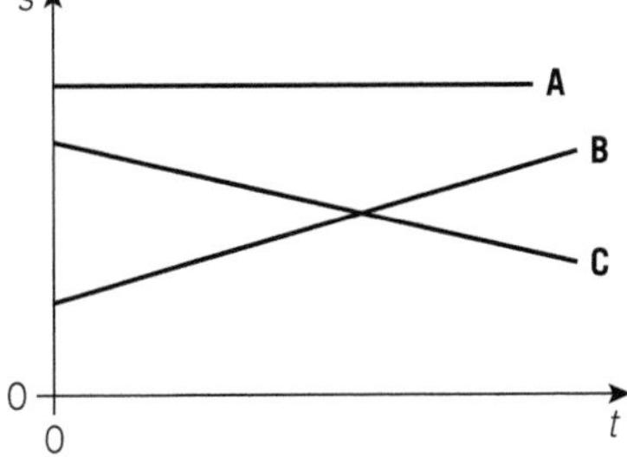

▲ Figure 2

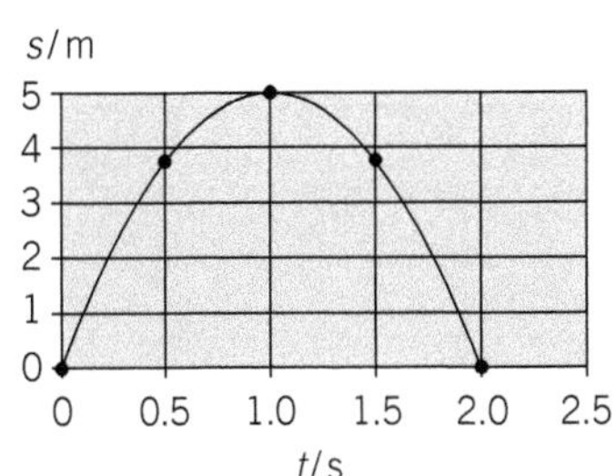

▲ Figure 3

3.3 Acceleration
3.4 More on velocity–time graphs

Specification reference: 3.1.1

3.3 Acceleration

Acceleration is a vector quantity – it has both magnitude and direction. An object is accelerating when its speed is increasing and it is decelerating when its speed decreases. An object will also have an acceleration when its direction of travel changes.

Defining acceleration

Acceleration is defined as the rate of change of velocity. In SI units, acceleration is measured in metre per second squared ($m\,s^{-2}$). The word equation for acceleration is

$$\text{acceleration} = \frac{\text{change in velocity}}{\text{time taken}}$$

The equation for acceleration a is

$$a = \frac{\Delta v}{\Delta t}$$

where Δv is the change in velocity in $m\,s^{-1}$ and Δt is the time taken in s.

You can also write this as

$$a = \frac{v - u}{t}$$

where u is the initial velocity, v is the final velocity, and t is the time taken for the change in velocity.

Common misconception

Do not define acceleration as 'the rate of change of speed'.

Maths: Change in velocity

The Greek delta Δ is shorthand for 'a change in'. So Δv means a change in velocity.

Revision tip: Signs

A positive acceleration means increasing velocity.

A negative acceleration means decreasing velocity or a deceleration.

Worked example: Deceleration

A car is travelling on a straight road. The driver applies the brakes. The velocity of the car changes from $30\,m\,s^{-1}$ to $12\,m\,s^{-1}$ in a time of 5.0 s. Calculate the acceleration of the car.

Step 1: Write down the known and unknown quantities.

$u = 30\,m\,s^{-1}$ $\quad v = 12\,m\,s^{-1}$ $\quad t = 5.0\,s$ $\quad a = ?$

Step 2: Write down the equation, substitute and solve.

$$a = \frac{v-u}{t} = \frac{12 - 30}{5.0} = -3.6\,m\,s^{-2}\ \text{(2 s.f.)}$$

The minus sign means the car is decelerating.
The magnitude of the deceleration is $3.6\,m\,s^{-2}$ (2 s.f.).

3.4 Velocity–time graphs

You can interpret the motion of an object from its velocity v against time t (or v–t) graph.

Information from v–t graphs

- The gradient of a v–t graph is equal to the acceleration a of the object.
- The area under a v–t graph is equal to the displacement s of the object.

A positive gradient shows that the velocity of the object is increasing. A negative gradient shows that the object is decelerating. You can calculate the instantaneous acceleration of an object by drawing a tangent to the graph and then determining the gradient of this tangent, see Figure 1.

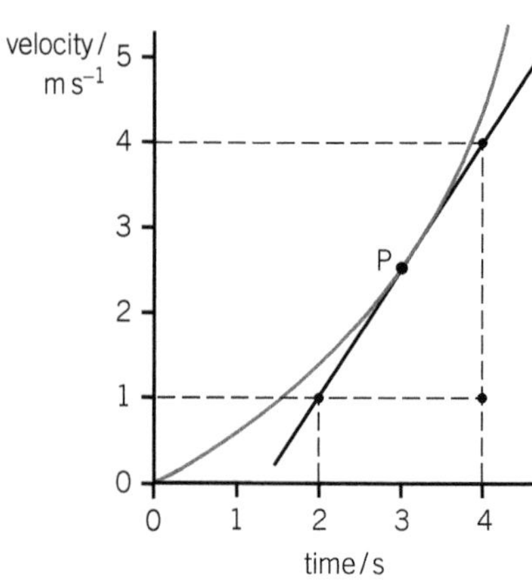

▲ **Figure 1** *The acceleration a can be determined from the gradient of the tangent drawn at point P*

Model answers: Analysing motion

Figure 2 shows the v–t graph for an object travelling in a straight line.

Determine:

a the acceleration of the object at time $t = 3.0$ s;

b the total distance travelled in 6.0 s.

▲ **Figure 2**

Answer

a The acceleration is the gradient at $t = 3.0$ s

$$a = \frac{\Delta v}{\Delta t} = \frac{10.0 - 0}{4.0} = 2.5\,\mathrm{m\,s^{-2}}$$

> The object has uniform acceleration from $t = 0$ to $t = 4.0$ s.
>
> This is a well-structured answer – the equation first, then the substitution and finally the answer quoted to 2 significant figures.

b The area under the graph is equal to the displacement s, which is the same as the distance travelled.

> This opening statement makes the physics very clear.

s = area of triangle

$$s = \frac{1}{2} \times 6.0 \times 10.0 = 30\,\mathrm{m}$$

> You can also calculate the total area by adding the areas of the two smaller triangles as follows:
>
> $$s = \left(\frac{1}{2} \times 4.0 \times 10.0\right) + \left(\frac{1}{2} \times 2.0 \times 10.0\right) = 30\,\mathrm{m}$$
>
> The answer is the same.
>
> The method shown in this model answer is shorter and neater.

Maths: Improving accuracy

On a y–x graph, the gradient is calculated by dividing Δy by Δx.

To improve accuracy, always use a large 'triangle' to determine the gradient.

Revision tip

Uniform acceleration means *constant* acceleration.

Summary questions

1 Describe the motion of an object with the v–t graph shown in Figure 3a, 3b, and 3c. *(3 marks)*

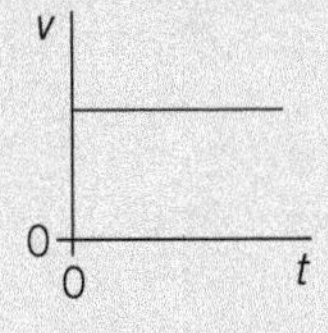

▲ **Figure 3a**

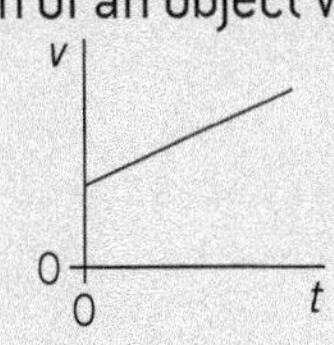

▲ **Figure 3b**

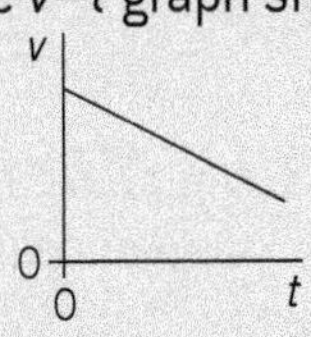

▲ **Figure 3c**

2 An aircraft can change its velocity from 100 m s^{-1} to 320 m s^{-1} in 20 s. Calculate the acceleration of the aircraft. *(2 marks)*

3 A rocket has a constant acceleration of 3.0 m s^{-2}. Calculate the change in its velocity in a time interval of 1.5 minutes. *(2 marks)*

4 A ball bounces off a wall. The change in velocity of the ball is 10 m s^{-1} and acceleration of the ball when in contact with the wall is 400 m s^{-2}. Calculate the time for which the ball is in contact with the wall. *(2 marks)*

5 Use Figure 1 to determine the instantaneous acceleration of the object at time $t = 3.0$ s. *(2 marks)*

6 A cyclist, initially at rest, accelerates uniformly for 10 s and reaches a velocity of 4.0 m s^{-1} on a straight road. She then travels at this constant velocity for a further period of 35 s. Finally she decelerates uniformly to rest in 5.0 s.

a Sketch a velocity–time graph for her journey. *(3 marks)*

b Calculate the total distance travelled by the cyclist. *(3 marks)*

c Calculate the magnitude of the deceleration. *(2 marks)*

3.5 Equations of motion
3.6 Car stopping distances

Specification reference: 3.1.2

3.5 Equations of motion

The four equations of motions (also known as the '*suvat*' equations) are expressions that can be applied to the motion of an object travelling in a straight line with a constant (or uniform) acceleration. The following labels are used for the quantities:

$s \rightarrow$ displacement (or distance travelled when the direction of travel remains the same)

$u \rightarrow$ initial velocity

$v \rightarrow$ final velocity

$a \rightarrow$ acceleration

$t \rightarrow$ time taken for the change in velocity

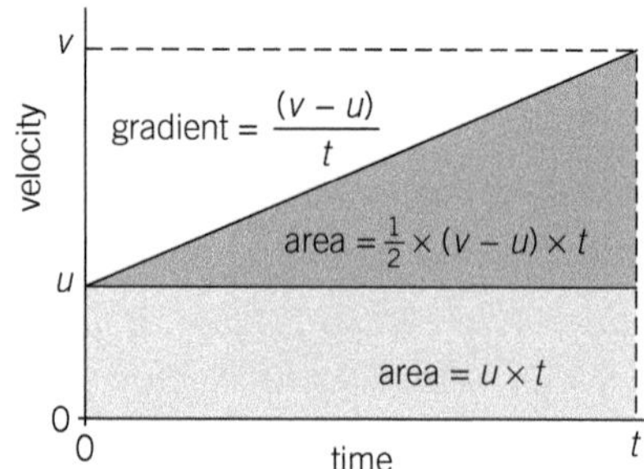

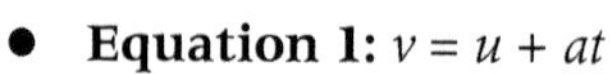

▲ **Figure 1** *The gradient of the graph is equal to acceleration and the area under the graph is equal to displacement*

The four equations of motion

There are four equations of motion that can be derived from the velocity against time graph shown in Figure 1.

- **Equation 1:** $v = u + at$

  The gradient of the graph is equal to a.

  Therefore $a = \frac{v - u}{t}$. When rearranged, this gives

  $$\mathbf{v = u + at}$$

- **Equation 2:** $s = \frac{1}{2}(u + v)t$

  The area under the line graph is equal to the displacement s. This is equal to the area of a trapezium of 'height' t and 'parallel lengths' u and v. Therefore

  $$\mathbf{s = \frac{1}{2}(u + v)t}$$

- **Equation 3:** $s = ut + \frac{1}{2}at^2$

  The displacement is equal to the sum of the area of the rectangle of 'height' u and 'length' t and area of triangle of 'height' $(v - u)$ and 'length' t.

  Therefore $s = [u \times t] + \left[\frac{1}{2}(v - u)t\right]$

  However, $v - u = at$ from equation 1. Therefore $s = ut + \frac{1}{2}(at)t$, or simply

  $$\mathbf{s = ut + \frac{1}{2}at^2}$$

- **Equation 4:** $v^2 = u^2 + 2as$

  This final equation can be derived from equations 1 and 2.

  From equation 1 we have $t = \frac{v - u}{a}$. When this is substituted into equation 2, we have

  $$s = \frac{1}{2}(v + u) \times \frac{(v - u)}{a} = \frac{v^2 - u^2 - uv + uv}{2a}$$

  Therefore $2as = v^2 - u^2$

  This simplifies to

  $$\mathbf{v^2 = u^2 + 2as}$$

Revision tip: Get the signs right

The acceleration a is a positive number for acceleration and a negative number for deceleration. For an object projected vertically upwards, $a = -9.81\ \mathrm{m\,s^{-2}}$.

Worked example: Braking car

A car is travelling at a velocity of 31 m s^{-1} on a straight road. The driver applies the brakes. The car comes to rest after travelling a distance of 18 m. Calculate the magnitude of the deceleration of the car.

Step 1: Write down all the quantities given and include the unknown quantity.

$s = 18$ m $u = 31$ m s^{-1} $v = 0$ $a = ?$

Step 2: Identify the correct *suvat* equation, rearrange the equation, substitute the values, and calculate a.

$$v^2 = u^2 + 2as$$

$$a = \frac{v^2 - u^2}{2s} = \frac{0^2 - 31^2}{2 \times 18} = -27 \text{ m s}^{-2} \text{ (2 s.f.)}$$

The magnitude of the deceleration is 27 m s^{-2} (2 s.f.).

3.6 Car stopping distances

The stopping distance of a car is made up of two components – thinking distance and braking distance.

Stopping distance is defined as the sum of thinking distance and braking distance.

Thinking distance is defined as the distance travelled between the moment the driver sees a reason to stop and the moment when the driver applies the brake.

The **braking distance** is the distance travelled by the car from the instant the brake is applied until the car stops.

- Thinking distance depends on the initial speed u of the car and the 'reaction' time of the driver – thinking distance $\propto u$.
- Braking distance depends on a number of factors, including the initial speed u of the car, the condition of the tyres, brakes and road, and the alertness of the driver – braking distance $\propto u^2$.

Summary questions

1 A stone is thrown vertically upwards at a speed of 15 m s^{-1}. Calculate its velocity after 1.0 s. *(2 marks)*

2 A car is travelling at 5.0 m s^{-1}. It suddenly speeds up for 7.0 s with a constant acceleration of 3.0 m s^{-2}. Calculate the distance travelled by the car in this interval of time. *(2 marks)*

3 Figure 2 shows the v–t graph for a car. The driver starts to react at time $t = 0$.
Calculate the thinking distance, braking distance, and the stopping distance of the car. *(3 marks)*

4 The distance travelled by a car is 80 m as its speed increases from 10 m s^{-1} to 25 m s^{-1}.
Calculate the time taken to travel this distance. *(3 marks)*

5 A metal ball is dropped from rest from the top of a building 100 m high. Calculate its speed just before it hits the ground below. *(3 marks)*

6 A golf ball is dropped on a hard floor. It hits the floor at a speed of 14 m s^{-1} and bounces back up at a speed of 12 m s^{-1}. It is in contact with the ground for 4.0 ms.
Calculate the magnitude of the average acceleration of the ball. *(3 marks)*

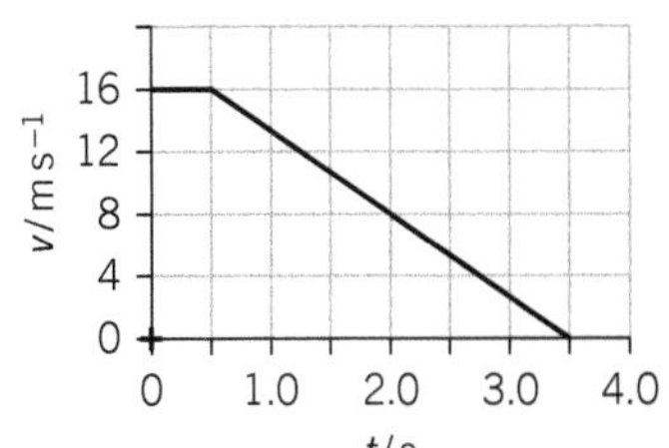

▲ **Figure 2**

3.7 Free fall and g

Specification reference: 3.1.2

3.7 Free fall and g

All objects on the surface of the Earth fall with the same acceleration, as long as air resistance is negligible. This acceleration of free fall g is equal to 9.81 m s^{-2} and is independent of the mass of an object. You can use the equations of motion (see Topic 3.5), to analyse the motion of objects falling freely in the Earth's gravitational field.

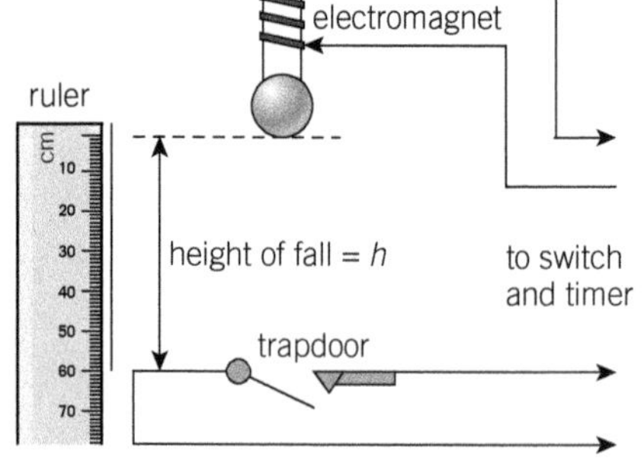

▲ **Figure 1** *Arrangement used to determine the acceleration of free fall*

Revision tip: Gradients

Use a large 'triangle' to determine the gradient of a straight line. This will improve the accuracy.

Determining acceleration of free fall g

Figure 1 shows an arrangement that can be used in the laboratory to determine the acceleration of free fall. A heavy steel ball is dropped from rest from an electromagnet. The distance h of fall is measured using a metre rule and the time t of fall is determined using a trapdoor and timer arrangement. The experimental value for g can be determined by plotting a graph of h against t^2.

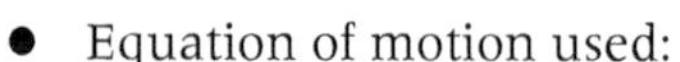

- Equation of motion used: $s = ut + \frac{1}{2}at^2$, with $u = 0$, $a = g$, and $s = h$
- Comparison with equation for a straight line: $s = \left(\frac{g}{2}\right)t^2$, with $s \rightarrow y$ and $t^2 \rightarrow x$

A graph of h against t^2 will be a straight line with a gradient equal to $\frac{g}{2}$. Therefore $g = 2 \times$ gradient

Using light gates and taking pictures

You can determine the speed of a falling object, for example, a metal plate, using a light gate connected to either a timer or a data-logger. This can then be used to calculate a value for g. This method of analysis is illustrated in the worked example below.

You can also take stroboscopic images of a falling object. The images of the falling object are taken at regular time intervals. The equation of motion $s = ut + \frac{1}{2}at^2$ can be used to determine g.

Worked example: Analysing data from a light gate and timer

A metal plate of length 9.00 cm is dropped from a height of 1.00 m above a light gate connected to a timer. The metal plate falls vertically through the light gate. The timer records 0.020 s.

a Calculate the acceleration of free fall.

b Determine the % difference between the experimental value and the accepted value for the acceleration of free fall.

Step 1: Calculate the final velocity of the metal plate.

$$v = \frac{\text{length of plate}}{\text{time taken}} = \frac{0.090}{0.020} = 4.50\ \text{m s}^{-1}$$

Step 2: Identify the correct *suvat* equation, rearrange the equation, substitute the values, and calculate a.

$v^2 = u^2 + 2as$, with $u = 0$ and $a = g$

$$g = \frac{v^2}{2s} = \frac{4.50^2}{2 \times 1.00}$$

$g = 10.1$ m s^{-2}

Step 3: Determine the % difference.

$$\%\ \text{difference} = \frac{10.1 - 9.81}{9.81} \times 100 = 3.0\%$$

Summary questions

1 A rock and a metal ball are dropped on the surface of the Earth. State the acceleration of free fall of each object. *(1 mark)*

2 A heavy metal ball is dropped from rest. Calculate the distance travelled by the ball after 0.30 s. *(2 marks)*

3 The acceleration of free fall on the surface of the Moon is 1.6 m s^{-2}. Calculate the distance travelled by a ball dropped on the surface of the Moon after 0.30 s. *(2 marks)*

4 An apple falls off a tree from a height of 2.5 m. It takes 0.70 s before it hits the ground below. Estimate a value for the acceleration of fall. *(3 marks)*

5 A student conducts a free fall experiment in the laboratory using the arrangement shown in Figure 1. The results are shown below.
height $h = 800 \pm 3$ mm $t = 0.403 \pm 0.002$ s
Determine a value for the acceleration of free fall. *(3 marks)*

6 Figure 2 shows a graph of s against t^2 plotted by a student following an experiment. The distance of fall is s and the time taken is t. Use Figure 2 to determine a value for g. *(3 marks)*

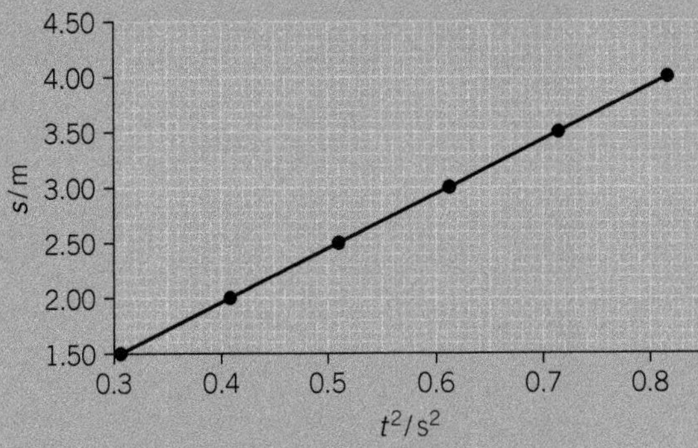

▲ **Figure 2**

7 Determine the absolute uncertainty in the value for the acceleration of free fall in **Q5**. Write the acceleration of free fall in the format $g = \ldots\ldots \pm \ldots\ldots$ m s^{-2}. *(4 marks)*

3.8 Projectile motion

An object thrown vertically on the Earth will describe a straight-line path. An object thrown at an angle to the vertical will follow a curved (parabolic) path. You can analyse the motion of projectiles using the equations of motion.

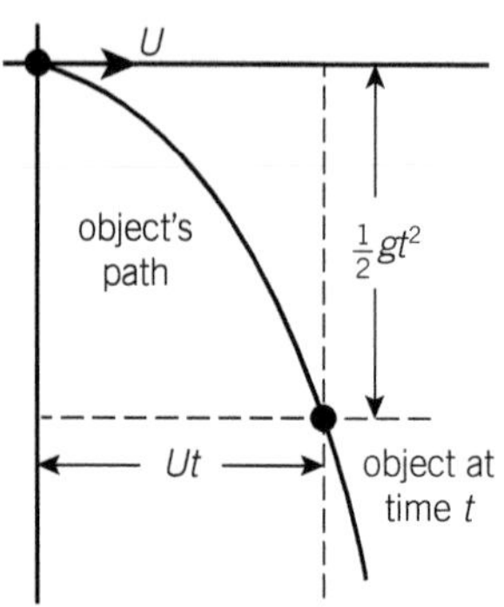

▲ **Figure 1** *An object projected horizontally*

Independent horizontal and vertical motions

When the effects of air resistance are ignored, all projectiles on the Earth will have an acceleration of 9.81 m s^{-2} vertically downwards. The component of g in the horizontal direction is zero because $g\cos 90° = 0$. Therefore

- Horizontally: The horizontal velocity remains constant.
- Vertically: The velocity changes and the motion can be described using the equations of motion.

Figure 1 shows the path described by an object projected horizontally with an initial velocity equal to U.

Table 1 lists the equations for horizontal and vertical motion.

Revision tip

You can use the equations of motion to analyse the *vertical* motion of a projectile. The horizontal velocity of a projectile stays the same.

▼ **Table 1**

	Horizontal motion	Vertical motion
Acceleration	zero	$+g$
Displacement	horizontal displacement $= x$ $x = Ut$	vertical displacement $= y$ initial velocity $= U\cos 90° = 0$ Using $s = ut + \frac{1}{2}at^2 \Rightarrow y = \frac{1}{2}gt^2$
Velocity	U (remains constant)	Using '$v = u + at$' $\Rightarrow v = gt$

Worked example: Projected horizontally

A metal ball is projected horizontally with a velocity of 3.0 m s^{-1} from the end of a flat table of height 75 cm. Calculate the time it takes before it hits the floor below.

Step 1: Write down the important quantities for the vertical motion of the ball. Assume downward direction is positive.

$s = 0.75$ m $u = 0$ $a = +9.81$ m s^{-2} $t = ?$

Step 2: Identify the correct *suvat* equation, rearrange the equation, substitute the values, and calculate *a*.

$$s = ut + \frac{1}{2}at^2 = \frac{1}{2}gt^2$$

$$t = \sqrt{\frac{2s}{g}} = \sqrt{\frac{2 \times 0.75}{9.81}} = 0.39 \text{ s}$$

$t = 0.39$ s (2 s.f.)

Projected at an angle

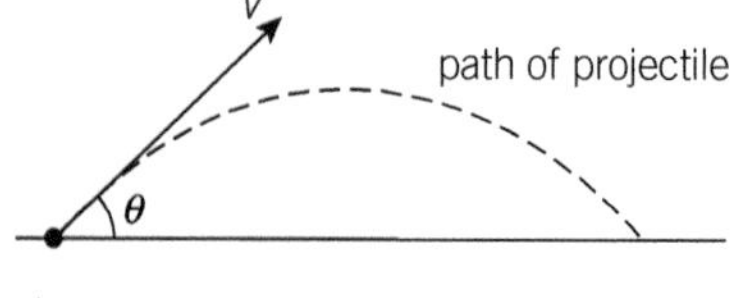

▲ **Figure 2** *An object projected at an angle to the horizontal*

Figure 2 shows a projectile fired at a velocity v at an angle θ to the horizontal. You can analyse the motion of this projectile by independently examining the horizontal and vertical motions.

- Horizontally: The velocity remains constant at $v\cos\theta$.
- Vertically: The initial velocity is $v\sin\theta$.

The acceleration is $-g$. (The vertical downward direction is taken as *positive*.)

Worked example: At an angle

A heavy ball is projected at a velocity of $12\ \mathrm{m\,s^{-1}}$ at an angle of $40°$ to the horizontal. Calculate the maximum height of the ball.

Step 1: Write down the important quantities for the vertical motion of the ball.

Note: At the maximum height, the ball stops momentarily in the vertical direction.

$s = ?$ $u = 12 \sin 40°$ $v = 0$ $a = -9.81\ \mathrm{m\,s^{-2}}$

Step 2: Identify the correct *suvat* equation, rearrange the equation, substitute the values, and calculate *s*.

$$v^2 = u^2 + 2as$$

$$s = \frac{v^2 - u^2}{2a} = \frac{0 - (12\sin 40)^2}{2 \times -9.81}$$

$$s = 3.0\ \mathrm{m}\ (2\ \text{s.f.})$$

Summary questions

1 A ball is projected horizontally at a velocity of $5.0\ \mathrm{m\,s^{-1}}$ from a tall platform.
State and explain the value of the horizontal velocity. *(2 marks)*

2 Calculate the horizontal displacement of the ball in **Q1** after 0.25 s. *(1 mark)*

3 The ball in **Q1** takes 0.70 s before it lands on the ground below.
Calculate the height of the platform. *(3 marks)*

4 A bullet is fired from a gun at a horizontal velocity of $250\ \mathrm{m\,s^{-1}}$.
Calculate the horizontal and vertical displacements of the bullet after a time of 0.30 s in flight. *(4 marks)*

5 A golf ball has a velocity of $30\ \mathrm{m\,s^{-1}}$ at an angle of $45°$ to the horizontal ground.
Calculate its maximum height. *(3 marks)*

6 For the ball in **Q5**, calculate its horizontal displacement when it lands on the ground. *(3 marks)*

Chapter 3 Practice questions

1 A student plots a graph to determine the distance travelled by an object. The area of which graph is equal to the distance travelled?

A Force against time.

B Acceleration against time.

C Displacement against time.

D Velocity against time. *(1 mark)*

2 A ball is dropped from rest at time $t = 0$. Air resistance has negligible effect on the motion of the ball.

What is the distance travelled by the ball between $t = 2.0$ s and $t = 3.0$ s?

A 4.9 m

B 20 m

C 25 m

D 44 m *(1 mark)*

3 A student conducts an experiment on the acceleration of free fall g. Her lab book has the following information:

equation: $g = \frac{2s}{t^2}$% uncertainty in s = 3.0% % uncertainty in t = 1.2%

What is the percentage uncertainty in the value for g?

A 0.6%

B 1.8%

C 4.2%

D 5.4% *(1 mark)*

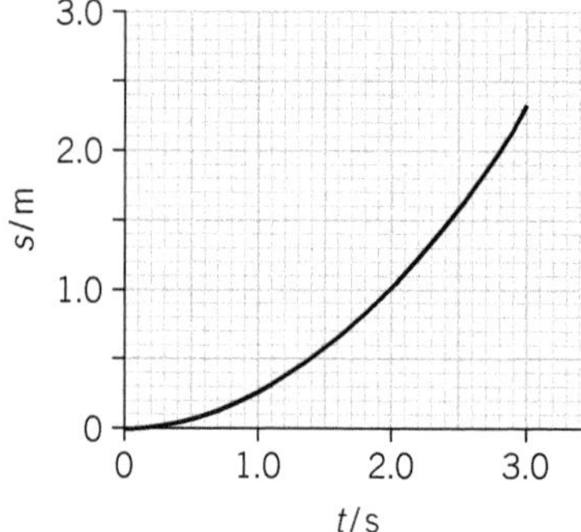

▲ **Figure 1**

4 **a** Define acceleration. *(1 mark)*

b Figure 1 shows the displacement–time graph plotted by a student for a ball rolling down a ramp.

i Use Figure 1 to describe and explain the motion of the ball. *(2 marks)*

ii Use Figure 1 to determine the velocity of the ball at time $t = 2.0$ s. *(3 marks)*

iii The ball was at rest at time $t = 0$.

Use your answer to (ii) to calculate the acceleration of the ball. *(2 marks)*

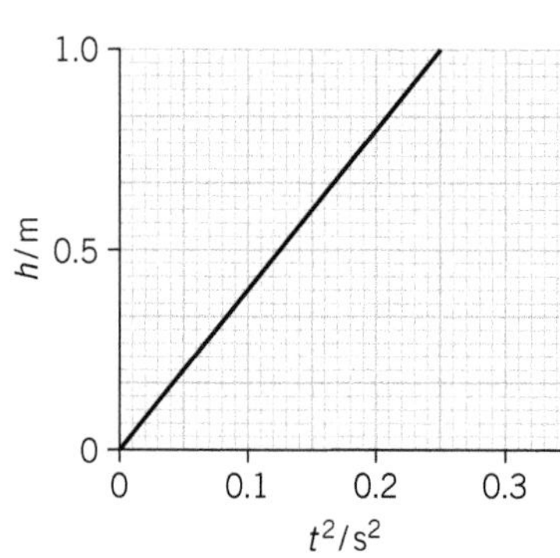

▲ **Figure 2**

5 A heavy metal ball is dropped from different heights. The time t taken to fall to the ground is recorded for each height h. Figure 2 shows a graph plotted to determine the acceleration of free fall g.

a Explain why plotting h against t^2 produces a straight-line graph through the origin. *(2 marks)*

b Use Figure 2 to determine a value for g. Explain your answer. *(3 marks)*

c State and explain the change, if any, to the graph in Figure 2 when the experiment is conducted with a much heavier metal ball. *(1 mark)*

Chapter 4 Forces in action

In this chapter you will learn about ...

- ☐ Mass and weight
- ☐ Centre of mass and centre of gravity
- ☐ Free body diagram
- ☐ Drag and terminal velocity
- ☐ Moment and torque
- ☐ Equilibrium
- ☐ Triangle of forces
- ☐ Density
- ☐ Pressure
- ☐ Archimedes' principle

FORCES IN ACTION
4.1 Force, mass, and weight
4.2 Centre of mass
4.3 Free-body diagrams

Specification reference: 3.2.1, 3.2.3

4.1 Force, mass, and weight

A resultant force F acting on an object of mass m will produce acceleration a. The resultant force, mass, and acceleration are related by the following equation:

$$\text{resultant force} = \text{mass} \times \text{acceleration}$$

or

$$F = ma$$

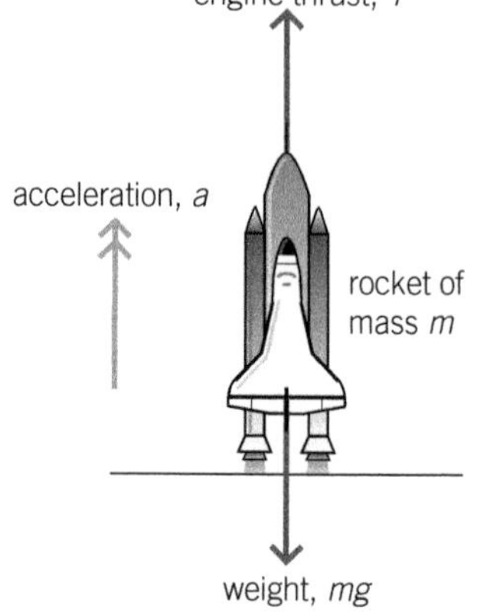

▲ **Figure 1** *What is the resultant force acting on this rocket?*

Force and acceleration are both vectors. The direction of the acceleration is in the same direction as the net force. The SI unit for force is the newton (N).

- For a constant mass, $F \propto a$
- For a constant acceleration, $F \propto m$
- For a constant force, $a \propto \frac{1}{m}$

The **newton** is defined as the resultant force that will give a mass of 1 kg an acceleration of 1 m s^{-2} in the direction of the force.

Weight

Mass and weight are not the same.

The **weight** of an object is defined as the gravitational force acting on the object.

An object in free fall on the surface of the Earth has an acceleration g. According to Newton's second law ($F = ma$ above), the weight W of the object must be equal to the product of the mass m of the object and the acceleration of free fall g. Therefore

$$W = mg$$

Figure 1 shows a rocket about to lift off. In order to determine its vertical acceleration, you need to determine the resultant force F on this rocket, where

$$F = \text{thrust} - \text{weight}$$

Revision tip

In the equation $F = ma$, F is the resultant force acting on the object. When $F = 0$, $a = 0$, the object can either be stationary or travelling at a constant velocity.

4.2 Centre of mass

On the Earth, the centre of gravity and the centre of mass are at the same point. In deep space, where there is no gravitational field, there is no centre of gravity, but the object has a centre of mass.

Centre of gravity is defined as a point where the entire weight of the object appears to act.

Centre of mass is defined as a point through which any externally applied force produces straight-line motion without any rotation.

Experiment to find the centre of gravity

You can determine the centre of gravity of a flat object using the following steps:

1. Suspend the object freely from a point (see Figure 2).
2. Use a plumb-line to draw a straight vertical line on the object.

3. Repeat step 1 using different suspension points.
4. Determine the point of intersection of these straight lines.
5. The intersection point is the centre of gravity.

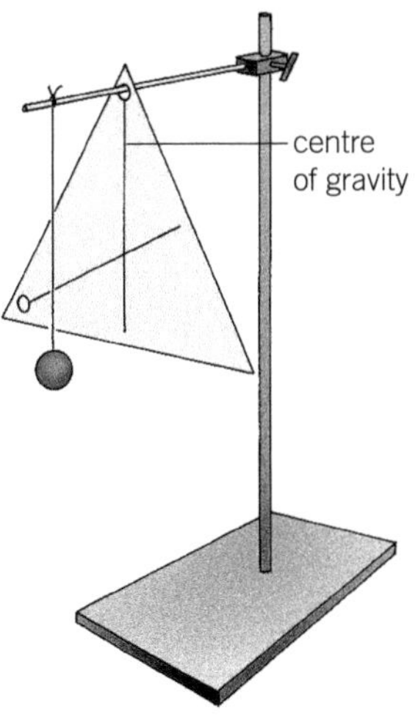

▲ **Figure 2** *Determining centre of gravity*

4.3 Free-body diagram

A free-body diagram shows all the forces acting on a particular object. The following named forces are very useful when drawing free-body diagrams:

thrust	weight	friction	normal contact force
drag	tension	upthrust	

Figure 3 shows a block on a ramp and all the forces acting on it.

Worked example: Block on a ramp

Use the free-body diagram shown in Figure 3 to determine the frictional force F and the normal contact force N.

Step 1: Always start with a neat free-body diagram – this has already been done for you.

Step 2: The block is stationary therefore the resultant force is zero. Determine F by resolving the 1.20 N force along the ramp.

force up the ramp = force down the ramp

$F = 1.20\cos 60°$ (The angle between the 1.20 N force and the ramp is 60°.)

$F = 0.60\,\text{N}$ (2 s.f.)

Step 3: Determine N by resolving the 1.20 N force perpendicularly to the ramp.

$N = 1.20\cos 30°$

$N = 1.0\,\text{N}$ (2 s.f.)

Synoptic link

The equation $F = ma$ is often referred to as Newton's second law. As you will see in Topic 7.3, Newton's second law of motion, this law is defined in terms of rate of change of momentum. $F = ma$ is just a special case for constant mass.

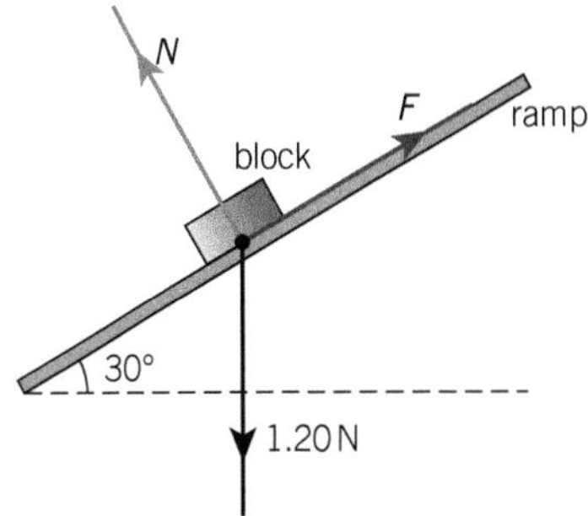

▲ **Figure 3** *Forces acting on a stationary block on a ramp*

Summary questions

1. A car of 1000 kg starts from rest. The constant forward force acting on the car is 400 N.
   Calculate the initial acceleration of the car. *(1 mark)*
2. After some time, the air resistance acting on the car in **Q1** is 150 N.
   Calculate the instantaneous acceleration of the car. *(2 marks)*
3. A car is travelling along a straight level road at a constant velocity. State and explain the resultant force acting on the car. *(2 marks)*
4. A 20 N force acting on an object produces an acceleration of $4.0\,\text{m s}^{-2}$. Predict its acceleration when the force is 65 N. *(2 marks)*
5. An electron experiences two forces acting at right angles. The magnitude of each force is 2.0×10^{-15} N. The mass of the electron is 9.1×10^{-31} kg. Calculate the acceleration of the electron. *(3 marks)*
6. The electron in **Q5** is initially at rest. Calculate the distance travelled after 20 ns of acceleration. *(2 marks)*

4.4 Drag and terminal velocity

An object falling in a vacuum on the Earth will have an acceleration of 9.81 $m\,s^{-2}$ because there is only one force acting on the object – its weight. Objects moving through the air will also experience a resistive force (drag) and an upthrust. The resultant force on the object is no longer its weight and therefore its acceleration cannot be 9.81 $m\,s^{-2}$.

Drag

Resistive forces originate when two materials rub against each other. The term drag is used for the resistive force experienced by an object moving through a fluid (liquid or gas). The factors affecting drag are:

- The speed of the object (the greater the speed, the greater is the drag).
- The cross-sectional area at right angles to the direction of travel (the larger the area, the greater is the drag).
- The type of fluid or its viscosity.

For many objects, drag force is directly proportional to speed^2. Table 1 shows how you can test the relationship between drag and speed – the ratio of force to speed^2 is a constant.

▼ **Table 1** *Drag is proportional to speed^2. A ten-fold increase in the speed of an object will increase the drag by a factor of 10^2 or 100*

speed / $m\,s^{-1}$	2.0	3.0	4.0	5.0	6.0
drag / N	3.2	7.2	12.8	20	28.8
$\frac{drag}{speed^2}$ / $N\,s^2\,m^{-2}$	0.80	0.80	0.80	0.80	0.80

Terminal velocity

For a given object of fixed area, the drag will simply depend on the speed of the falling object. Figure 1 shows a skydiver of mass m at three stages in his downward motion through the air.

Just before falling ⇒	During the flight ⇒	At terminal velocity ⇒
• speed = 0 • drag = 0 • resultant force F = weight • acceleration = g	• speed > 0 • drag > 0 • resultant force $F = mg -$ drag • acceleration = $\frac{mg - \text{drag}}{m}$	• speed = constant • drag = weight • resultant force = 0 • acceleration = 0

▲ **Figure 1** *Falling vertically through the air*

At the **terminal velocity**, the speed of the skydiver is constant and his acceleration is zero. The net force on the skydiver is zero therefore drag = weight (if we assume the upthrust to be negligible).

Figure 2 shows the velocity–time graph for an object falling through a fluid. Remember that the gradient of the graph is equal to acceleration. The acceleration is maximum and equal to g at time = 0 and the acceleration is zero at terminal velocity.

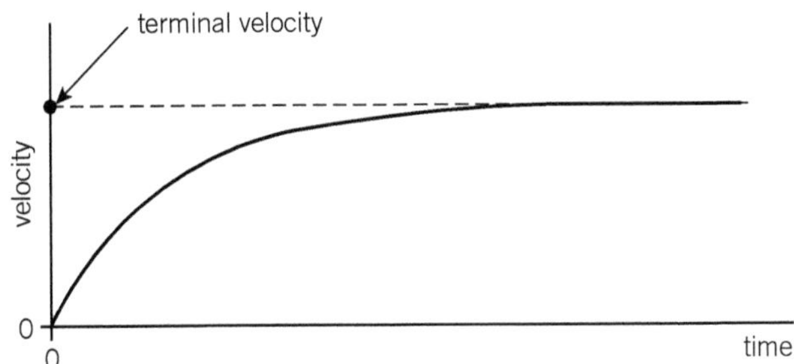

▲ **Figure 2** *Velocity–time graph for an object falling through a fluid*

Investigating drag

Figure 3 shows one possible arrangement for investigating the factors affecting drag.

Revision tip

Do not use the term 'gravity' to mean 'weight'.

Worked example: Drag

A paper cone of mass 5.0 g is dropped from a height of 1.00 m. It reaches terminal velocity almost immediately after release and spends 1.4 s in flight. Calculate the terminal velocity of the cone and the maximum drag.

Step 1: Calculate the terminal velocity v of the paper cone.

$$v = \frac{\text{distance}}{\text{time}} = \frac{1.00}{1.4} = 0.71\ \text{m s}^{-1}$$

Step 2: Use your knowledge of terminal velocity to determine drag.

At terminal velocity, resultant force = 0.

drag = weight = mg

drag = $5.0 \times 10^{-3} \times 9.81 = 4.9 \times 10^{-2}$ N (2 s.f.)

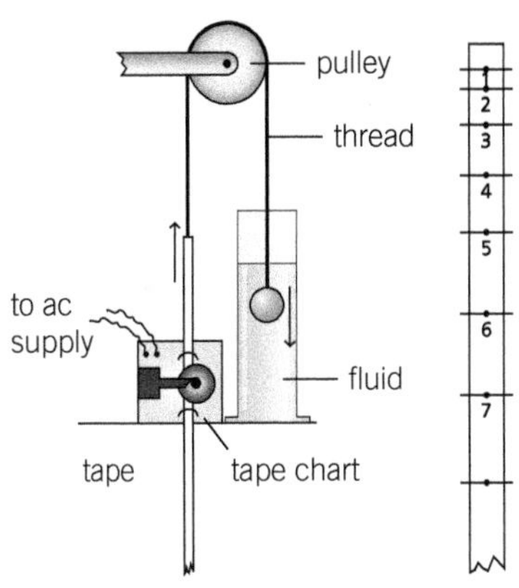

▲ **Figure 3** *The falling object is connected to a tape that passes through a ticker timer. The ticker timer produces 50 dots per second on the moving tape*

Summary questions

1 A small car and a large truck are travelling at the same speed. State and explain which of these will experience a greater drag force. *(1 mark)*

2 State the resultant force on an object travelling at terminal velocity. *(1 mark)*

3 A 2.0 kg mass is dropped from rest. Calculate the initial resultant force and acceleration of this mass. *(2 marks)*

4 A skydiver of total mass 70 kg is falling through air. The drag force acting on the skydiver is 300 N. Calculate the acceleration of the skydiver. *(3 marks)*

5 Use the data in Table 1 to derive a relationship between drag D and speed v. *(1 mark)*

6 The results shown in Table 1 are for an object of mass 100 g. Predict the terminal velocity of this object when falling through air. *(3 marks)*

4.5 Moments and equilibrium
4.6 Couples and torques

Specification reference: 3.2.3

4.5 Moments and equilibrium

A body is in equilibrium when the forces acting on the object do not produce any rotation or any acceleration. No acceleration means that the resultant force on the object must be zero. The object can either have a constant velocity or be stationary. As you will see later, the total moment acting on the object must also be zero.

Moment of a force

The 'turning effect' of a force acting on an object is called a moment. A moment is applied to a door about its hinges when you pull the door at its handle.

▲ **Figure 1** *Moment of the force is the product of the force F and the perpendicular distance x*

Moment of a force is defined as the product of the force applied and the perpendicular distance of the line of action of the force from the axis or point of rotation.

Therefore

moment = Fx

where F is the force and x is the perpendicular distance.

A moment can either be clockwise or anticlockwise. The SI unit for moment is N m, but you can also use N cm.

The principle of moments

The two conditions for an object in equilibrium are:

- resultant force = 0
- resultant moment = 0

The second condition is the same as the principle of moments.

Principle of moments: The sum of the clockwise moments about a point or an axis is equal to the sum of the anticlockwise moments about the same point or axis.

Figure 2 shows a uniform beam pivoted at its centre of gravity. It is in equilibrium and balanced horizontally by two weights W_1 and W_2. When taking moments about the pivot, we get

clockwise moment = anticlockwise moment

$$W_2d_2 = W_1d_1$$

The force at the pivot must act vertically upwards and equal to $W_1 + W_2$; this is because the net force on the beam must be zero.

Revision tip
For equilibrium, both the net force and net moment acting on the object must be zero.

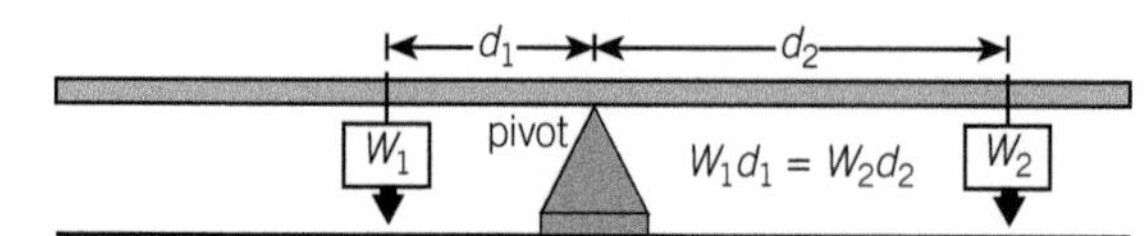

▲ **Figure 2** *A uniform beam in equilibrium*

Worked example: Equilibrium

Figure 3 shows a uniform shelf of weight 28 N held in equilibrium by a cord. Calculate the tension T in the cord.

Step 1: Determine the perpendicular distance d of the line of action of T from the left hand side of the shelf.

$\tan\theta = \frac{0.40}{0.60}$ $\quad\theta = \tan^{-1}(0.666...) = 33.7°$

$\sin 33.7° = \frac{d}{0.60}$ $\quad d = 0.6 \times \sin 33.7° = 0.33\,\text{m}$

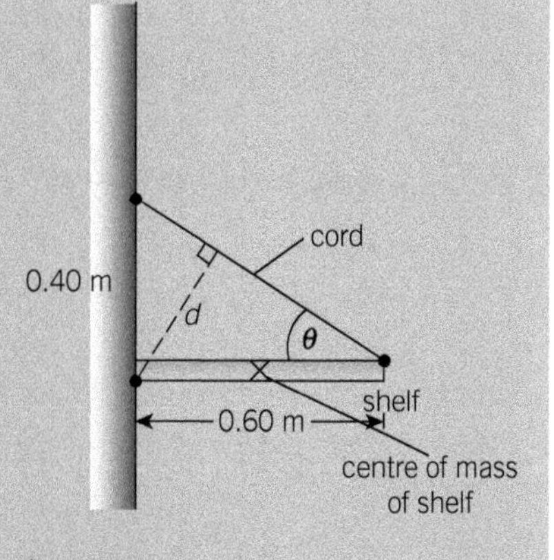

▲ **Figure 3**

Step 2: Use the principle of moments to determine T. The centre of gravity of the shelf is at its midpoint – it is 0.30 m from the end of the shelf.

Take moments about the left hand side of the shelf.

clockwise moment = anticlockwise moment

$28 \times 0.30 = 0.33 \times T$

$T = 25\,\text{N}$ (2 s.f.)

4.6 Couples and torques

A couple in physics means two equal but opposite forces acting on an object. The forces must be parallel and along different lines. See Figure 4.

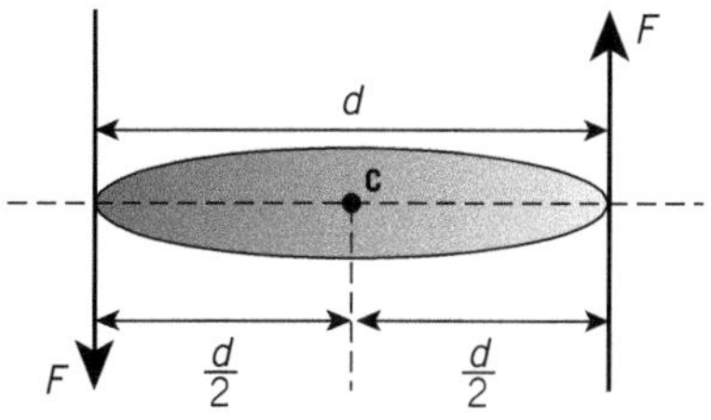

Torque of a couple

The **torque** of a couple is defined as the product of one of the forces and the perpendicular separation between the forces.

$$\text{torque} = Fd$$

The torque of a couple is the same as the total moment of the two forces about their midpoint, c. Torque is also measured in N m.

▲ **Figure 4** *A couple gives rise to a torque. The magnitude of the torque is Fd.*

Summary questions

1. A 2.0 N force acts at a perpendicular distance of 10 cm from the pivot. Calculate the moment of the force. *(1 mark)*
2. The moment of a force is 20 N m. The magnitude of the force is 400 N. Calculate the perpendicular distance of the force from the pivot. *(2 marks)*
3. A couple has two 3.0 N forces with a perpendicular separation of 15 cm. Calculate the torque of the couple. *(1 mark)*
4. In Figure 2, $W_1 = 1.2$ N, $d_1 = 20$ cm and $W_2 = 0.80$ N. Calculate:
   a. the distance d_2 in cm; *(2 marks)*
   b. the force at the pivot. *(1 mark)*
5. Figure 5 shows a beam of weight W in equilibrium. Determine expressions for the vertical contact forces S_x and S_y in terms of d_x, d_y, and W. *(4 marks)*
6. The beam in Figure 5 has mass 100 N, $d_x = 0.80$ m, and $d_y = 1.20$ m. Calculate the values of the forces S_x and S_y. *(2 marks)*

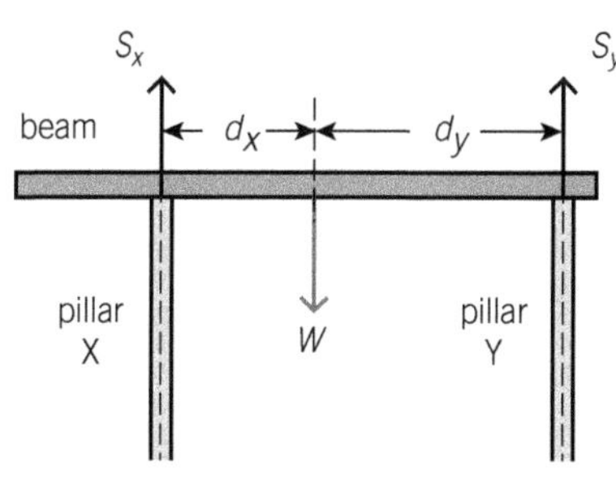

▲ **Figure 5**

4.7 Triangle of forces

For a point object to be in equilibrium the resultant force must be zero. In Topics 2.3, Scalar and vector quantities and 2.4, Adding vectors, you saw the rules for constructing a vector diagram to determine the resultant of two vectors. These rules can be extended to several forces. If an object is in equilibrium under the action of three coplanar (in the same plane) forces, then the vector diagram will form a closed triangle. This triangle is called a triangle of forces, see Figure 1.

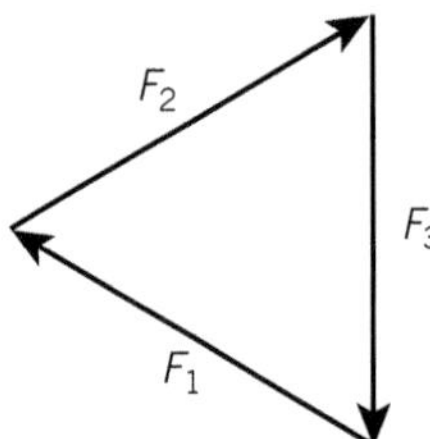

▲ **Figure 1** *Three forces acting on a point object in equilibrium will produce a closed vector triangle or a triangle of forces*

Equilibrium

Figure 2 shows the free-body diagram for a point object subjected to three forces. The object is in equilibrium. The resultant force on the object is zero. You can therefore construct a triangle of forces. Notice that the arrows on this diagram are all in a cyclic order.

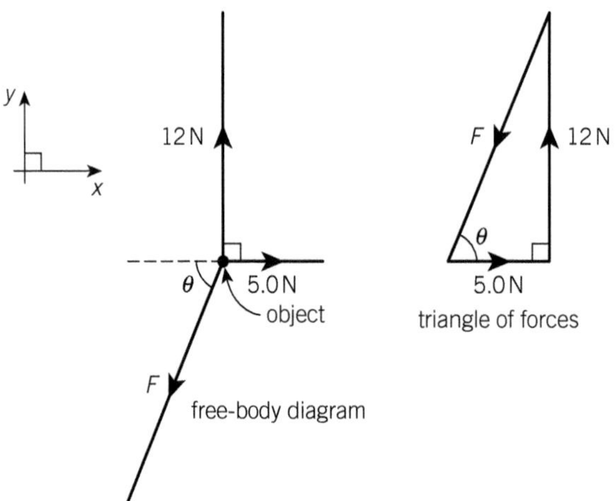

▲ **Figure 2** *A triangle of forces can be constructed from a free-body diagram. Notice that the resultant force of any two forces has a magnitude equal to the third force and has a direction opposite to this third force*

Resolving forces

There is an alternative to drawing a triangle of forces – you can resolve forces. When the resultant force is zero, then the total force in two mutually perpendicular directions must individually also be zero.

For the object in Figure 2 this implies:

In the x-direction the total force = 0 therefore $F\cos\theta = 5.0$

In the y-direction the total force = 0 therefore $F\sin\theta = 12$

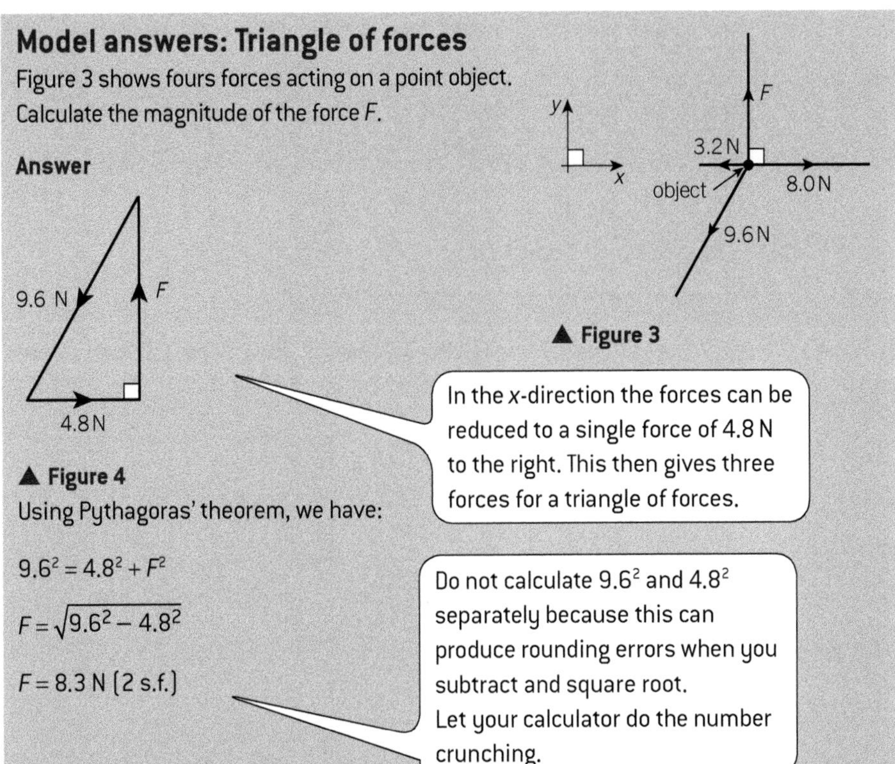

Model answers: Triangle of forces

Figure 3 shows fours forces acting on a point object. Calculate the magnitude of the force F.

▲ Figure 3

Answer

▲ Figure 4

In the x-direction the forces can be reduced to a single force of 4.8 N to the right. This then gives three forces for a triangle of forces.

Using Pythagoras' theorem, we have:

$9.6^2 = 4.8^2 + F^2$

$F = \sqrt{9.6^2 - 4.8^2}$

$F = 8.3$ N (2 s.f.)

Do not calculate 9.6^2 and 4.8^2 separately because this can produce rounding errors when you subtract and square root. Let your calculator do the number crunching.

Common misconception

Check the angles carefully and only use the Pythagoras theorem when the triangle of forces has a right angle.

Maths: Sine and cosine rules

Sine rule: $\dfrac{a}{\sin A} = \dfrac{b}{\sin B} = \dfrac{c}{\sin C}$

Cosine rule: $a^2 = b^2 + c^2 - 2bc \cos A$

Summary questions

1. Three coplanar forces act on an object. The object is moving at a constant velocity. What is the resultant of these three forces? *(1 mark)*
2. Figure 5 shows the forces S, F, and W acting on a block of wood resting on a ramp. Draw a triangle of forces. *(2 marks)*

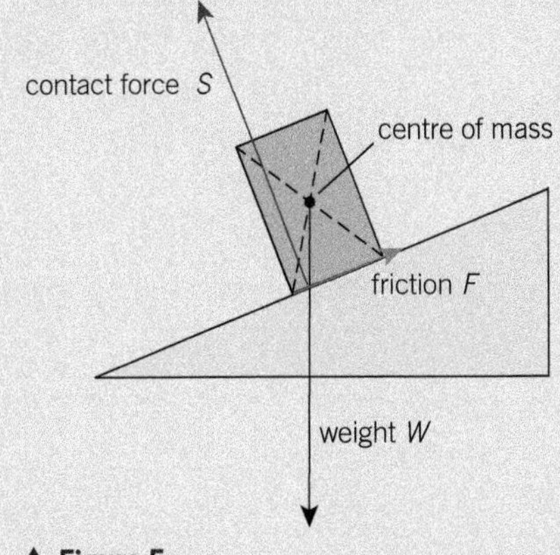

▲ Figure 5

3. Use Figure 2 to determine the magnitude of the force F and the angle θ. *(4 marks)*
4. Use Figure 4 to determine the angle between the 9.6 N force and the x-direction. *(2 marks)*
5. An object of 20 kg is supported by two cables. The angle made by each cable is 20° to the vertical. The object is at rest and the tension in each cable is T. Use your knowledge of triangle of forces to calculate the magnitude of the tension T. *(3 marks)*
6. Determine the tension T in **Q5** by resolving forces in the vertical direction. *(3 marks)*

4.8 Density and pressure
4.9 $p = h\rho g$ and Archimedes' principle

Specification reference: 3.2.4

4.8 Density and pressure

Lead is much denser than water, but water is much denser than air.

A book resting on a table exerts pressure on the table. This pressure is related to the weight of the book and its cross-sectional area.

Density

The **density** of a substance is defined as its mass per unit volume.

The word equation for density is:

$$\text{density} = \frac{\text{mass}}{\text{volume}}$$

This may be written as

$$\rho = \frac{m}{V}$$

where ρ is the density, m is the mass, and V is the volume. The SI unit for density is $\mathrm{kg\,m^{-3}}$.

Revision tip

$1\,\mathrm{cm^3} = (10^{-2}\,\mathrm{m})^3 = 10^{-6}\,\mathrm{m^3}$

$1\,\mathrm{mm^2} = (10^{-3}\,\mathrm{m})^3 = 10^{-9}\,\mathrm{m^3}$

Determining density

In order to calculate density, you need to determine the mass of the substance and its volume.

- The mass of a substance can be measured using a top pan balance.
- The volume of a regular solid can be calculated once its dimensions are measured using a ruler, a vernier calliper, or a micrometer. For an irregular solid, its volume can be determined from the volume of water displaced when it is completely submerged. The volume of the displaced water can be measured directly using a measuring cylinder.

Pressure

Pressure is the normal force exerted on a surface per unit cross-sectional area.

The word equation for pressure is:

$$\text{pressure} = \frac{\text{force}}{\text{cross-sectional area}}$$

This may be written as

$$p = \frac{F}{A}$$

where p is the pressure, F is the force, and A is the cross-sectional area. The SI unit for pressure is $\mathrm{N\,m^{-2}}$ or pascal (Pa). $1\,\mathrm{Pa} = 1\,\mathrm{N\,m^{-2}}$.

4.9 $p = h\rho g$ and Archimedes' principle

Imagine swimming under water. You will experience pressure from the water and also an upward force called upthrust.

$$p = h\rho g$$

Figure 1 shows a fluid column of height h and cross-sectional area A. The density of the fluid is ρ.

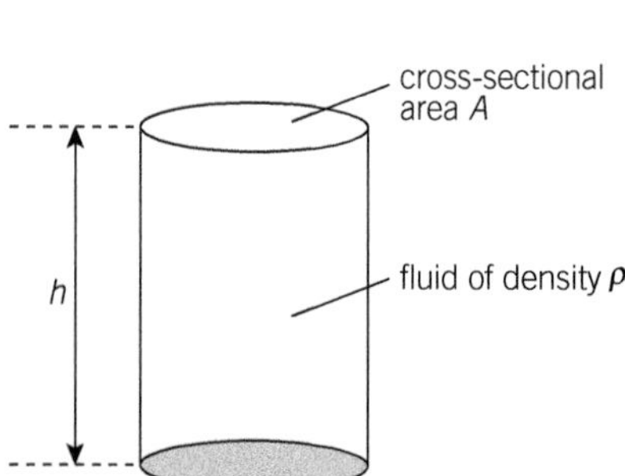

▲ **Figure 1** *A column of fluid exerts pressure at its base*

pressure at the base = weight of column/cross-sectional area

$p = \frac{mg}{A}$ (weight = mass × g = mg)

$p = \frac{(\rho V)g}{A}$ ($m = \rho$ × volume of column = ρV)

$p = \frac{V}{A}\rho g$

$p = h\rho g$ (height h of column = V/A)

The pressure p exerted by the column is independent of its cross-sectional area and $p \propto h$.

Archimedes' principle

An object submerged in a fluid (liquid or gas) experiences an upward force – the upthrust, because the pressure at the bottom surface of the object is greater than the pressure at its top surface.

Archimedes' principle: The upthrust exerted on a body immersed in a fluid, whether fully or partially submerged, is equal to the weight of the fluid that the body displaces.

Worked example: Upthrust

An empty balloon of weight 0.024 N is filled with helium until its volume is 2.0×10^{-2} m³. Calculate the net upward force on the balloon when it is released.

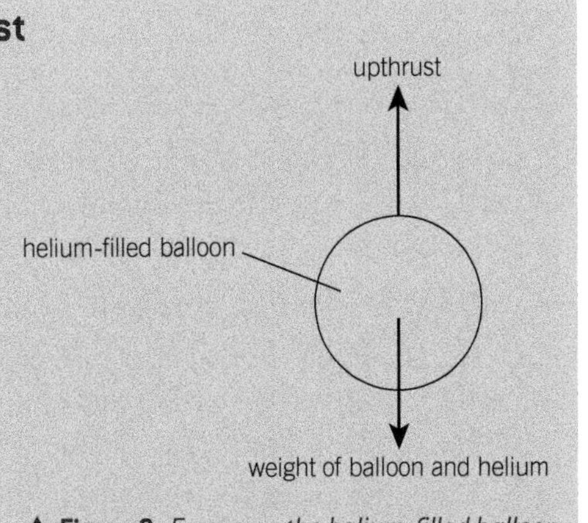

▲ **Figure 2** *Forces on the helium-filled balloon*

density of helium = 0.18 kg m^{-3}

density of air = 1.3 kg m^{-3}

Step 1: Calculate the total downward force on the balloon.

total downward force = weight of balloon + weight of helium

total downward force = 0.024 + $(2.0 \times 10^{-2} \times 0.18) \times 9.81$ = 0.0593 N

Step 2: Use Archimedes' principle to calculate the upthrust and therefore the net force.

upthrust = weight of air displaced = $(2.0 \times 10^{-2} \times 1.3) \times 9.81$ = 0.255 N

net upward force = 0.255 – 0.0593 = 0.20 N (2 s.f.)

Summary questions

Data required: density of water = 1.0×10^3 kg m^{-3} atmospheric pressure = 1.0×10^5 Pa

1 Calculate the density of a liquid of volume 0.50 m³ and mass 350 kg. *(1 mark)*

2 Calculate the pressure at the base of a 3.0×10^4 kg concrete column of cross-sectional area of 1.2 m². *(2 marks)*

3 Calculate the vertical force on a circular surface of radius 15 cm due to the atmosphere. *(2 marks)*

4 Calculate the total pressure at the bottom of a swimming pool where the depth of water is 3.0 m. *(3 marks)*

5 The mass of the Earth is 6.0×10^{24} kg and it has a radius of 6.4 Mm. Calculate the average density of the Earth. *(2 marks)*

6 Mercury is 13.6 times denser than water. Calculate the height of a mercury column that provides at its base a pressure equivalent to atmospheric pressure. *(2 marks)*

Chapter 4 Practice questions

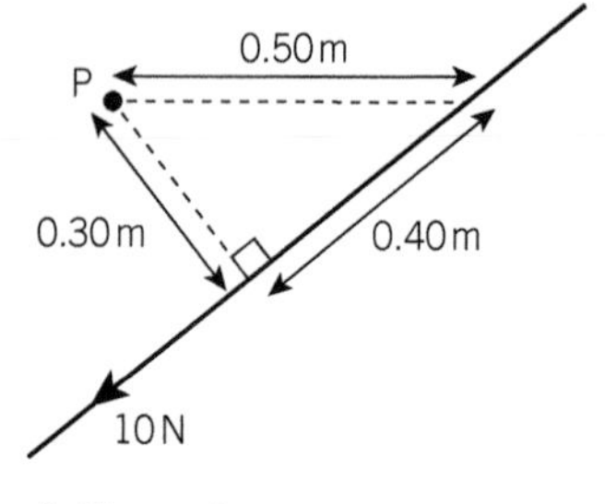

▲ **Figure 1**

1 Which is the correct definition for the centre of gravity of an object?

The centre of gravity is the point where ...

A the middle of the object happens to be

B all the mass appears to be

C the weight of the objects appears to be

D the net force is zero. *(1 mark)*

2 What is the moment of the force about the pivot P in Figure 1?

A 3.0 N

B 4.0 N

C 5.0 N

D 12 N *(1 mark)*

3 Which base units are used to measure pressure?

A $kg\,m^{-2}$

B $kg\,m\,s^{-1}$

C $kg\,m^{3}\,s^{-2}$

D $kg\,m^{-1}\,s^{-2}$ *(1 mark)*

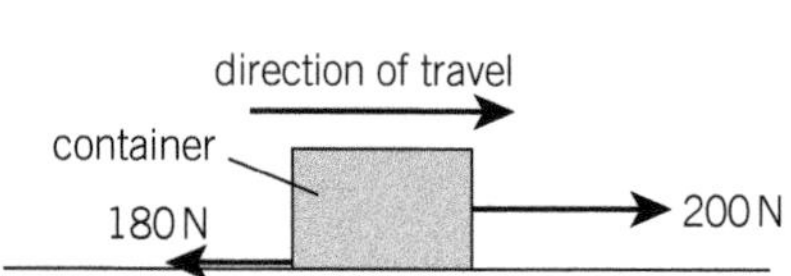

▲ **Figure 2**

4 A container is pulled along horizontal ground. Figure 2 shows the pulling force of 200 N and the frictional force of 180 N acting on the container.

The mass of the container is 400 kg.

a Calculate its horizontal acceleration. *(2 marks)*

b Explain why the weight and normal contact force have no effect on the horizontal acceleration of the container. *(1 mark)*

c The surface area of the container is $1.8\,m^2$. Calculate the pressure it exerts on the ground. *(2 marks)*

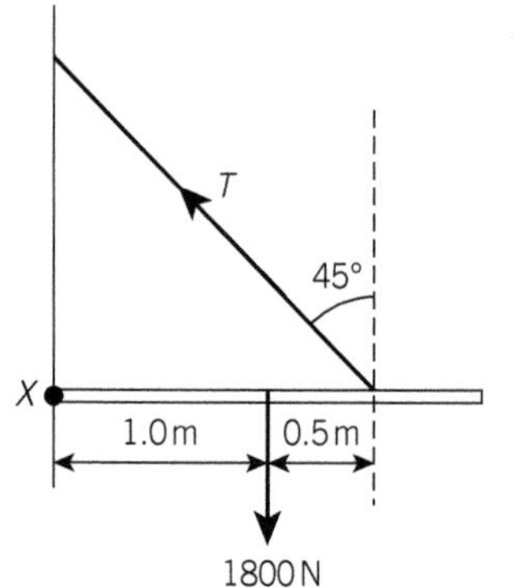

▲ **Figure 3**

5 **a** State the principle of moments. *(1 mark)*

b Figure 3 shows two of the forces acting on a uniform board.

The tension *T* in the cable is at an angle of 45° to the vertical and the weight of the board is 1800 N.

i The density of the material used to make the board is $900\,kg\,m^{-3}$. Calculate the volume of the board. *(3 marks)*

ii Calculate the tension *T* in the cable. *(3 marks)*

iii Explain why the force at **X** must have a component in the horizontal direction. *(1 mark)*

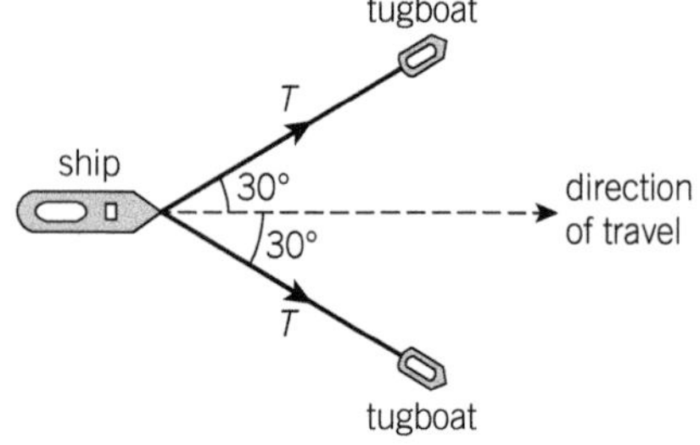

▲ **Figure 4**

6 A ship is towed at **constant** speed by two tugboats, see Figure 4.

The tension *T* in each cable attached to the ship is 8.6 kN.

a State and explain the resultant force acting on the ship. *(1 mark)*

b Calculate the total force due to the tensions in the cables. *(3 marks)*

c State and explain the direction and magnitude of the drag acting on the ship. *(2 marks)*

Chapter 5 Work, energy, and power

In this chapter you will learn about ...

- ☐ Work done
- ☐ Energy
- ☐ Conservation of energy
- ☐ Kinetic energy
- ☐ Gravitational potential energy
- ☐ Power
- ☐ Efficiency

5 WORK, ENERGY, AND POWER
5.1 Work done and energy
5.2 Conservation of energy

Specification reference: 3.3.1

5.1 Work done and energy

Work is done when the point of application of a force moves. Work done on an object is linked to energy transfer of the object. A force does no work if there is no motion.

Work done

Work done is defined as the product of force F and the distance x moved in the direction of the force.

The word equation for work done is:

work done = force × distance moved in the direction of the force

The equation for work done is $W = F \times x$

Work done is a scalar quantity. The SI unit for work done is N m or joule (J).

$$1\,\text{J} = 1\,\text{N m}.$$

1 **joule** is defined as the work done by a force of 1 N when the point of application of the force moves a distance of 1 m in the direction of the force.

If the force acts at an angle θ to the distance moved (see Figure 1), then the work done by the force is:

work done = component of force in direction of travel × distance moved

$$W = F\cos\theta \times x \qquad \text{or} \qquad W = Fx\cos\theta$$

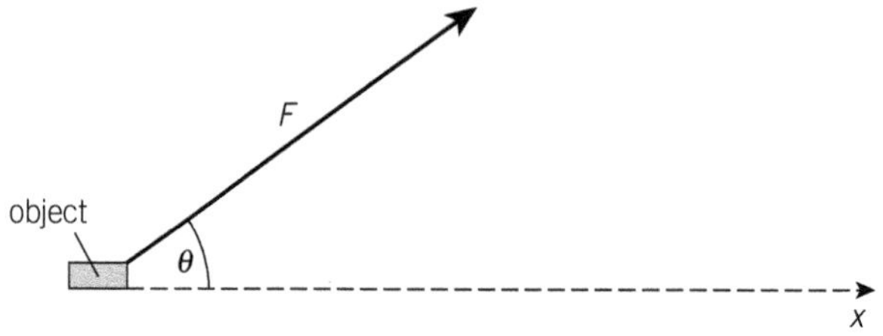

▲ **Figure 1** *Work done is given by the equation $W = Fx\cos\theta$*

5.2 Conservation of energy

Energy can take many different forms – kinetic energy, gravitational potential energy, elastic potential energy, thermal (heat) energy, sound energy, and so on. A car moving on a hill possesses both kinetic energy and gravitational potential energy. The chemical energy in the fuel produces lots of thermal energy.

Energy

Energy is the capacity to do work.

The work done on an object is equal to the energy transferred, that is,

work done = energy transferred

This means that 600 J of work done in lifting a crate vertically will increase the gravitational potential energy of the crate by 600 J.

Revision tip

The term 'potential' in physics is used to mean 'stored'.

Principle of conservation of energy

Energy can be converted from one form to another. For example, an object falling through the air on the Earth converts gravitational potential energy to kinetic energy and thermal energy. However, the total energy of the system remains the same.

The principle of **conservation of energy** states that the total energy of a closed system remains constant – energy can never be created or destroyed, but it can be transferred from one form to another.

Revision tip

A closed system is one where there are no external forces acting.

Worked example: Rough surface

A rope is used to drag a heavy box along a horizontal surface at a constant speed. The tension in the rope is 38 N. The rope makes an angle of 42° to the horizontal. The crate moves a horizontal distance of 15 m. Calculate the work done on the box.

Step 1: Write down the important quantities given in the question.

$F = 38$ N $\quad x = 15$ m $\quad \theta = 42°$

Step 2: Use the equation $W = Fx\cos\theta$ to calculate the work done by the force.

$W = Fx\cos\theta$

$W = 38 \times 15 \times \cos 42° = 420$ N (2 s.f.)

Summary questions

1 An object of weight 20 N falls through a vertical distance of 5.0 m. Calculate the work done by the weight. *(1 mark)*

2 The falling object in **Q1** transfers 30 J of energy to the surrounding air. Determine the kinetic energy of the object after falling 5.0 m. *(1 mark)*

3 A 100 N force acting at an angle of 45° to the horizontal moves an object through a horizontal distance of 20 m. Calculate the work done by the force. *(2 marks)*

4 A ball is rolling along on a smooth horizontal surface at a constant velocity. Explain why no work is done by the weight of the ball. *(2 marks)*

5 A force of 20 N acts on an object. The work done by the force is 100 J. The distance moved by the object is 8.0 m. Calculate the angle between the force and direction of motion of the object. *(2 marks)*

6 A 2.0 kg metal sphere falls through a vertical distance of 10 m. The kinetic energy of the ball just before it hits the ground is 120 J. Calculate the average drag force acting on the falling sphere. *(4 marks)*

5.3 Kinetic energy and gravitational potential energy

Specification reference: 3.3.2

5.3 KE and GPE

In mechanics, you need to be familiar with both kinetic energy (KE) and gravitational potential energy (GPE). KE is associated with a moving object and GPE is linked to the position of an object in a gravitational field. GPE is energy 'stored' by an object. In Topic 6.2, you will come across elastic potential energy – energy 'stored' when the shape of an object is changed.

Kinetic energy

The kinetic energy E_k of a moving object is given by the equation

$$E_k = \frac{1}{2}mv^2$$

where m is the mass of the object and v is its speed.

Kinetic energy is a scalar quantity. The base units of kinetic energy are $kg\,m^2\,s^{-2}$, which is the same as joule (J). Kinetic energy of an object is directly proportional to speed^2 – increasing the speed by a factor of 10 will increase its kinetic energy by a factor of $10^2 = 100$.

Common misconception

In the equation $E_k = \frac{1}{2}mv^2$, only the speed v is squared.
A common error made is squaring the term '$\frac{1}{2}mv$'. Just be vigilant.

Maths: Standard form or scientific notation

For an object with initial speed u and final speed v, the change in kinetic energy is $\frac{1}{2}m(v^2 - u^2)$.

This is not the same as $\frac{1}{2}m(v - u)^2$.

Model answer: KE of an accelerating car

A car of mass 900 kg is travelling along at a velocity of $5.0\,m\,s^{-1}$. The car accelerates at $3.0\,m\,s^{-2}$ for 4.0 s. Calculate the change in its kinetic energy in this period of time.

Answer

Final speed:

$u = 5.0\,m\,s^{-1}$ $a = 3.0\,m\,s^{-2}$ $t = 4.0\,s$

$v = u + at = 5.0 + 3.0 \times 4.0 = 17\,m\,s^{-1}$

You need to calculate the final speed of the car using one of the equations of motion.

Initial KE: $E_k = \frac{1}{2}mv^2 = \frac{1}{2} \times 900 \times 5.0^2$

Final KE: $E_k = \frac{1}{2}mv^2 = \frac{1}{2} \times 900 \times 17^2$

Change in kinetic energy = ΔE_k

$\Delta E_k = \frac{1}{2} \times 900 \times (17^2 - 5.0^2)$

$\Delta E_k = 1.2 \times 10^5$ J (2 s.f.)

You can calculate the individual kinetic energies and then subtract. This can lead to rounding and transfer errors.

It is much neater to do the calculation as a single step on your calculator.

Gravitational potential energy

Figure 1 shows an object of mass m falling through a vertical distance h in a vacuum. The work done by the weight must equal the change in the gravitational potential energy E_p. Therefore

$$E_p = \text{force} \times \text{distance moved in the direction of the weight}$$

$$E_p = (mg) \times h$$

$$E_p = mgh$$

Gravitational potential energy is a scalar quantity. The SI unit is joule (J).

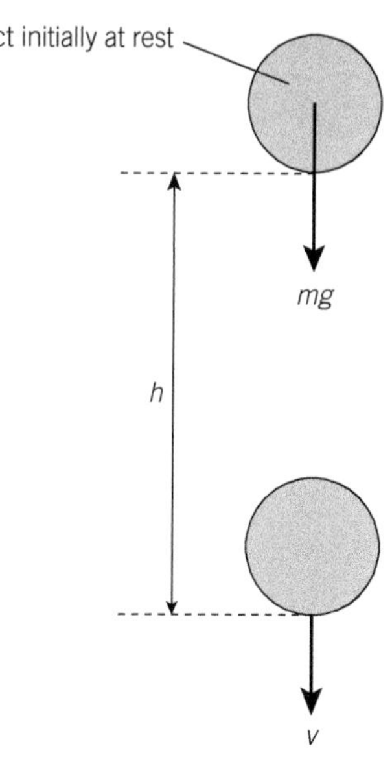

▲ **Figure 1** *The work done by the falling weight is mgh*

KE and GPE changes

There are many situations where energy is transferred between kinetic energy and gravitational potential energy. Examples include a rollercoaster, a swinging pendulum, and a ball rolling up or down a slope.

In Figure 1, the object is initially at rest and has speed v after falling through the vertical distance h.

According to the principle of conservation of energy, the sum of the KE and GPE of the object at any instant must be constant, as long as there are no thermal losses. This is equivalent to:

change in GPE = change in KE

$$mgh = \frac{1}{2}mv^2$$

The mass m cancels on both sides of the equation. Therefore

$$v^2 = 2gh$$

Summary questions

1 Calculate the change in the gravitational potential energy of a 1500 kg lift travelling through a vertical distance of 310 m. *(1 mark)*

2 Calculate the kinetic energy of a dust particle of mass 1.0 g travelling at $120\ \text{km s}^{-1}$ in deep space. *(2 marks)*

3 A stone falls vertically. The change in its GPE is 20 J. Explain the change in the KE of the stone. State any assumption made. *(2 marks)*

4 A 1000 kg car has kinetic energy 50 kJ. Calculate its speed. *(2 marks)*

5 An electron is travelling at a speed of $1.2 \times 10^5\ \text{m s}^{-1}$. Its kinetic energy is increased by 8.0×10^{-21} J. The mass of an electron is 9.1×10^{-31} kg. Calculate its final speed. *(4 marks)*

6 Figure 2 shows a simple pendulum. The object is released from rest from a height of h_0. At a height h it has speed v.

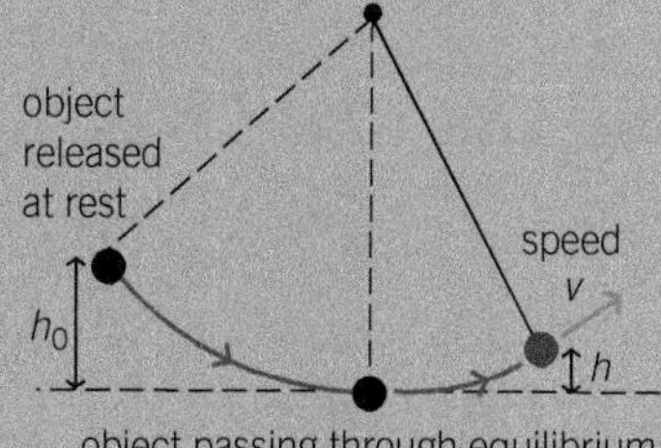

▲ **Figure 2**

a Describe the energy changes taking place as the object travels from its maximum height to the height h. *(2 marks)*

b Calculate v given $h_0 = 3.0$ m and $h = 0.20h_0$. *(4 marks)*

5.4 Power and efficiency

Specification reference: 3.3.3

5.4 Power and efficiency

The term power is the most frequently misused term in physics examinations. Though linked to energy or work done, power has a precise definition, so you must use it carefully. The efficiency of a device or a mechanical system cannot be greater than 100%. If you get an answer greater than 100% in a calculation, then you have definitely gone wrong and need to double-check your working.

Power

Power is defined as the rate of work done or the rate of energy transfer.

The word equation for power is

$$\text{power} = \frac{\text{work done}}{\text{time}} \quad \text{or} \quad \text{power} = \frac{\text{energy transfer}}{\text{time}}$$

This may be written as

$$P = \frac{W}{t}$$

where P is the power, W is the work done (or energy transfer), and t is the time taken. The SI unit for power is J s^{-1} or watt (W).

1 **watt** is defined as 1 joule of work done per second.

The equation above may also be written as $W = Pt$

A 24 W table lamp will transfer 24 J of energy per second. In 2 s, the energy transfer is $24 \times 2 = 48$ J, and so on.

Maths: Rate

The term 'rate' in physics means 'per unit time' or simply 'divided by time'. So rate of work done can be written as $\frac{\text{work done}}{\text{time}}$

Common misconception

The letter W is used for work done (or energy transfer). It must not be confused with W (watt), which is the unit for power.

Worked example: Car power

A 1000 kg electric car, starting from rest, can reach a top speed of $30\ \text{m s}^{-1}$ in 4.5 s. Calculate the average output power of the car in kW.

Step 1: Calculate the change in the kinetic energy of the car.

$$E_k = \frac{1}{2}mv^2 = \frac{1}{2} \times 1000 \times 30^2 = 4.5 \times 10^5\ \text{J}$$

Step 2: Use the equation for power to calculate the average output power of the car in watts.

P = energy transfer/time

$$P = \frac{4.5 \times 10^5}{4.5}$$

$P = 1.0 \times 10^5$ W

Step 3: Convert the power from W to kW using 1 kW = 10^3 W.

$$\text{power} = \frac{1.0 \times 10^5}{10^3} = 100\ \text{kW (2 s.f.)}$$

Power and motion

Consider a car travelling at a constant speed v on a level road. The constant speed means that the car has zero acceleration and the net force on the car must be equal to zero. The forward force F acting on the car must be equal to the total resistive force, see Figure 1. In this situation all the work done by the force F is transferred not to the kinetic energy of the car, but to thermal losses.

▲ **Figure 1** *The output power of the car is given by the equation $P = Fv$*

The output power P of the car can be determined as follows:

$$\text{power} = \frac{\text{work done}}{\text{time}}$$

$$\text{power} = \text{force} \times \frac{\text{distance}}{\text{time}}$$

$$\text{power} = \text{force} \times \left(\frac{\text{distance}}{\text{time}}\right)$$

or

$$P = Fv$$

where v is the speed of the car.

Efficiency

The efficiency of a system or a device is defined by the following equations:

$$\text{efficiency} = \frac{\text{useful output energy}}{\text{total input energy}} \times 100\%$$

or

$$\text{efficiency} = \frac{\text{useful output power}}{\text{total input power}} \times 100\%$$

It is important to remember the following:

- Efficiency can be expressed either as a percentage or as a number between 0 and 1.0.
- The efficiency of a system cannot exceed 100% because this would violate the principle of conservation of energy.
- The efficiency of most mechanical systems is less than 100% because of thermal losses caused by frictional losses.

Summary questions

1 The total input energy to a motor is 56 J and the useful output energy is 2.4 J. Calculate the efficiency of the motor. *(1 mark)*

2 The input power to a table lamp is 24 W. Calculate the input energy to the lamp in a period of 1.0 hour. *(2 marks)*

3 The output power of a car is 6.0 kW. The car is travelling on a level road at a constant speed of 20 m s^{-1}. Calculate the forward force acting on the car. *(2 marks)*

4 The useful output power to a device is 20 W. The efficiency of the device is 0.25. Calculate the total input power to the device. *(2 marks)*

5 A motor can lift a 100 g mass through a height of 120 cm in a time of 2.4 s.

a Calculate the output power of the motor. *(3 marks)*

b The motor has an efficiency of 12%. Calculate the input power to the motor. *(2 marks)*

6 The drag force acting on a car is directly proportional to speed^2. The speed of the car is doubled. Calculate the factor by which the output power of the car would increase. *(3 marks)*

Chapter 5 Practice questions

1 Which is the correct unit for power?

A J s
B J s^{-1}
C N m
D N m^{-1} *(1 mark)*

2 The mass of a truck is doubled and its speed is also doubled.

What is the factor by which its kinetic energy increases?

A 2
B 4
C 8
D 16 *(1 mark)*

3 A load of mass 40 kg is lifted vertically at a constant speed of 1.5 m s^{-1}.

What is the rate of work done in lifting the load?

A 45 W
B 60 W
C 440 W
D 590 W *(1 mark)*

4 A pendulum bob is released from rest from a vertical height h. The speed of the bob at the bottom of the swing is v.

What is the correct equation for v and h?

A $v = gh$
B $v = 2gh$
C $v^2 = 2gh$
D $v^2 = gh$ *(1 mark)*

5 **a** State the principle of conservation of energy. *(1 mark)*

b A metal ball of mass 0.040 kg is dropped into a viscous liquid. After travelling a vertical distance of 0.080 m the ball has a velocity of 0.90 m s^{-1}. Calculate:

i the loss in the gravitational potential energy of the ball; *(1 mark)*

ii the kinetic energy of the ball; *(1 mark)*

iii the average work done against drag; *(2 marks)*

iv the average value for drag. *(2 marks)*

6 Figure 1 shows a trolley of mass 0.80 kg being pulled along a horizontal table by a falling object of mass 0.21 kg.

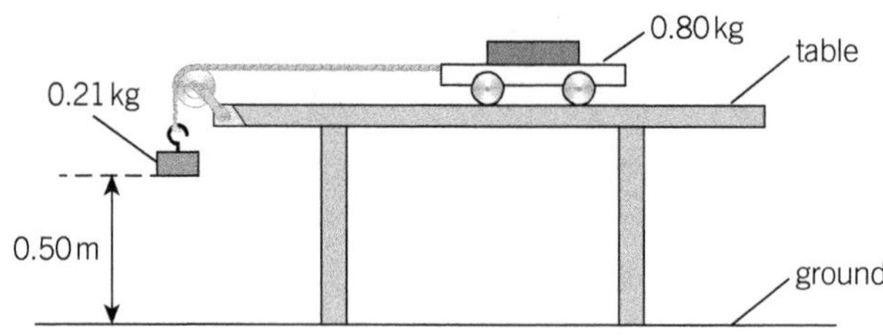

▲ **Figure 1**

The trolley is held at rest. The string attached to the trolley is taut. The 0.21 kg mass is 0.50 m above the ground.

a Show that the acceleration of the trolley is about 2 m s^{-2}. *(2 marks)*

b Calculate the kinetic energy of the trolley just before the 0.21 kg mass hits the ground. *(3 marks)*

c Calculate the change in the gravitational potential energy of the 0.21 kg mass. *(1 mark)*

d Explain why the answers in (b) and (c) are not the same. *(1 mark)*

Chapter 6 Materials

In this chapter you will learn about ...

- [ ] Hooke's law
- [ ] Force constant
- [ ] Elastic potential energy
- [ ] Force–extension graphs
- [ ] Stress and strain
- [ ] Elastic and plastic behaviour
- [ ] Stress-strain graphs
- [ ] The Young modulus

6 MATERIALS
6.1 Springs and Hooke's law
6.2 Elastic potential energy

Specification reference: 3.4.1, 3.4.2

6.1 Springs and Hooke's law

A spring or a wire can be stretched by a pair of equal and opposite forces; such forces are called tensile forces. Forces that squash a spring or an object are called compressive forces. Hooke's law can be used to model the behaviour of springs and wires.

Hooke's law

Hooke's law: The extension of the spring is directly proportional to the force applied. This is true as long as the elastic limit of the spring is not exceeded.

The extension x is the difference between the final length of the spring and its original length.

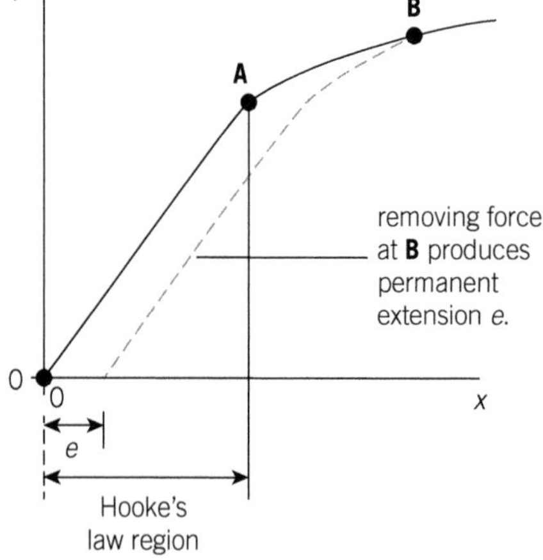

▲ **Figure 1** *Force F against extension x graph for a spring*

Force constant

Figure 1 shows the force against extension graph for a spring (or a wire).

- The spring shows elastic behaviour up to point A (elastic limit).
  - **Elastic:** Removal of the force will return the spring to its original length.
- The spring obeys Hooke's law in the linear region of the graph.

$$F \propto x$$

or

$$F = kx$$

In the equation above, k is a constant for the spring known as the force constant. The SI unit for force constant is N m^{-1}.

- The gradient of the linear section of the graph is equal to force constant.
- Beyond the elastic limit the spring shows plastic behaviour.
  - **Plastic:** Removal of the force will not return the spring to its original length – it will have a permanent deformation.

Revision tip

Stiffer springs have larger values of force constant.

6.2 Elastic potential energy

Energy is stored in a spring when it is extended or compressed elastically. This stored energy is called elastic potential energy.

Elastic potential energy equations

The area under a force–extension graph is equal to the work done by the force, see Figure 2.

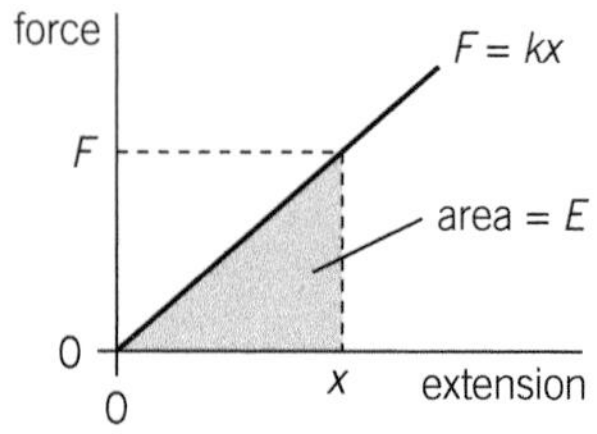

▲ **Figure 2** *The area under a straight-line force–extension graph is equal to elastic potential energy, E*

For a spring (or wire) with extension x and force F, the work done on the spring is equal to the elastic potential energy E.

E = area of the triangle of 'height' F and 'base length' x

$$E = \frac{1}{2}Fx$$

Since $F = kx$, we also have

$$E = \frac{1}{2}(kx)x = \frac{1}{2}kx^2$$

For a given spring, $E \propto x^2$; doubling the extension will quadruple the energy stored by the spring.

Worked example: Energy stored in a spring

A compressible spring of force constant 60 N m^{-1} has its compression changed from 10.0 cm to 18.0 cm. Calculate the change in the energy stored in the spring.

Step 1: Write down the quantities given in the question.

$k = 60\,\text{N m}^{-1}$ $\quad x_1 = 0.100\,\text{m}$ $\quad x_2 = 0.180\,\text{m}$

Step 2: Use the equation $E = \frac{1}{2}kx^2$ to calculate the change in the energy stored.

initial stored energy $= \frac{1}{2} \times 60 \times 0.100^2$

final stored energy $= \frac{1}{2} \times 60 \times 0.180^2$

change in stored energy $= \frac{1}{2} \times 60 \times (0.180^2 - 0.100^2)$

change in stored energy = 0.67 J (2 s.f.)

Common misconception

The work done by a constant force is given by the equation $W = Fx$, but for a spring or wire, the work done (or the elastic potential energy) is given by $E = \frac{1}{2}Fx$.

Summary questions

1 An object of weight 1.2 N is hung from the bottom of a spring. The spring has a force constant of 25 N m^{-1}.
Calculate the extension of the spring. *(2 marks)*

2 A student records the following results from an experiment on a wire.
force = 10 N and extension = 2.5 mm
force = 30 N and extension = 9.2 mm
Discuss whether or not the wire obeys Hooke's law. *(3 marks)*

3 A spring has force constant 120 N m^{-1}.
Calculate the elastic potential energy in the spring when it has extension 20 mm. *(2 marks)*

4 The energy stored in the spring in **Q3** is 5.4 J for a new extension.
Calculate this new extension. *(3 marks)*

5 The extension of a spring is increased by a factor of 5.
Calculate the factor by which the elastic potential energy of the spring would increase. *(2 marks)*

6 A spring is used to propel a small metal ball of mass 20 g.
The spring has force constant 150 N m^{-1} and its maximum compression is 9.0 cm.
Calculate the maximum speed of the metal ball. *(4 marks)*

6.3 Deforming materials
6.4 Stress, strain, and the Young modulus

Specification reference: 3.4.1, 3.4.2

6.3 Deforming materials

Not all materials deform in the same way. A ductile material can be drawn into a wire and it stretches a lot beyond its elastic limit. Metals such as gold and copper are both ductile. A brittle material snaps when it reaches its elastic limit. Glass and cast iron are both brittle. A polymeric material such as rubber deforms elastically.

Loading and unloading

Figure 1 shows the force–extension graphs for a metal wire, rubber, and polythene.

- **Metal wire:** For forces less than the elastic limit, the wire obeys Hooke's law and it behaves elastically. It deforms plastically for forces greater than the elastic limit.
- **Rubber:** Rubber has long chains of molecules. It behaves elastically and does not obey Hooke's law.
- **Polythene:** Polythene does not obey Hooke's law and it undergoes plastic deformation for very small values of force.

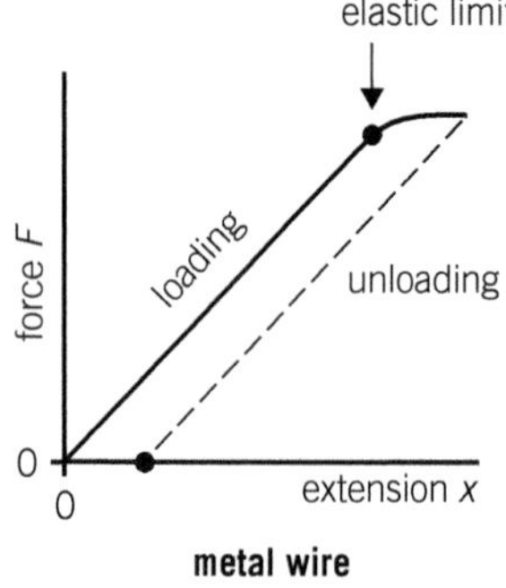

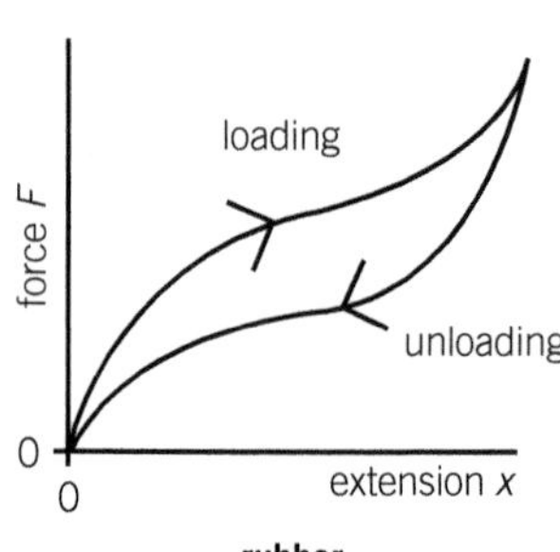

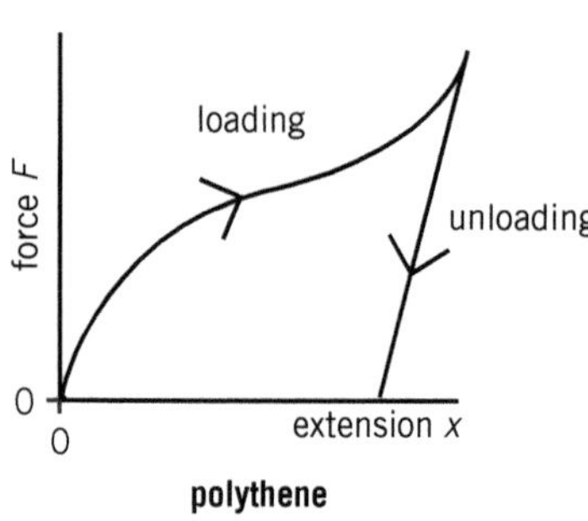

▲ **Figure 1** *Force–extension graphs*

6.4 Stress, strain, and the Young modulus

The factors that affect the extension x of a wire are: its original length L, its cross-sectional area A, the force F applied to the wire and the material of the wire.

Stress and strain

Tensile stress is defined as the force per unit cross-sectional area.

The word equation for stress σ is

$$\text{stress} = \frac{\text{force}}{\text{cross-sectional area}} \quad \text{or} \quad \sigma = \frac{F}{A}$$

The unit of stress is the same as that for pressure: N m^{-2} or pascal (Pa).

Tensile strain is defined as the extension (or compression) per unit original length.

The word equation for strain ε is

$$\text{strain} = \frac{\text{extension}}{\text{original length}} \quad \text{or} \quad \varepsilon = \frac{x}{L}$$

Strain has no unit. However, sometimes it is written as a percentage. For example, a strain of 0.012 may also be written as 1.2%.

Breaking stress is the value of the stress at which the material snaps or breaks.

The **ultimate tensile strength** is the maximum stress the material can withstand.

elastic limit

stress

strain

0

0

▲ **Figure 2** *Stress–strain graph for a ductile material*

Young modulus

Figure 2 shows the stress–strain graph for a ductile material.

When the material does not exceed its elastic limit,

$$\text{stress} \propto \text{strain}$$

or

$$E = \frac{\sigma}{\varepsilon}$$

where E is the Young modulus of a material. The unit for Young modulus is $N\,m^{-2}$ or Pa.

Young modulus of a material is defined as the ratio of stress to strain.

Revision tip

Stiffer materials have larger values of Young modulus.

Stress–strain graphs

The force–extension and stress–strain graphs have the same shapes for rubber and polythene.

The stress–strain graph for a brittle material is the same as Figure 2, except there is no plastic deformation.

Determining Young modulus

To determine the Young modulus of a material in the form of a long wire:

- Plot a graph of stress against strain. The Young modulus is the gradient of the straight-line section of the graph.
- Each value of stress is determined by hanging a mass m from the end of the wire and measuring the diameter d of the wire using a micrometer.

$$\text{stress} = \sigma = \frac{mg}{\left(\frac{\pi d^2}{4}\right)}$$

- Each value of strain is determined by measuring the extension x and the original length L using a metre rule. To improve the accuracy and precision, the extension should be measured using a device with a vernier scale (±0.1 mm or better).

Worked example: Force constant and the Young modulus

Show that the force constant k of a wire is equal to $\frac{EA}{L}$, where E is the Young modulus, A is the cross-sectional area of the wire, and L is the original length of the wire.

Step 1: Write the equations for stress, strain, and the Young modulus.

$$\text{stress} = \sigma = \frac{F}{A} \qquad \text{strain} = \varepsilon = \frac{x}{L} \qquad \text{Young modulus} = E = \frac{\sigma}{\varepsilon}$$

Step 2: Substitute the stress and strain labels into the equation for Young modulus.

$$E = \frac{\sigma}{\varepsilon} = \frac{\left(\frac{F}{A}\right)}{\left(\frac{x}{L}\right)} = \frac{FL}{Ax}$$

Step 3: Rearrange the equation with F being the subject.

$$F = \frac{EA}{L}x$$

Comparing the equation above with $F = kx$, the force constant k must be $\frac{EA}{L}$

Summary questions

1 State the units of stress, strain, and Young modulus. *(1 mark)*

2 A rubber band of original length 10.0 cm is extended to a length of 15.0 cm. Calculate the strain. *(2 marks)*

3 A 12 N weight is hung from a wire of radius 1.0 mm. Calculate the stress in the wire. *(2 marks)*

4 The material of the wire in **Q3** has Young modulus 1.8×10^{11} Pa. Calculate the strain of the wire. *(3 marks)*

5 Use the results recorded by a student to calculate the Young modulus of the material.
mass supported by wire = 5.0 kg original length = 3.2 m
extension of wire = 3.0 mm diameter of wire = 1.4 mm *(4 marks)*

6 For the wire in **Q5**, calculate the energy stored in the wire per unit volume. *(3 marks)*

Chapter 6 Practice questions

1 Which of these is the correct unit for the force constant of a spring?

A J s **C** N m

B J s^{-1} **D** N m^{-1} *(1 mark)*

2 Which of these statements is correct?
Rubber is elastic because...

A it obeys Hooke's law

B it has no plastic deformation

C it does not return all the energy stored

D it returns to its original shape when forces are removed. *(1 mark)*

3 The extension of a spring is x and the energy stored (elastic potential energy) is E. The spring is pulled until its new extension is $3x$.

What is the *change* in the energy stored in the spring if its extension is altered from x to $3x$?

A $2E$ **C** $8E$

B $3E$ **D** $9E$ *(1 mark)*

4 A 1.5 kg load is suspended from the end of a wire of diameter 1.2 mm.

What is the tensile stress of this wire?

A 3.9×10^3 Pa

B 1.2×10^4 Pa

C 3.3×10^6 Pa

D 1.3×0^7 Pa *(1 mark)*

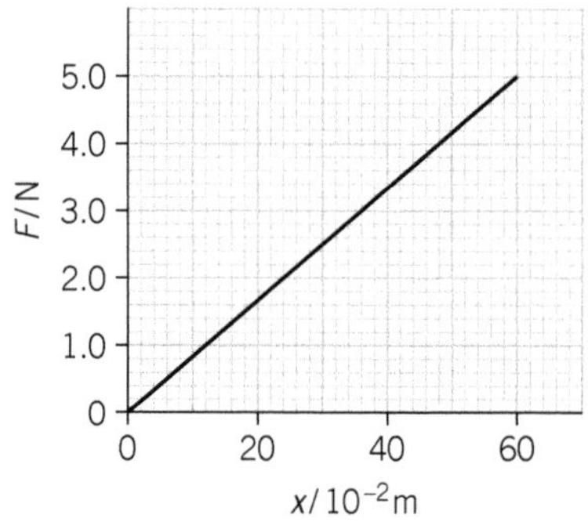

5 **a** Figure 1 show the force F against extension x graph for a spring.

i Calculate the force constant of the spring. *(2 marks)*

ii Calculate the extension of the spring when the applied force is 9.0 N.

State any assumption made. *(3 marks)*

b Two identical springs are connected in series and support a load of weight 10 N. The force constant of each spring is 40 N m^{-1}. Calculate the total energy stored in the springs. *(3 marks)*

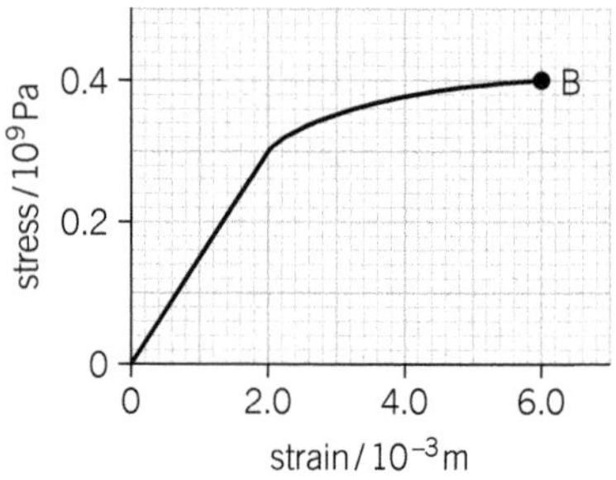

▲ **Figure 1**

6 **a** Define the Young modulus of a material. *(1 mark)*

b Figure 2 shows the stress against strain graph for a metal in the form of a wire being investigated by a student.

The material breaks at point B when the applied force is 16 N.

i Calculate the cross-sectional area of the wire when it breaks. *(2 marks)*

ii Calculate the Young modulus of the metal. *(3 marks)*

iii State and explain how the shape of the graph would change when a thicker wire of the same metal is investigated. *(2 marks)*

▲ **Figure 2**

7 A metal wire of length 1.20 m is fixed at its top end and an object is hung from its lower end. The extension of the wire is 0.54 mm. The cross-sectional area of the wire is 2.1×10^{-7} m^2 and the Young modulus of the wire is 8.0×10^{10} Pa.

a Calculate the strain of the wire. *(2 marks)*

b Calculate the mass of the object hung from the wire. *(4 marks)*

Chapter 7 Laws of motion and momentum

In this chapter you will learn about ...

- ☐ Newton's first, second, and third laws
- ☐ Momentum
- ☐ Rate of change of momentum
- ☐ Impulse
- ☐ Force–time graph
- ☐ Conservation of momentum
- ☐ Elastic collision
- ☐ Inelastic collision

7

LAWS OF MOTION AND MOMENTUM
7.1 Newton's first and third laws of motion
7.2 Linear momentum

Specification reference: 3.1.2, 3.5.1, 3.5.2

7.1 Newton's first and third laws of motion

Newton's laws are universal laws that can be used to model the motion of planets and gas atoms. It is important to understand these laws.

First law

Newton's first law of motion: An object will remain at rest or continue to move with constant velocity unless acted upon by a resultant force.

When the resultant force on an object is zero, it will either be motionless or travel at a constant velocity. A non-zero resultant force will accelerate the object.

Third law

Newton's third law of motion: When two objects interact, they exert equal and opposite forces on each other.

This is one of the most misunderstood laws. It is worth noting the following points:

- The forces cannot act on a single object.
- The forces are in opposite directions.
- The forces have the same magnitude.
- Both forces must be of the same type (e.g., gravitational, electrostatic, etc.).

All interactions can be explained in terms of four fundamental forces – gravitational, electromagnetic, strong nuclear, and weak nuclear.

7.2 Linear momentum

Knowledge of momentum is essential in Newton's second law (see Topic 7.3) and collisions. Momentum is a vector quantity. In one dimension, we can use positive and negative signs to denote the directions.

Momentum

The **linear momentum** p, or just momentum, is defined as the product of the mass m of an object and its velocity v.

$$\text{momentum} = \text{mass} \times \text{velocity}$$

or

$$p = mv$$

The SI unit of momentum is $\mathrm{kg\,m\,s^{-1}}$.

Common misconception

Momentum is defined as the product of mass (a scalar) and velocity (a vector). A common mistake made in exams is to define it as: momentum = mass × speed. This must be avoided.

Conservation of momentum

For a system of interacting objects, the total momentum in a specified direction remains constant, as long as no external forces act on the system.

In a closed system, where no external forces act, the total initial momentum is equal to the total final momentum. It is important to take account of the vector nature of velocity and momentum when using this law.

- In all collisions, the total energy and the total momentum remain constant.
- In a perfectly elastic collision the total kinetic energy remains constant.
- In an inelastic collision there is a change in the kinetic energy, with transformation of energy to other forms (e.g., heat, sound, etc.).

Revision tip

Momentum is a vector quantity and therefore can be resolved.

Worked example: Explosive

An explosive is used to break up a stationary large rock. After the explosion the rock breaks up into two fragments. The mass of the smaller fragment is 30% of the total mass of the rock. The immediate velocity of the larger fragment is 15 m s^{-1}. Calculate the ratio:

$$\frac{\text{kinetic energy of smaller fragment}}{\text{kinetic energy of larger fragment}}$$

Step 1: Use the principle of conservation of momentum to determine the velocity of the smaller fragment.

The initial momentum is zero – therefore, the final momentum must also be zero.

final momentum = initial momentum

$(0.30M \times v) + (0.70M \times 15) = 0$ $\quad M$ = total mass of the rock

$$v = \frac{0.70M \times 15}{0.30M} = -35 \text{ m s}^{-1}$$

The minus sign signifies the fragments flying off in opposite directions. The speed of the smaller fragment is 35 m s^{-1}.

Step 2: Calculate the ratio of kinetic energies.

$$\text{ratio} = \frac{\frac{1}{2} \times 0.30M \times 35^2}{\frac{1}{2} \times 0.70M \times 15^2} = 2.3 \text{ (2 s.f.)}$$

Summary questions

1 State and explain the gravitational force exerted by a person of weight 700 N on the Earth. *(1 mark)*

2 Calculate the momentum of a 900 kg car travelling at a speed of 30 m s^{-1}. *(1 mark)*

3 Calculate the momentum of a 1.2 kg object dropped from rest after falling for 3.0 s. *(3 marks)*

4 A cannon of mass 1200 kg fires a 20 kg shell at a speed of 240 m s^{-1}. Calculate the magnitude of the recoil velocity of the cannon. *(3 marks)*

5 Figure 1 shows a ball of mass m before and after hitting a wall. Calculate the change in momentum of the ball. *(2 marks)*

6 A 1100 kg car is travelling at 30 m s^{-1}. The momentum of the car decreases by 1.3×10^4 kg m s^{-1}. Calculate the final speed of the car. *(3 marks)*

7 A 40 g metal ball travelling at 80 m s^{-1} hits a stationary lump of clay of mass 300 g. The ball becomes embedded in the clay. Calculate:

a the common speed of the clay and metal ball immediately after the impact; *(3 marks)*

b the loss in kinetic energy. *(2 marks)*

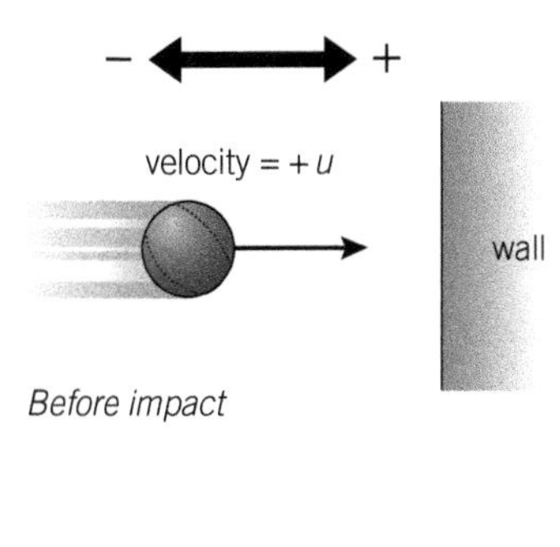

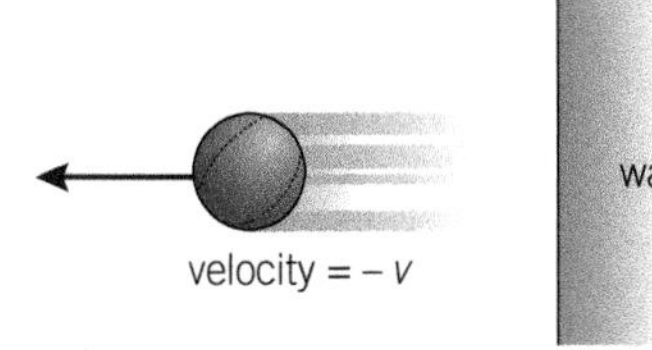

▲ **Figure 1**

7.3 Newton's second law of motion
7.4 Impulse
7.5 Collisions in two dimensions

Specification reference: 3.5.1, 3.5.2

7.3 Newton's second law of motion

You have already met the equation $F = ma$; this is a special case of Newton's second law of motion. This law relates the rate of change of momentum of an object to the resultant force acting on the object.

It is important that you do not state this law as $F = ma$.

Newton's second law

Newton's second law: The net (resultant) force acting on an object is directly proportional to the rate of change of its momentum, and is in the same direction.

resultant force ∝ rate of change of momentum
or
resultant force = k × rate of change of momentum.

In SI units, the constant k is taken as equal to 1, so this law can be written as resultant force = rate of change of momentum

$$F = \frac{\Delta p}{\Delta t}$$

where F is the resultant force and Δp is the change in momentum in a time Δt.

Units: $F \rightarrow$ (N) $\Delta p \rightarrow$ (kg m s^{-1}) $\Delta t \rightarrow$ (s)

$F = ma$

The equation $F = ma$ can be derived from Newton's second law.

For an object of constant mass m with initial velocity u and final velocity v after a time t, the resultant force F is given by

$$F = \frac{\Delta p}{\Delta t} = \frac{mv - mu}{t} = m\left(\frac{v - u}{t}\right) = ma$$

A constant force will produce a constant acceleration a.

▲ **Figure 1** *The area under a force–time graph is equal to impulse*

Revision tip
Momentum has the same unit as impulse; kg m s^{-1} or N s.

7.4 Impulse

Impulse of a force is defined as the product of force and time.

According to Newton's second law, the change in momentum Δp of an object is given by $\Delta p = F \times \Delta t$.

Therefore

- change in momentum = impulse

Force–time graph

For a constant force F acting for a time t, the impulse is equal to Ft. This is the same as the area under the force–time graph, see Figure 1. Therefore

- area under force–time graph = impulse = change in momentum

Worked example: *F–t* graph

Figure 2 shows a force F against time t graph for an object. The change in momentum of the object is 2.0 kg m s^{-1} in the first 1.0 ms.

Use the graph to determine the total change in momentum of the object.

Step 1: Determine the total area under the graph.

$$\text{total area under the graph} = (4 \times 1) + \left(\frac{1}{2} \times 4 \times 3\right) = 10 \text{ 'units'}$$

Step 2: Determine the change in momentum.

area under graph = impulse = change in momentum

An area of 4 'units' is equal to 2.0 kg m s^{-1}.

$$\text{Therefore, total change in momentum} = \frac{10}{4} \times 2.0 = 5.0 \text{ kg m s}^{-1} \text{ (2 s.f.)}$$

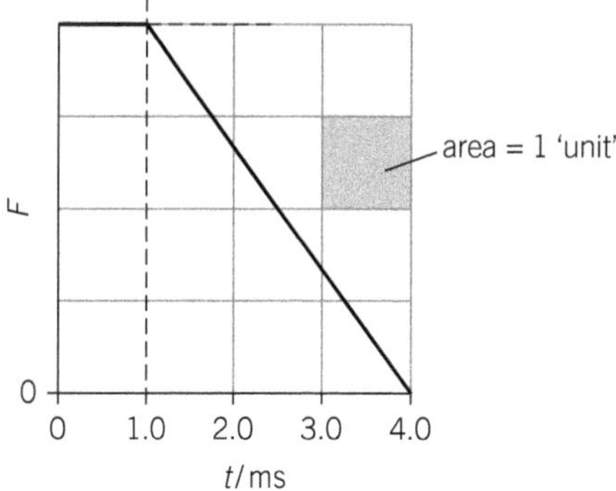

▲ **Figure 2**

7.5 Collisions in two dimensions

For collisions in two dimensions:

- total momentum in the x-direction remains constant
- total momentum in the y-direction remains constant.

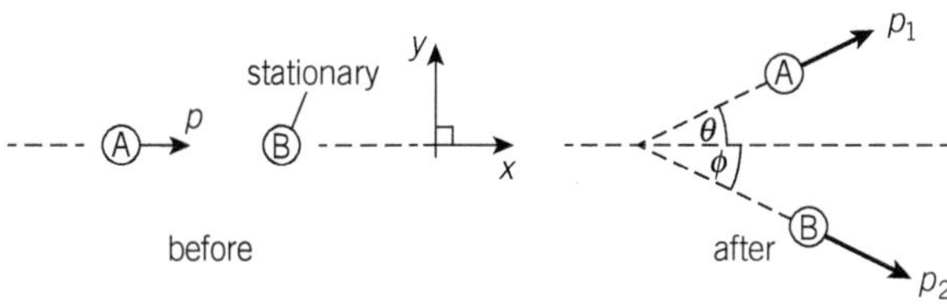

▲ **Figure 3** *There is no change in the total momentum*

Figure 3 shows the collision between objects A and B.

x-direction: initial momentum = final momentum

$$p = p_1 \cos\theta + p_2 \cos\phi$$

y-direction: total momentum is zero

$$p_1 \sin\theta = p_2 \sin\phi$$

Summary questions

1. The change in momentum of a toy car is 0.12 kg m s^{-1}. What is the impulse acting on the toy car? *(1 mark)*
2. The velocity of a 900 kg mass car changes from 30 m s^{-1} to 10 m s^{-1} in 5.0 s. Use the change in momentum of the car to calculate the magnitude of the force acting on the car. *(3 marks)*
3. Figure 4 shows the force–time graph for a car of mass 1000 kg. Calculate the impulse acting on the car. *(2 marks)*
4. Calculate the change in the velocity of the car in **Q3**. *(2 marks)*
5. A metal ball of mass 40 g hits a solid floor with a speed of 30 m s^{-1}. It rebounds with the same speed. It is in contact with the floor for 2.0 ms. Calculate the impulse of the force acting on the ball during its impact with the floor. *(3 marks)*
6. Draw a labelled vector diagram for Figure 3 to show that momentum is conserved. Explain your answer. *(3 marks)*

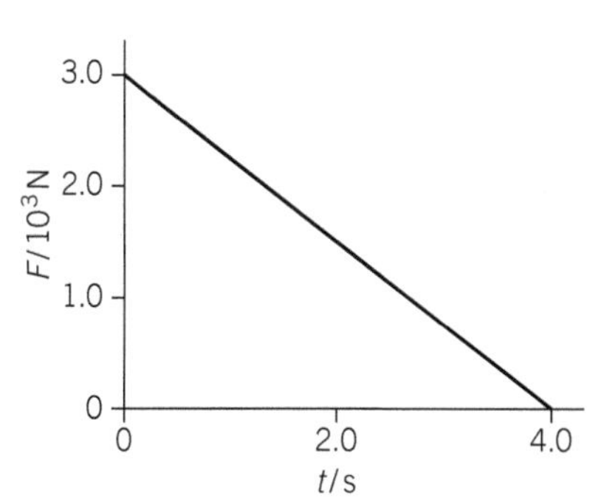

▲ **Figure 4**

Chapter 7 Practice questions

1 What is the area under a force against time graph equal to?

A energy
B impulse
C acceleration
D velocity *(1 mark)*

2 What quantity is conserved in all collisions?

A impulse
B force
C kinetic energy
D momentum *(1 mark)*

3 An object is travelling to the right with a momentum of $20\,kg\,m\,s^{-1}$. It collides head-on with another object moving to the left with a momentum of $30\,kg\,m\,s^{-1}$. The objects join together during the collision.

What is the final momentum of the objects?

A $10\,kg\,m\,s^{-1}$ to the left
B $10\,kg\,m\,s^{-1}$ to the right
C $50\,kg\,m\,s^{-1}$ to the left
D $50\,kg\,m\,s^{-1}$ to the right *(1 mark)*

4 The momentum of a car is $2000\,kg\,m\,s^{-1}$. The speed of the car increases when a force of 200 N acts on the car for a period of 3.0 s.

What is the final momentum of the car?

A $600\,kg\,m\,s^{-1}$
B $1400\,kg\,m\,s^{-1}$
C $2000\,kg\,m\,s^{-1}$
D $2600\,kg\,m\,s^{-1}$ *(1 mark)*

5 a Define linear momentum. *(1 mark)*

b State Newton's second law of motion. *(1 mark)*

c An object of mass 800 g falls vertically and hits the ground with a speed of $6.0\,m\,s^{-1}$. It rebounds with a speed of $4.0\,m\,s^{-1}$. The contact time with the ground is 25 ms.

i Calculate the change in the momentum of the object. *(1 mark)*

ii Calculate the magnitude of the average force exerted by the ground on the object. *(2 marks)*

iii State and explain the direction of the force experienced by the ground. *(2 marks)*

6 a Define impulse of a force. *(1 mark)*

b Figure 1 shows the variation of the force F experienced by a ball with time t.

The mass of the ball is 200 g. The ball was travelling at a constant initial speed of $3.0\,m\,s^{-1}$ before the force shown in Figure 1 was applied.

i Describe and explain how the acceleration of the ball varies between $t = 0$ and $t = 0.4$ s. *(2 marks)*

ii Calculate the final speed v of the ball. Explain your answer. *(4 marks)*

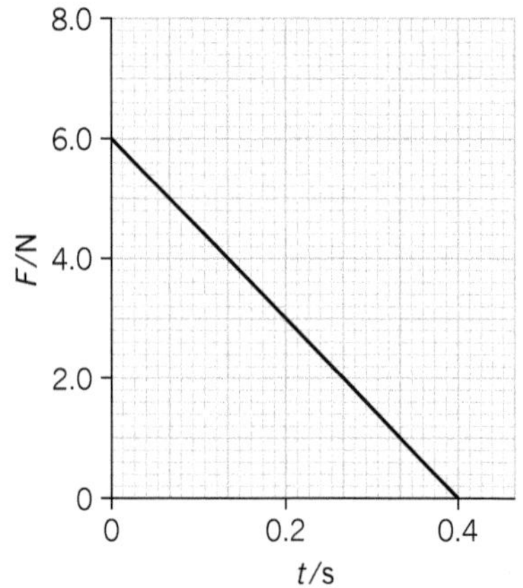

▲ **Figure 1**

7 a Explain what is meant by an inelastic collision. *(1 mark)*

b A railway engine of mass 8.0×10^4 kg travelling at a constant speed of $0.50\,m\,s^{-1}$ collides with a stationary carriage of mass 4.0×10^4 kg. The engine and the carriage couple together and move off at a constant speed v.

i Calculate the initial momentum of the engine. *(1 mark)*

ii Calculate the speed v. Explain your answer. *(3 marks)*

iii Calculate the change in the kinetic energy of:

1 the engine; *(2 marks)*

2 the carriage. *(1 mark)*

iv Explain why the answers to (iii)1 and (iii)2 are not the same. *(1 mark)*

Module 4 Electrons, waves, and photons

Chapter 8 Charge and current

In this chapter you will learn about ...

- [ ] Charge
- [ ] Quantisation of charge
- [ ] Charge carriers
- [ ] Current
- [ ] Conventional current
- [ ] Kirchhoff's first law
- [ ] Mean drift velocity
- [ ] Number density
- [ ] Insulators, metals, and semiconductors

8 CHARGE AND CURRENT
8.1 Current and charge
8.2 Moving charges

Specification reference: 4.1.1

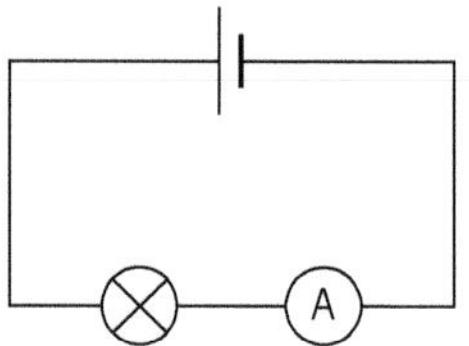

▲ **Figure 1** *The ammeter is connected in series to the lamp*

8.1 Current and charge

Charged particles can either be positive or negative. Charge is measured in coulomb (C). An electron has a negative charge and a proton has a positive charge. In a conducting liquid (an electrolyte) there are negative and positive ions.

An electric current is the flow of charged particles. Current is measured in ampere (A). The ampere is one of the SI base units. In the laboratory you can measure the current in a component using an ammeter connected in series with the component, see Figure 1.

Electric current

Electric current is defined as the rate of flow of charge.

Electric current I can be calculated using the equation

$$I = \frac{\Delta Q}{\Delta t}$$

where ΔQ is the charge transferred in a time Δt.

Units: $I \rightarrow \text{A}$ $\quad \Delta Q \rightarrow \text{C}$ $\quad \Delta t \rightarrow \text{s}$

A current of 1 ampere is the same as a charge of 1 coulomb transferred per second; $1\,\text{A} = 1\,\text{C}\,\text{s}^{-1}$.

A charge of 1 **coulomb** is defined as the flow of charge in a time of 1 second when the current is 1 ampere.

Revision tip

Tiny currents are measured in pA, nA, μA, and mA.

$1\,\text{pA} = 10^{-12}\,\text{A}$ $\quad 1\,\text{nA} = 10^{-9}\,\text{A}$

$1\,\mu\text{A} = 10^{-6}\,\text{A}$ $\quad 1\,\text{mA} = 10^{-3}\,\text{A}$

Worked example: Tiny current

The current in a resistor is 3.2 pA. Calculate the number of electrons travelling through the resistor in a time of 1.0 minutes.

Step 1: Write down the quantities given in the question.

$I = 3.2 \times 10^{-12}\,\text{A}$ $\quad \Delta t = 60\,\text{s}$ $\quad \Delta Q = ?$

Step 2: Calculate the charge flow ΔQ.

$$I = \frac{\Delta Q}{\Delta t}$$

$$\Delta Q = I\Delta t = 3.2 \times 10^{-12} \times 60$$

$$\Delta Q = 1.92 \times 10^{-10}\,\text{C}$$

Step 3: The magnitude of the charge on each electron is 1.60×10^{-19} C. Use this to calculate the number of electrons responsible for the charge.

$$\text{number of electrons} = \frac{1.92 \times 10^{-10}}{1.60 \times 10^{-19}} = 1.2 \times 10^{9}\ (2\text{ s.f.})$$

The number of electrons passing through the resistor is 1.2 billion.

Quantisation of charge

All objects can be charged by removing or depositing electrons. The magnitude of the charge on a single electron is given the letter e. At a microscopic level, a single electron removed from a neutral atom will give the remaining atom (ion) a positive charge of $+e$, removing two electrons will result in a positive ion of charge of $+2e$, and so on. Adding electrons to a neutral atom can result in a negative ion of charge $-e$, $-2e$, and so on.

- The charge Q on particles is quantised, with $Q = \pm ne$ (n = integer).

8.2 Moving charges

In a metal, the current is due to the movement of electrons. In an electrolyte, the current is due to the simultaneous movement of positive and negative ions.

Electrons and ions

Figure 2 shows the negative electrons in a metal wire moving away from the negative side of the cell. The convention for the direction of current in a circuit is opposite in direction to the flow of electrons.

- The direction of conventional current I is from positive to negative.
- The direction of flow of electrons is from negative to positive.

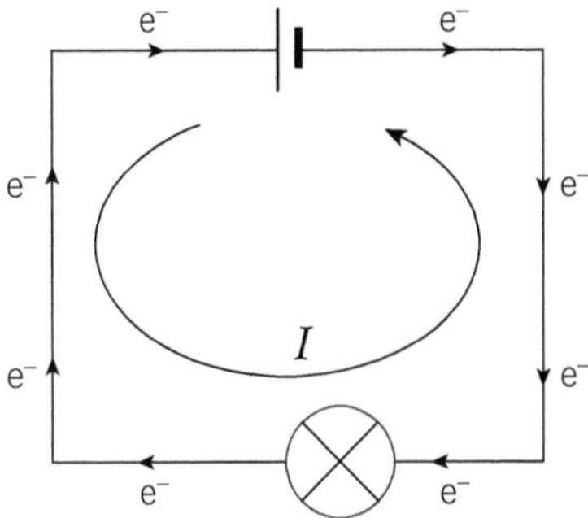

▲ **Figure 2** *The direction of conventional current I is in the opposite direction to the flow of electrons*

Summary questions

1. Name the charged particles responsible for current in a metal wire and in an electrolyte. *(1 mark)*
2. **a** According to a student the charge of an ion can be $+8.0 \times 10^{-20}$ C. Explain whether or not the student is correct. *(1 mark)*

   b Calculate the current in a diode given that the charge flow is 12 C in a time of 60 s. *(2 marks)*
3. An oil droplet has an excess of 8 electrons. Calculate the charge on the oil droplet. *(2 marks)*
4. A solar cell delivers an average current of 20 mA in a period of 4.0 hours. Calculate the total flow of charge through the solar cell. *(2 marks)*
5. Calculate the total number of electrons flowing through the solar cell in **Q4**. *(2 hours)*
6. Modern electronic circuits can measure tiny current equivalent to a million electrons per minute. Calculate this tiny current. *(2 hours)*

8.3 Kirchhoff's first law
8.4 Mean drift velocity

Specification reference: 4.1.1, 4.1.2

8.3 Kirchhoff's first law

In an electrical circuit, the charge carriers responsible for electric current are the electrons. Unlike the positive ions, which are fixed, electrons are free to move within a circuit. In a series circuit, you would expect all the electrons entering a particular component to come out. This is because charge cannot be created or destroyed – it is conserved.

The first law

Kirchhoff's first law states that for any point (junction) in an electrical circuit, the sum of currents into that point (junction) is equal to the sum of currents out of that point.

This can be written as

$$\Sigma I_{\text{in}} = \Sigma I_{\text{out}}$$

where sum ΣI_{in} is the sum of the currents into a point and ΣI_{out} is the sum of the currents out of the same point. Kirchhoff's first law is a natural consequence of the conservation of charge.

Maths: Sum of

The Greek letter sigma Σ is used as a shorthand for 'sum of...'.

Figure 1 shows some examples of Kirchhoff's first law.

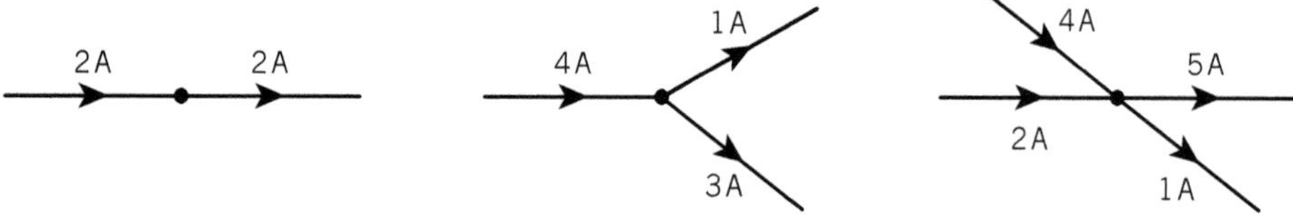

▲ **Figure 1** *Examples of the first law*

Revision tip

The current in a series circuit is constant because of conservation of charge.

8.4 Mean drift velocity

The electrons in a metal move around at high speeds and make frequent collisions with the vibrating fixed positive metal ions. The net result is that the electrons move in random directions. When the wire is connected to a battery or a power supply, the electrons gain an additional velocity towards the positive electrode of the external supply. The mean drift velocity v is the average velocity gained by the electrons along the length of the wire.

Figure 2 shows a cross-section through a current-carrying wire.

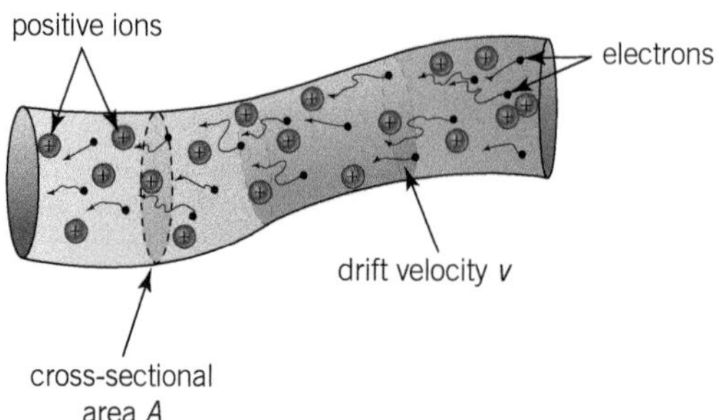

▲ **Figure 2** *Electrons have a mean drift velocity along the length of the wire because they make repeated collisions with the vibrating positive ions*

$I = Anev$

The **number density** is the number of charge carriers (e.g. electrons) per unit volume of the material.

We can classify the conduction properties of materials by their number density.

- Metals are good conductor with n about $10^{28}\,m^{-3}$.
- Insulators such as plastic have much smaller number density than metals.
- Semiconductors such as silicon have a number density between insulators and metals, about $10^{18}\,m^{-3}$.

The mean drift velocity v depends on the current I in the wire, the cross-sectional area A of the wire, the number density n of the free electrons, and the elementary charge e.

The equation for the current is $I = Anev$.

Worked example: Lamp filament

The current in a tungsten filament lamp is 0.33 A. The filament has a radius of 2.4×10^{-4} m. The mean drift velocity of the electrons is $3.0 \times 10^{-4}\,m\,s^{-1}$.

Calculate the number density of the free electrons for tungsten.

Step 1: Write down the quantities given in the question.

$I = 0.33\,A \quad r = 2.4 \times 10^{-4}\,m \quad v = 3.0 \times 10^{-4}\,m\,s^{-1} \quad e = 1.6 \times 10^{-19}\,C \quad n = ?$

Step 2: Rearrange the equation $I = Anev$, substitute the values, and calculate n.

$$n = \frac{I}{Aev} = \frac{0.33}{[\pi \times (2.4 \times 10^{-4})^2] \times 1.6 \times 10^{-19} \times 3.0 \times 10^{-4}}$$

$n = 3.8 \times 10^{28}\,m^{-3}$ (2 s.f.)

Summary questions

1. In an electric circuit, 2 billion electrons enter a point. State and explain how many electrons will leave the same point. *(1 mark)*
2. Determine the currents in the part of the circuit shown in Figure 3. *(3 marks)*
3. The mean drift velocity in a resistor is $2.0\,mm\,s^{-1}$ when the current is 4.0 A.
   Calculate the mean drift velocity when the current is reduced to 1.0 A. *(2 marks)*
4. Figure 4 shows a cross-section through a current-carrying conductor. Describe and explain how the mean drift velocity of the free electrons changes from B to A. *(2 marks)*
5. The number density of free electrons for copper is $8.5 \times 10^{28}\,m^{-3}$. Estimate the number of free electrons in a copper wire of length 1.0 cm and diameter 1.0 mm. *(2 marks)*
6. The number density of free electrons in a metal is $5.0 \times 10^{28}\,m^{-3}$. Calculate the mean drift velocity of the electrons in a metal wire of cross-sectional area $1.0\,mm^2$ when the current is $3.0\,\mu A$. *(3 marks)*

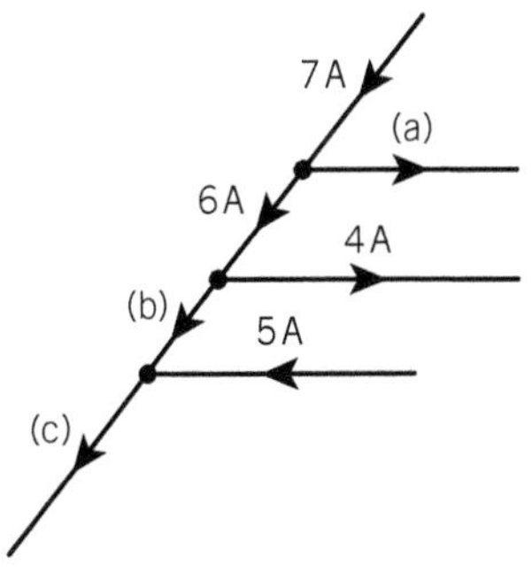

▲ Figure 3

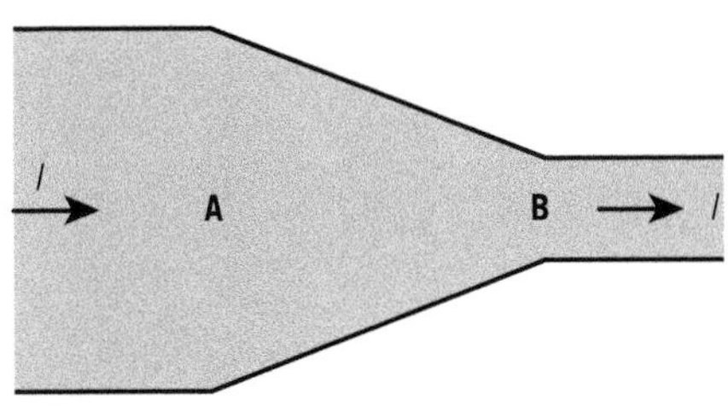

▲ Figure 4

Chapter 8 Practice questions

1 Which quantity is conserved according to Kirchhoff's first law?

A energy

B charge

C potential difference

D electromotive force *(1 mark)*

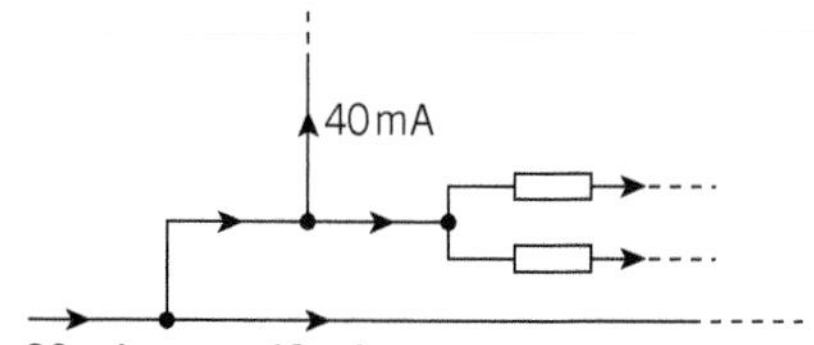

▲ **Figure 1**

2 Figure 1 shows parts of an electrical circuit.
What is the current in one of the identical resistors?

A 5 mA **C** 20 mA

B 10 mA **D** 40 mA *(1 mark)*

3 How many electrons represent a charge of $-6.4\,\mu C$?

A 4.0×10^{7} **C** 4.0×10^{13}

B 4.0×10^{10} **D** 4.0×10^{16} *(1 mark)*

4 The diameter of a current-carrying wire gets thinner in the direction of electron flow.
Which statement is correct about the mean drift velocity of the charge carriers in the direction of electron flow?
The mean drift velocity ...

A decreases

B increases

C stays the same

D is equal to $3 \times 10^{8}\,m\,s^{-1}$. *(1 mark)*

positive ion

negative ion

▲ **Figure 2**

5 **a** Figure 2 shows the conduction of ions within a liquid. State and explain the direction of the conventional current in the liquid. *(1 mark)*

b The current in a wire is 40 mA. Calculate:

i the total charge passing through a point in the wire in a period of 30 s; *(2 marks)*

ii the total number of electrons responsible for the charge in (i). *(2 marks)*

6 A metal wire of diameter 1.2 mm is connected to an electrical appliance. The current in the wire is 9.5 A and the appliance is operated for a time of 2.0 minutes. The metal has 8.5×10^{22} free electrons per cm^3 of the wire. Calculate:

a the number density n of the free electrons within the wire in m^{-3}; *(1 mark)*

b the mean drift velocity of the electrons; *(3 marks)*

c the mean distance travelled along the length of the wire by each electron in 2.0 minutes. *(1 mark)*

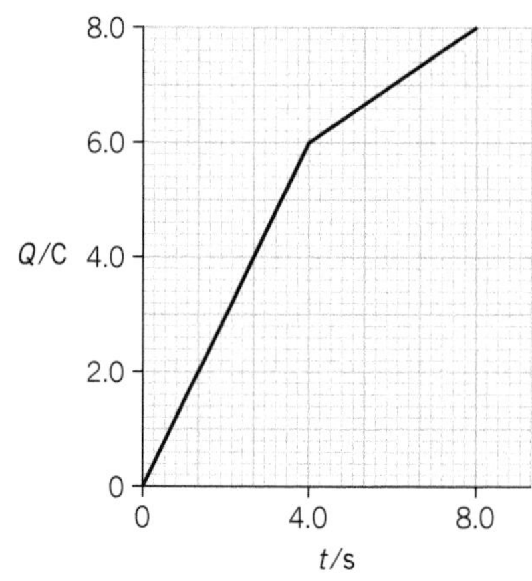

▲ **Figure 3**

7 Figure 3 shows variation of charge Q passing though a point in a circuit with time t.

a State and explain what physical quantity is equal to the gradient of the graph. *(1 mark)*

b Calculate the current at time $t = 3.0\,s$. *(1 mark)*

c Describe the variation of current I with time t. *(3 marks)*

Chapter 9 Energy, power, and resistance

In this chapter you will learn about ...

- ☐ Circuit symbols
- ☐ Potential difference (p.d.)
- ☐ Electromotive force (e.m.f.)
- ☐ The electron gun
- ☐ Resistance
- ☐ *I*–*V* characteristics
- ☐ Resistors and filament lamps
- ☐ Diodes and light-emitting diodes
- ☐ Resistivity
- ☐ Thermistor and light-dependent resistors (LDR)
- ☐ Electrical power
- ☐ Kilowatt-hour

9 ENERGY, POWER, AND RESISTANCE
9.1 Circuit symbols
9.2 Potential difference and electromotive force
9.3 The electron gun

Specification reference: 4.2.1, 4.2.2

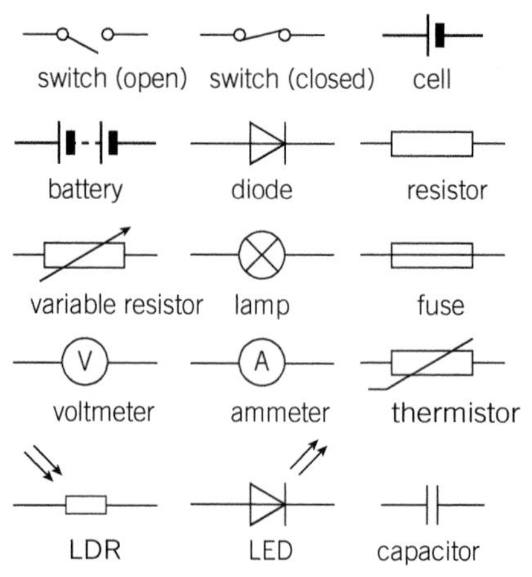

▲ **Figure 1** *Circuit symbols*

9.1 Circuit symbols

Figure 1 shows the circuit symbols used in electrical circuits – use them with great care when drawing circuit diagrams.

9.2 Potential difference and electromotive force

When analysing circuits, it is important that you distinguish the terms potential difference (p.d.) and electromotive force (e.m.f.). A voltmeter is a device calibrated to measure the p.d. across a component – a voltmeter is placed in parallel with the component.

Potential difference

The term potential difference is used when the charge carriers (electrons) lose energy in a component.

Potential difference is defined as the energy transferred from electrical energy to other forms (heat, light, etc.) per unit charge.

The word equation for p.d. V is

$$\text{p.d.} = \frac{\text{energy transferred}}{\text{charge}} \quad \text{or} \quad V = \frac{W}{Q}$$

where W is the energy transferred and Q is the charge flow.

Revision tip

The term 'voltage' is often used instead of potential difference.

Electromotive force

The term electromotive force is used when charge carriers (electrons) gain energy from a source. A chemical cell, a dynamo, a solar cell, a thermocouple, and a power supply are all examples of sources of e.m.f.

Electromotive force is defined as the energy transferred from chemical energy (or another form) to electrical energy per unit charge.

The word equation for e.m.f. $\mathcal{E}$ is

$$\text{e.m.f.} = \frac{\text{energy transferred}}{\text{charge}} \quad \text{or} \quad \mathcal{E} = \frac{W}{Q}$$

where W is the energy transferred and Q is the charge flow.

- p.d. and e.m.f. are both measured in volts (V).
- 1 volt = 1 joule per coulomb ($1\,\text{V} = 1\,\text{J}\,\text{C}^{-1}$).
- A potential difference of 1 **volt** is defined as 1 joule of energy transferred per unit coulomb.

Revision tip

$1\,\text{V} = 1\,\text{J}\,\text{C}^{-1}$

Think of a voltmeter as an instrument calibrated to show 'joules per coulomb'.

Analysing a circuit

Figure 2 shows a simple electrical circuit. A battery of e.m.f. 3.0 V is connected in series with a filament lamp and a resistor. The p.d. across the lamp is 2.4 V and the p.d. across the resistor is 0.6 V.

- Battery: 3.0 J of chemical energy is transferred into electrical energy per unit coulomb of charge.
- Lamp: 2.4 J of electrical energy is transferred to thermal energy and radiant energy per unit coulomb of charge.
- Resistor: 0.6 J of electrical energy is transferred to thermal energy per unit coulomb of charge.

▲ **Figure 2** *Imagine what happens to the energy of 1 C of charge travelling from the battery, around the circuit, and then back to the battery*

9.3 The electron gun

Figure 3 shows an electron gun – a device used to accelerate electrons to high speeds.

The p.d. between the hot filament and the anode is V. The electrons have negligible kinetic energy at the filament.

work done on an electron = final kinetic energy of an electron

$$Ve = \frac{1}{2}mv^2$$

where e is the elementary charge, m is the mass of the electron, and v is the final speed of the electron.

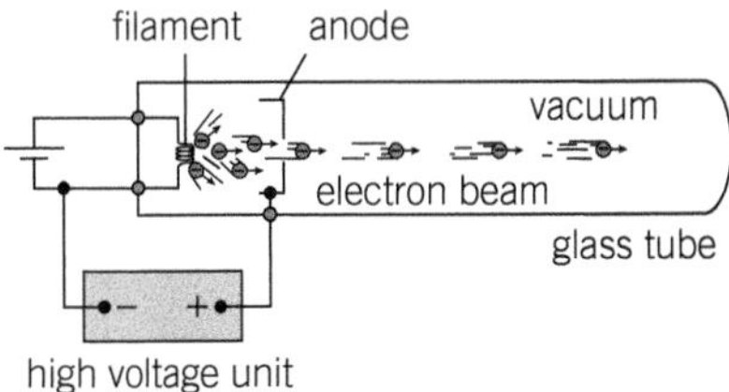

▲ **Figure 3** *An electron gun*

Worked example: High-speed electrons

An electron gun produces a beam of electrons of speed $2.0 \times 10^6\ \mathrm{m\,s^{-1}}$.

Calculate the accelerating p.d. V.

Step 1: Select the equation relating p.d. and the speed of the electron.

$$Ve = \frac{1}{2}mv^2$$

Step 2: Rearrange to make V the subject.

$$V = \frac{mv^2}{2e}$$

Step 3: Substitute the values and calculate V.

$$V = \frac{9.11 \times 10^{-31} \times (2.0 \times 10^6)^2}{2 \times 1.60 \times 10^{-19}}$$

$V = 11.4\ \mathrm{V} \approx 11\ \mathrm{V}$ (2 s.f.)

Summary questions

1. Name two sources of e.m.f. *(1 mark)*
2. Calculate the potential difference across a resistor that transfers 20 J of energy when a charge of 4.0 C flows through it. *(2 marks)*
3. A charge of 60 C travels through a power supply of e.m.f. 100 V. Calculate the energy transferred. *(2 marks)*
4. Calculate the energy transferred by an electron travelling through a p.d. of 1.0 kV. *(2 marks)*
5. The current in a filament lamp is 0.25 A and the p.d. across it is 6.0 V. The lamp is operated for 30 s. Calculate the rate of energy transfer in the lamp. *(3 marks)*
6. Calculate the speed of protons accelerated from rest through a potential difference of 10 kV. *(3 marks)*

9.4 Resistance
9.5 *I–V* characteristics
9.6 Diodes

Specification reference: 4.2.3

9.4 Resistance

The term resistance has a precise meaning in physics. You can use a multimeter set on 'ohms' to directly measure the resistance of some components (e.g., resistor, thermistor, etc.).

Defining resistance

Revision tip

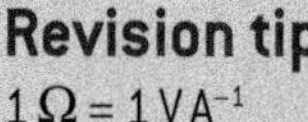

The **resistance** of a component in a circuit is defined by the following word equation:

$$\text{resistance} = \frac{\text{potential difference across component}}{\text{current in the component}}$$

or

$$R = \frac{V}{I}$$

where R is the resistance, V is the potential difference, and I is the current. The SI unit for resistance is ohm, Ω.

Units: $V \rightarrow \text{V}$ $\quad I \rightarrow \text{A}$ $\quad R \rightarrow \Omega$

A resistance of 1 **ohm** is defined as 1 volt per unit ampere.

Determining resistance

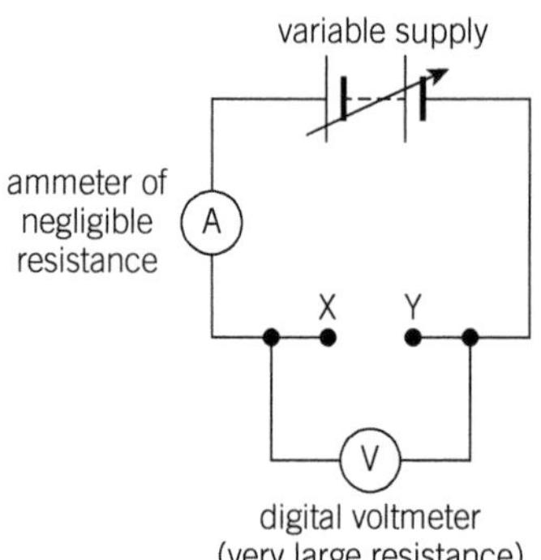

▲ **Figure 1** *The component is connected between X and Y*

Figure 1 shows a circuit that may be used to determine the resistance R of any component. The voltmeter measures the p.d. V across the component and the ammeter measure the current I in the component. The resistance is calculated using the equation $R = \frac{V}{I}$

Ohm's law: For a metallic conductor kept at a constant temperature, the current in the wire is directly proportional to the p.d. across its ends.

9.5 *I–V* characteristics

The *I–V* characteristic of a component is a graph of current against p.d.

Resistor and filament lamp

Figure 2 shows the *I–V* characteristics of a resistor (a) and a filament lamp (b).

Resistor:

- $I \propto V$ therefore its resistance remains constant.
- A resistor is an ohmic component.

Filament lamp:

- The resistance of the lamp is not constant – the *I–V* graph is not a straight line through the origin.
- The resistance of a filament lamp increases as the current, or the p.d., increases. At a microscopic level, the metal ions of the filament vibrate more quickly and with increased amplitudes. The conducting electrons makes more frequent collisions with these vibrating ions and transfer greater amounts of energy to these ions.
- A filament lamp is a non-ohmic component.

(a)

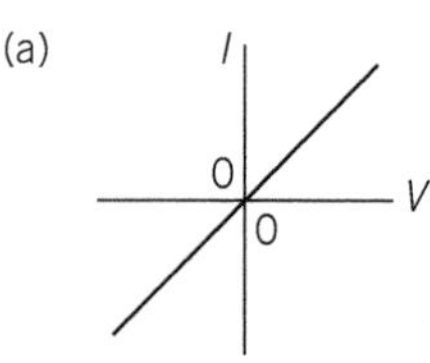

(b)

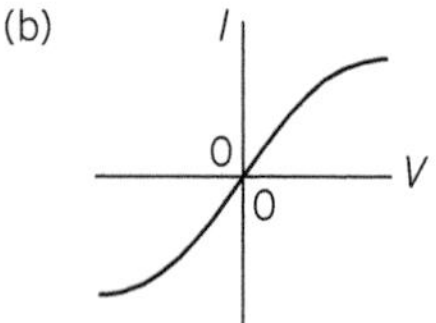

▲ **Figure 2** *I–V characteristics*

9.6 Diodes

Diodes are components made from semiconductors that conduct in one direction. A light-emitting diode emits light of a specific colour when it conducts. See Figure 3.

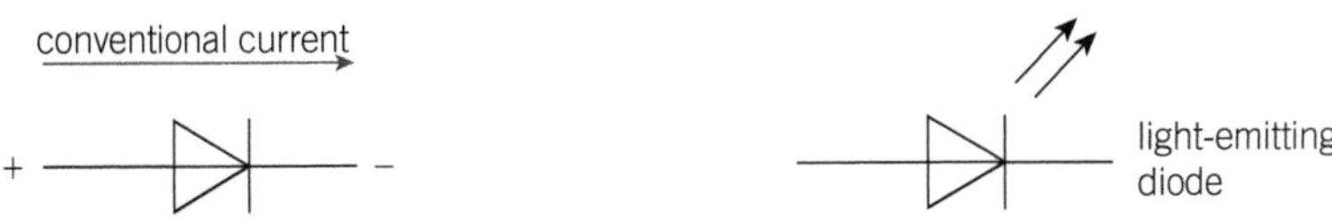

▲ **Figure 3** *Symbols for a diode and LED*

Light-emitting diodes

Figure 4 shows the *I*–*V* characteristics of an ordinary diode (e.g., silicon) and a light-emitting diode, LED.

- The diode does not conduct when it is reverse biased.
- The resistance is infinite for negative p.d.s.
- The diode starts to conduct when it is forward biased and the p.d. is greater than the threshold p.d. V_c.
- The resistance of the diode decreases significantly as the p.d. is increased beyond V_c. This is because of the increase in the number density of the charge carriers.

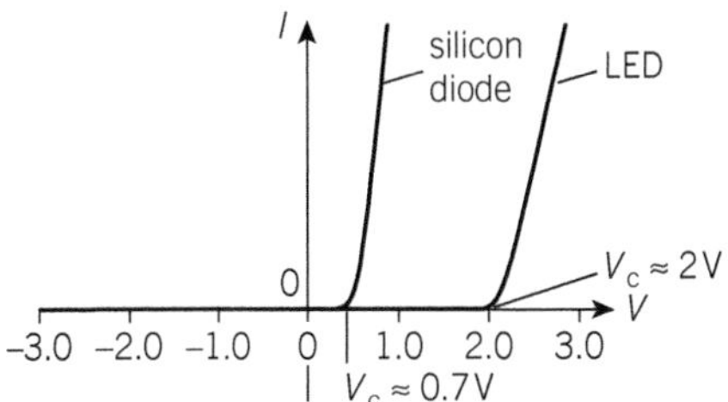

▲ **Figure 4** *I–V characteristics of diodes*

Worked example: What is it?

A student is investigating a component and records the results below.

Identify the component.

I/A	−20 mA	52 mA
V/V	−1.2 V	4.7 V

Step 1: Eliminate the diode.

The component cannot be a diode because it conducts in 'both directions'.

Step 2: Calculate the resistance at each p.d.

negative values: $R = \frac{V}{I} = \frac{1.2}{0.02} = 60\,\Omega$ (2 s.f.)

positive values: $R = \frac{V}{I} = \frac{4.7}{0.052} = 90\,\Omega$ (2 s.f.)

Step 3: Identify the component from the resistance values.

The resistance is not constant, so it cannot be a resistor.

The resistance increases as the p.d. increases, so it must be a filament lamp.

Summary questions

1. Name one non-ohmic component. *(1 mark)*
2. Calculate the p.d. across a 120 Ω resistor with a current of 30 mA. *(1 mark)*
3. The current in a resistor is doubled. State and explain what happens to the resistance of the resistor. *(1 mark)*
4. Show that the resistance of a reverse biased diode is infinite. *(1 mark)*
5. The resistance of an LED decreases from 200 Ω to 20 Ω when the p.d. across it is increased from 1.9 V to 2.1 V. Calculate the change in current in the LED. *(3 marks)*
6. For the LED in **Q5**, calculate the percentage change in the number density of the charge carriers. *(2 marks)*

9.7 Resistance and resistivity
9.8 The thermistor
9.9 The LDR

Specification reference: 4.2.3, 4.2.4

9.7 Resistance and resistivity

The resistance R of a conductor depends on:

- its length L
- its cross-sectional area A
- its temperature
- the material of the conductor.

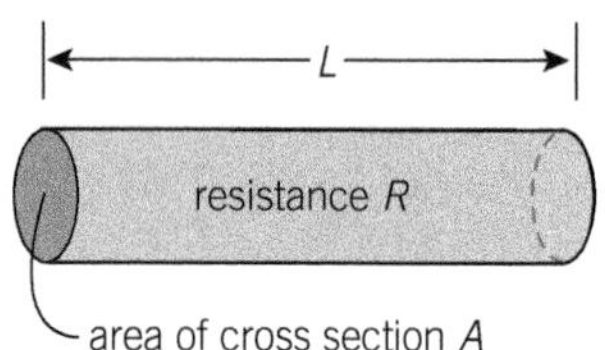

▲ **Figure 1** *Factors affecting resistance*

Resistivity equation

Figure 1 shows a conductor of length L and cross-sectional area A. For a given temperature, $R \propto L$ and $R \propto \frac{1}{A}$. Therefore:

$$R \propto \frac{L}{A}$$

or

$$R = \frac{\rho L}{A}$$

where ρ is a constant for the material known as its **resistivity**. The SI unit for resistivity is $\Omega\,\text{m}$. Metals have much smaller values of resistivity compared with insulators – metals have a greater number density of free electrons. For example: $\rho_{\text{silver}} = 1.6 \times 10^{-8}\,\Omega\,\text{m}$ and $\rho_{\text{glass}} \approx 10^{12}\,\Omega\,\text{m}$.

Revision tip

Resistance and resistivity are very different.

Resistivity is a property of a *material*, whereas resistance depends on the *physical dimensions* of the material.

Resistivity and temperature

Metals: Increasing the temperature of a metal increases the frequency and amplitude of vibration of the fixed metal ions. Therefore, the conducting electrons collide more frequently with the vibrating ions and this subsequently leads to greater resistance. The temperature of a metal has almost negligible effect on the number density of the electrons.

Semiconductors: Temperature affects the number density of charge carriers in a semiconductor. The number density increases with temperature and this leads to a dramatic decrease in resistance of the semiconductor.

Worked example: Very long

A student decides to make a '100 Ω resistor' from a length of manganin wire of cross-sectional area 0.70 mm². The resistivity of manganin is $4.8 \times 10^{-7}\,\Omega\,\text{m}$.

Calculate the length of the wire and suggest if the student's idea is feasible.

Step 1: Write down all the quantities given and convert the area into m². $[1\,\text{mm}^2 = 10^{-6}\,\text{m}^2]$

$R = 100\,\Omega$ $\quad \rho = 4.8 \times 10^{-7}\,\Omega\,\text{m}$ $\quad A = 0.70 \times 10^{-6}\,\text{m}^2$ $\quad L = ?$

Step 2: Use the resistivity equation to calculate the length.

$$L = \frac{RA}{\rho} = \frac{100 \times 0.70 \times 10^{-6}}{4.8 \times 10^{-7}} = 146\,\text{m (3 s.f.)}$$

Step 3: Make a suitable comment.

A resistor, about 150 m long, is just not practical in circuits.

The resistor can be made from a material of higher resistivity.

9.8 The thermistor

The resistance of a thermistor depends on its temperature.

Symbol: 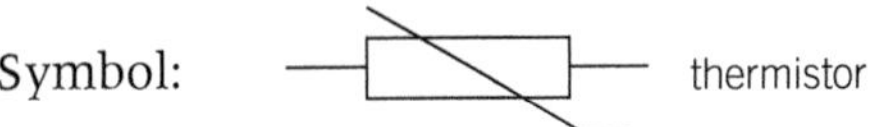

Properties of a thermistor

For an NTC thermistor, its resistance decreases as temperature increases. NTC stands for 'negative temperature coefficient', which means the resistance drops as temperature rises.

Figure 2 shows the variation in resistance of a typical thermistor with temperature.

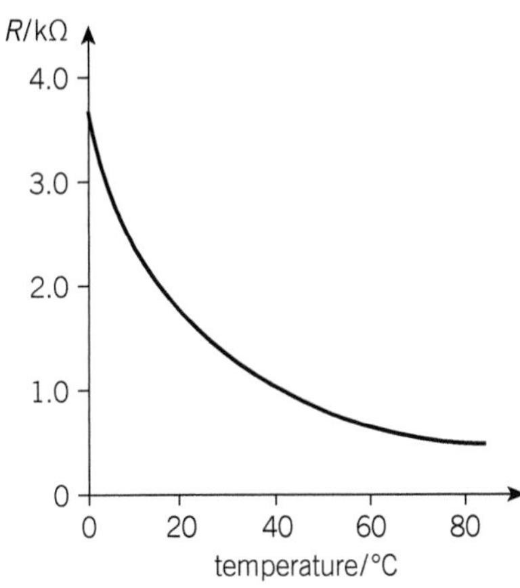

▲ **Figure 2** *Resistance–temperature graph for an NTC thermistor*

9.9 The LDR

The resistance of an LDR depends on the intensity of the incident light.

Symbol: 

Properties of a light-dependent resistor (LDR)

For an LDR, its resistance decreases as intensity of light incident on it increases. Figure 3 shows the variation of resistance of an LDR with light intensity.

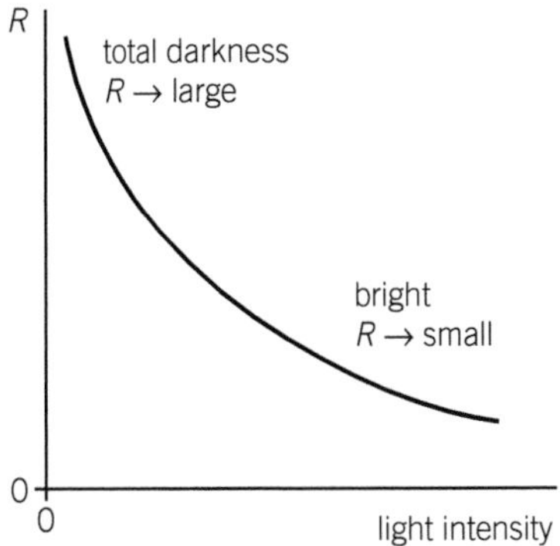

▲ **Figure 3** *Resistance–intensity graph for an LDR*

Summary questions

Data: resistivity of aluminium = $2.7 \times 10^{-8}\,\Omega\,m$
resistivity of silver = $1.6 \times 10^{-8}\,\Omega\,m$

1. The resistance of a length of wire is 2.3 Ω. Calculate the resistance of a similar wire that is 4 times longer. *(2 marks)*
2. Calculate the resistance of a cube of aluminium across its opposite ends when all the sides are 1.0 m. *(2 marks)*
3. Calculate the resistance of a 1.00 m long silver wire of cross-sectional area $1.2 \times 10^{-6}\,m^2$. *(2 marks)*
4. Calculate the resistance of an aluminium cable of radius 1.0 cm and 1.0 km long. *(3 marks)*
5. Use Figure 2 and the temperature ranges 0 °C to 20 °C and 20 °C to 40 °C to show the effect temperature has on the resistance of the thermistor. *(3 marks)*
6. A wire has resistance 12 Ω. It is pulled until its length is 10 times longer. Calculate its resistance now. State any assumption made. *(3 marks)*

Revision tip

For both thermistors and LDRs, the resistance decreases because of an increase in the number of charge carriers.

9.10 Electrical energy and power
9.11 Paying for electricity

Specification reference: 4.2.5

9.10 Electrical energy and power

A current-carrying component will transfer electrical energy into other forms. A resistor will transfer energy into thermal energy; whereas an LED will predominantly transfer energy into light. The term power is used to show how quickly this energy is transferred.

Electrical energy

Power is the rate of energy transfer.

For a current-carrying component, the power P is given by the equation

$$P = VI$$

where V is the p.d. across the component and I is the current in the component. The SI unit for power is $\mathrm{J\,s^{-1}}$ or watt (W). Therefore, $1\,\mathrm{W} = 1\,\mathrm{J\,s^{-1}}$.

The energy W transferred by a component is the product of power P and the time t. Therefore

$$W = VIt$$

In a period of 1 hour (3600 s), a 240 V lamp with a current of 0.25 A will transfer 216 kJ of electrical energy into heat and light. (Check this for yourself.)

Synoptic link

You have already met power in Topic 5.4, Power and efficiency.

Revision tip

$1\,\mathrm{W} = 1\,\mathrm{J\,s^{-1}}$

A power of 20 W just means $20\,\mathrm{J\,s^{-1}}$.

Common misconception

Do not confuse the unit watt (W) with the quantity work done (energy transfer) W – both are represented by the same letter.

Other equations for power

For a resistor, its constant resistance R is given by the equation $R = \frac{V}{I}$. Substituting for V and I into the power equation $P = VI$ gives two more useful equations

$$P = VI = (IR) \times I \qquad \therefore P = I^2R$$

and

$$P = VI = V \times \left(\frac{V}{R}\right) \qquad \therefore P = \frac{V^2}{R}$$

Worked example: Maximum current

A 12 Ω resistor can safely dissipate 0.50 W. Calculate the maximum current in mA for this resistor.

Step 1: Write down the quantities given in the question.

$R = 12\,\Omega \qquad P = 0.50\,\mathrm{W} \qquad I = ?$

Step 2: Select the correct power equation, rearrange the equation, substitute values, and calculate the current in ampere (A)

$P = I^2R$

$I = \sqrt{\frac{P}{R}} = \sqrt{\frac{0.50}{12}} = 0.20\,\mathrm{A}$ (2 s.f.)

Step 3: Change the current to mA. (1000 mA = 1 A)

current = 0.20 × 1000 = 200 mA (2 s.f.)

9.11 Paying for electricity

The joule (J) is the SI unit for energy. When considering energy used in domestic and industrial situations, the unit joule is too small. Electricity companies use the kilowatt-hour (kWh) as a convenient unit for billing customers.

Kilowatt-hour

The **kilowatt-hour** (kWh) is defined as the energy transferred by a device with a power of 1 kW operating for a time of 1 hour.

The amount of energy in kWh is calculated as follows

$$\text{number of kWh} = \text{power in kW} \times \text{time in hours}$$

The cost of operating a device is calculated as follows

$$\text{cost} = \text{number of kWh} \times \text{cost per kWh}$$

Revision tip

The kilowatt-hour is an alternative unit for **energy**.

1 kWh = 3.6 MJ

Worked example: Charging your phone

A 6.0 W phone charger is used on average for 3.0 hours every day. Calculate the annual cost of using the charger. Assume the cost of each kWh is 8.1p.

Step 1: Calculate the total energy in kWh.

number of kWh for 1 day = 0.006 kW × 3.0 h

number of kWh for 1 year = 0.006 × 3.0 × 365 = 6.57 kWh

Step 2: Calculate the cost.

cost = 6.57 × 8.1 = 53p (2 s.f.)

cost = £0.53 (2 s.f.)

Summary questions

1. State an alternative unit for W. *(1 mark)*
2. Calculate the current in a 2000 W, 240 V appliance. *(2 marks)*
3. Calculate the power dissipated by a 100 Ω resistor connected to a 240 V supply. *(2 marks)*
4. A 3000 W cooker is used for 4.0 hours. The cost of each kWh is 8.1p.
   Calculate the cost of using this cooker. *(2 marks)*
5. Calculate the heat energy in joules produced by a 60 W heater operated for 10 hours. *(2 marks)*
6. Calculate the resistance of the filament of a 12 V, 36 W lamp. *(2 marks)*
7. Use the definition of current and potential difference to show $P = VI$. *(3 marks)*
8. The power dissipated in a wire connected to a fixed power supply is 10 W.
   Calculate the power dissipated in the same wire connected to the same supply when it is pulled to double its original length. *(4 marks)*

Chapter 9 Practice questions

1 As the temperature of an NTC thermistor is increased, which statement is correct? Its resistance ...

A increases because its volume increases

B increases because it does not obey Ohm's law

C decreases because there are more free electrons

D decreases because the ions vibrate more. *(1 mark)*

2 The current in a $100\,\Omega$ resistor is 0.20 A.

What is the energy dissipated by the resistor in a period of 3.0 s?

A 4.0 J | **C** 20 J

B 12 J | **D** 60 J *(1 mark)*

3 A wire of length L and diameter d has resistance R.

Another wire made from the same metal has length $4L$ and diameter $\frac{d}{2}$.

What is the resistance of this new wire in terms of R?

A R | **C** $4R$

B $2R$ | **D** $16R$ *(1 mark)*

4 What are the base units for resistance?

A $\text{kg}\,\text{m}\,\text{s}^{-2}$

B $\text{kg}\,\text{m}^2\,\text{s}^{-1}$

C $\text{kg}\,\text{m}^2\,\text{A}^{-1}\,\text{s}^{-1}$

D $\text{kg}\,\text{m}^2\,\text{A}^{-2}\,\text{s}^{-3}$ *(1 mark)*

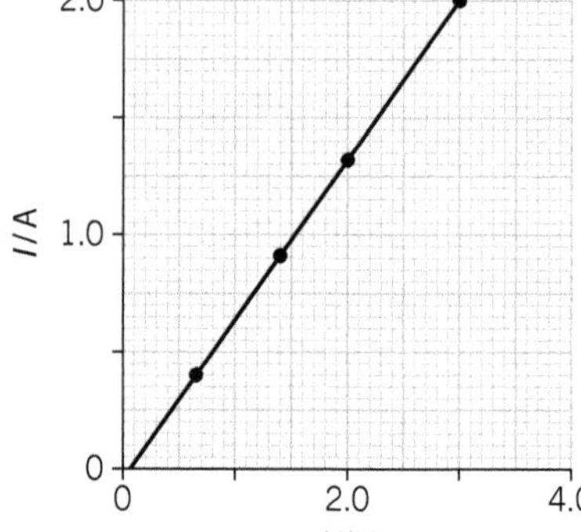

▲ **Figure 1**

5 a Define resistance. *(1 mark)*

b Figure 1 shows a graph of current I against potential difference V for a pencil 'lead' drawn by a student.

The pencil lead is 9.0 cm long and has diameter 0.82 mm.

i Use Figure 1 to determine the resistance of the pencil lead. *(1 mark)*

ii Calculate the resistivity of the material of the pencil lead. *(3 marks)*

iii Calculate the current in the pencil lead when its ends are connected to a supply of e.m.f. 2.5 V and negligible internal resistance. *(1 mark)*

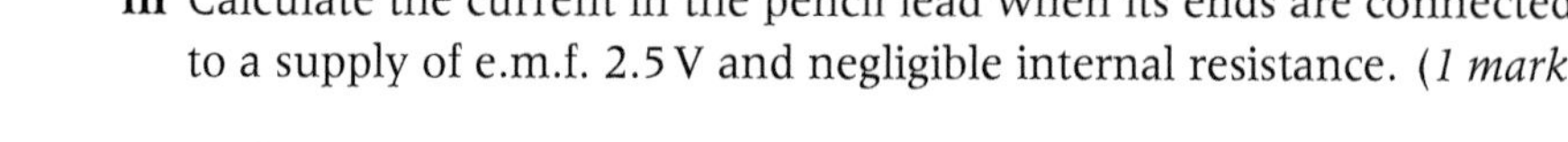

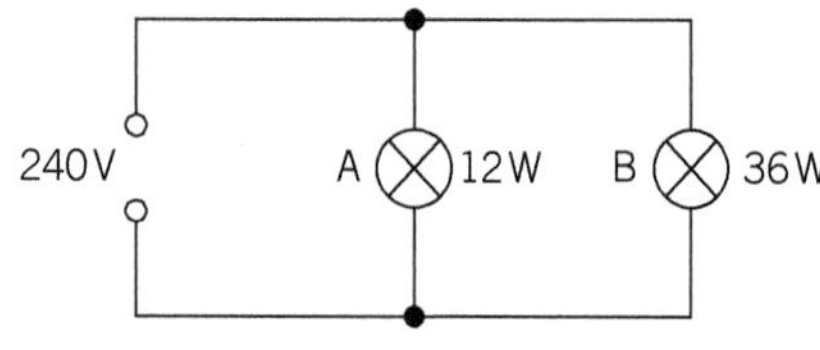

▲ **Figure 2**

6 Figure 2 shows two filament lamps A and B connected to a supply.

Lamps A and B are labelled as 230 V, 12 W and 230 V, 36 W, respectively.

a Explain what is meant by 12 W. *(2 marks)*

b Calculate the current in each lamp and therefore the total current delivered by the supply. *(3 marks)*

c The lamps shown in Figure 2 are operated for 24 hours.

Calculate the cost of operating the lamps given the cost of each kWh is 10.5p. *(3 marks)*

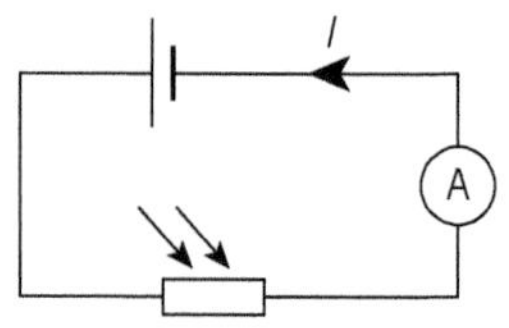

▲ **Figure 3**

7 Figure 3 shows a light-dependent resistor connected to a cell.

The cell has negligible internal resistance.

a Explain how the current I in the circuit changes as the intensity of the light falling on the LDR is decreased. *(2 marks)*

b Explain whether the LDR will dissipate greater power when in darkness or in sunlight. *(2 marks)*

Chapter 10 Electrical circuits

In this chapter you will learn about ...

- ☐ Kirchhoff's second law
- ☐ Series and parallel circuits
- ☐ Internal resistance
- ☐ Terminal p.d.
- ☐ Potential divider circuits
- ☐ Sensing circuits

10 ELECTRICAL CIRCUITS
10.1 Kirchhoff's laws and circuits
10.2 Combining resistors
10.3 Analysing circuits

Specification reference: 4.3.1

10.1 Kirchhoff's second law

You have already met Kirchhoff's first law in Topic 8.3. Kirchhoff's second law is a consequence of conservation of energy. Both laws are invaluable when analysing circuits.

The second law

Kirchhoff's second law: In any circuit, the sum of the electromotive forces is equal to the sum of the p.d.s around a closed loop.

This can be written mathematically as

$\Sigma\mathcal{E} = \Sigma V$ (Σ means 'sum of...')

or

$\Sigma\mathcal{E} = \Sigma IR$

where ε is the e.m.f., V is the p.d., and I is the current in a resistor of resistance R.

Revision tip

Kirchhoff's first law expresses conservation of *charge* and the second law expresses conservation of *energy*.

10.2 Combining resistors

Components can be connected in series (end-to-end), in parallel (across each other) or in a combination of series and parallel.

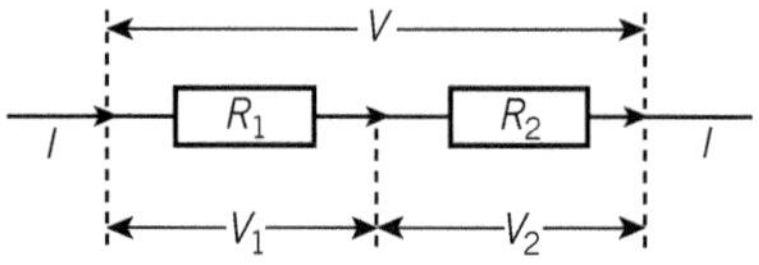

▲ **Figure 1** *Resistors in series*

Series

Figure 1 shows two resistors of resistance R_1 and R_2 connected in series.

Rules for a series circuit:

- The current I in each resistor is the same. (Kirchhoff's first law.)
- The total p.d. V across the resistors is the sum of the individual p.d.s. (Kirchhoff's second law.)

  $V = V_1 + V_2 + \dots$
- The total resistance R of the combination is given by $R = R_1 + R_2 + \dots$

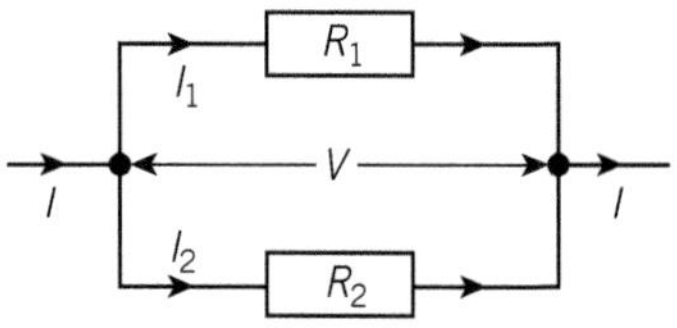

▲ **Figure 2** *Resistors in parallel*

Parallel

Figure 2 shows two resistors of resistance R_1 and R_2 connected in parallel.

Rules for a parallel circuit:

- The current I is the sum of the individual currents. (Kirchhoff's first law.)

  $I = I_1 + I_2 + \dots$
- The p.d. V across each resistor is the same. (Kirchhoff's second law.)
- The total resistance R of the combination is given by $\frac{1}{R} = \frac{1}{R_1} + \frac{1}{R_2} + \dots$

Revision tip

For resistors in parallel, the total resistance is always less than the smallest resistance value.

Maths: The parallel rule

The parallel rule for resistors may be written as $R = (R_1^{-1} + R_2^{-1} + \dots)^{-1}$. This is easier to do on a calculator.

10.3 Analysing circuits

There are some important equations and ideas you will need when analysing circuits.

Equations

- $I = \dfrac{\Delta Q}{\Delta t}$
- $W = VQ$
- $V = IR$
- $P = VI$ $\quad P = I^2R \quad$ $P = \dfrac{V^2}{R}$
- $R = R_1 + R_2 + \ldots$ $\quad \dfrac{1}{R} = \dfrac{1}{R_1} + \dfrac{1}{R_2} + \ldots$

Ideas

In a series circuit:

- the current in each component is the same
- the total p.d. across components is the sum of the p.d.s.

In a parallel circuit:

- the p.d. across each component is the same
- the total current is the sum of the currents.

Worked example: Predicting p.d.

Calculate the p.d. across the 10 Ω resistor in Figure 3.

Step 1: Calculate the total resistance of the resistors in parallel.

$$R = (R_1^{-1} + R_2^{-1})^{-1} = (12^{-1} + 18^{-1})^{-1} = 7.2\,\Omega$$

Step 2: Calculate the total resistance of the circuit.

$$R = R_1 + R_2 = 10 + 7.2 = 17.2\,\Omega$$

Step 3: Use Kirchhoff's laws to calculate the current in the circuit.

The current I is the same and $\Sigma\varepsilon = \Sigma Ir$

$$\therefore 1.5 + 1.5 = I \times 17.2$$

$$I = 0.174\text{ A}$$

Step 4: Use $V = IR$ to calculate the p.d. across the resistor.

$$V = 0.174 \times 10$$

$$V = 1.7\text{ V (2 s.f.)}$$

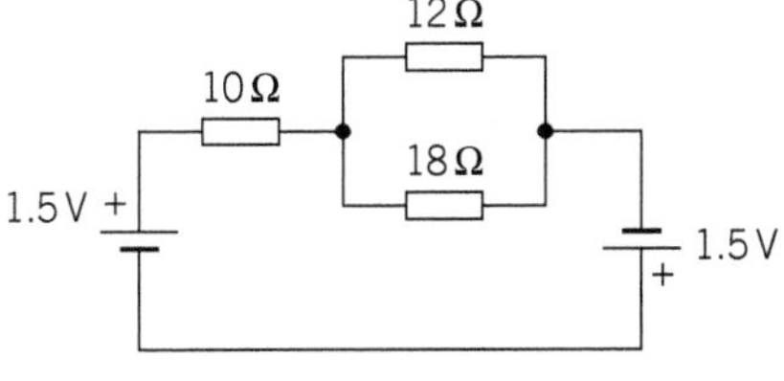

▲ Figure 3

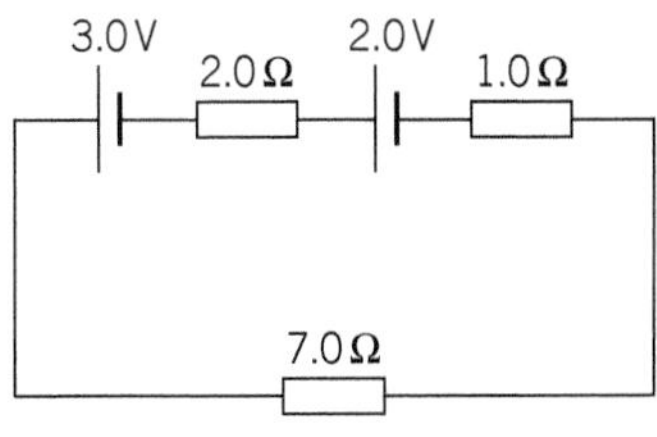

▲ Figure 4

Summary questions

1. You are given two 120 Ω resistors. Calculate the total resistance when these are connected in series and then in parallel. *(3 marks)*
2. A battery of e.m.f. 6.0 V is connected across 10 identical lamps. Calculate the p.d. across each lamp. *(1 mark)*
3. Calculate the total resistance and the e.m.f. for the circuit shown in Figure 4. *(2 marks)*
4. Calculate the current in the circuit shown in Figure 4 and the p.d. across the 7.0 Ω resistor. *(3 marks)*
5. A 100 Ω resistor is connected across a resistor of resistance R. A multimeter shows the total to be 70 Ω. Calculate the value of R. *(3 marks)*
6. A thermistor and a 100 Ω resistor are connected in series to a fixed power supply. The temperature of the thermistor is slowly decreased. Describe and explain the effect this has on the p.d. across the resistor. *(3 marks)*

10.4 Internal resistance
10.5 Potential divider circuits
10.6 Sensing circuits

Specification reference: 4.3.2, 4.3.3

10.4 Internal resistance

A source of e.m.f. has internal resistance. For example, a solar cell has internal resistance because the charges have to pass through the material of the cell.

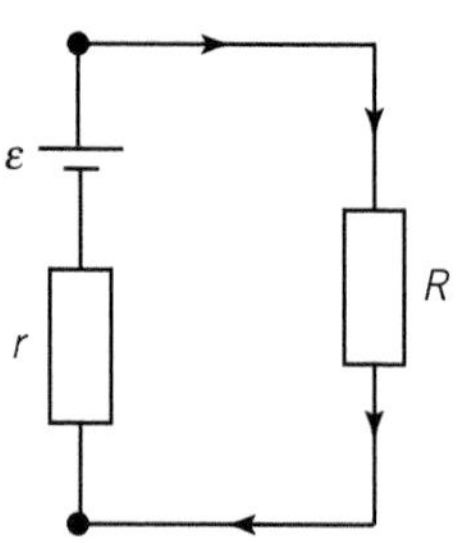

▲ **Figure 1** *A source of e.m.f. connected to an external 'load' of resistance R*

Terminal p.d. and 'lost volts'

A source of e.m.f. can be represented by an e.m.f. ε in series with its internal resistance r. Figure 1 shows a source of e.m.f. providing current I to an external resistor, of resistance R.

The current is the same in both resistors (Kirchhoff's first law). Using Kirchhoff's second law, we have

$\varepsilon = IR + Ir$ or $\varepsilon = I(R + r)$ or $\varepsilon = V + Ir$

where V is the terminal p.d., or simply the potential difference across the external resistor, and Ir is the p.d. across the internal resistor (which is also referred to as the 'lost volts').

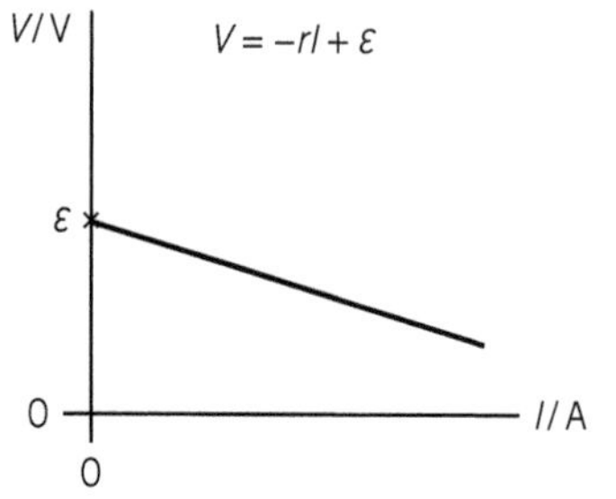

▲ **Figure 2** *The gradient of the graph is −r and the V-intercept is ε*

Determining e.m.f. and internal resistance

You can get different I and V values by varying the value of R. Since $\varepsilon = V + Ir$, we have

$V = -Ir + \varepsilon$

The equation for a straight line is $y = mx + c$

Therefore, a graph of V against I will be a straight line with:

- gradient = $-r$
- y-intercept = ε

Revision tip
The terminal p.d. is always less than, or equal to, the e.m.f.

10.5 Potential divider circuits

A potential divider circuit consists of two or more components connected to a battery or a power supply with the 'output' p.d. taken across one of the components, see Figure 3.

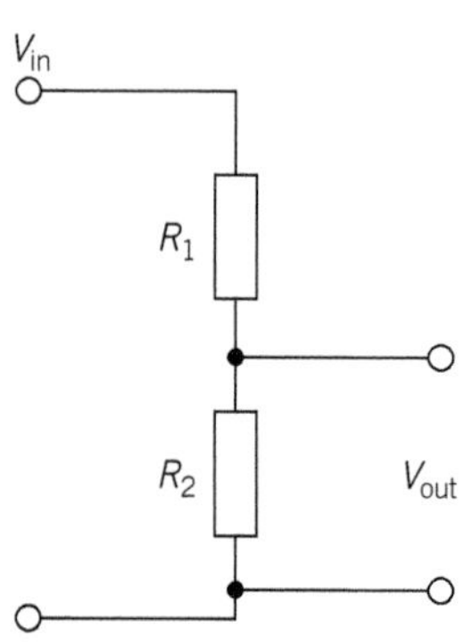

▲ **Figure 3** *A potential divider circuit*

Potential divider equations

The output p.d. V_{out} is taken across the resistor of resistance R_2 – this is a fraction of the total p.d. V_{in} across the resistors. The bigger the resistance R_2 the larger is V_{out}.

The output p.d. is given by the equation

$$V_{out} = \frac{R_2}{R_1 + R_2} \times V_{in}$$

The current in the resistors is the same therefore

$$I = \frac{V_1}{R_1} = \frac{V_2}{R_2} \quad \text{or} \quad \frac{V_1}{V_2} = \frac{R_1}{R_2}$$ (V_1 and V_2 are the p.d.s across the resistors)

10.6 Sensing circuits

You can use the potential divider equation to analyse a temperature-sensing circuit (Figure 4) and a light-sensing circuit (Figure 5).

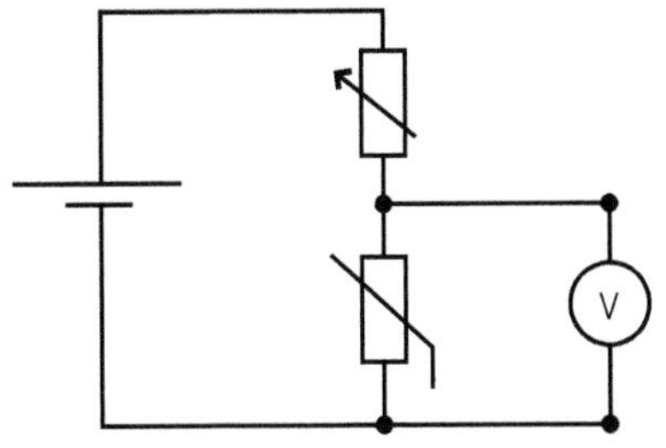

▲ **Figure 4** *Temperature-sensor has a thermistor*

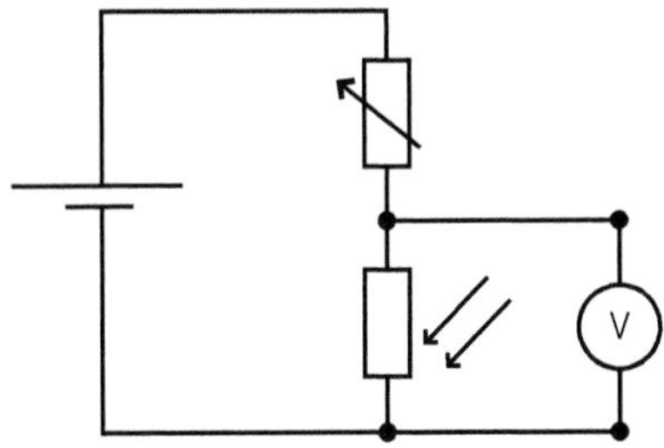

▲ **Figure 5** *Light-sensor has an LDR*

Worked example: Terminal p.d.

The battery in Figure 6 has e.m.f. 6.0 V and internal resistance 2.0 Ω. The maximum resistance of the variable resistor is 10 Ω. Calculate the maximum and minimum values of the terminal p.d.s.

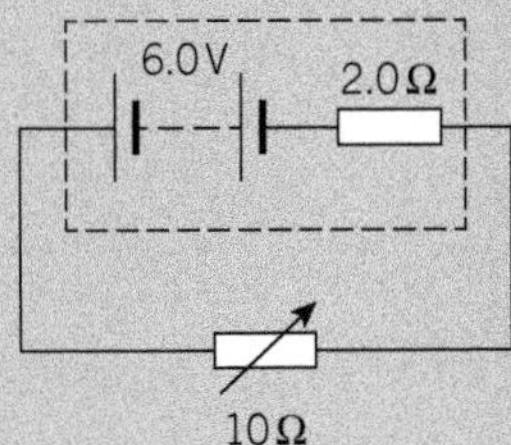

▲ **Figure 6**

Step 1: Calculate the maximum current and therefore the minimum terminal p.d.

$R = 0$; total resistance = 0 + 2.0 = 2.0 Ω

$$\text{current} = \frac{\mathcal{E}}{\text{total resistance}} = \frac{6.0}{2.0} = 3.0\,\text{A}$$

minimum terminal p.d. $V = IR = 3.0 \times 0 = 0\,\text{V}$

Step 2: Calculate the minimum current and therefore the maximum terminal p.d.

$R = 10\,\Omega$; total resistance = 10 + 2.0 = 12.0 Ω

$$\text{current} = \frac{\mathcal{E}}{\text{total resistance}} = \frac{6.0}{12.0} = 0.50\,\text{A}$$

maximum terminal p.d. $V = IR = 0.50 \times 10 = 5.0\,\text{V}$ (2 s.f.)

Summary questions

1 Explain what is meant by 'lost volts'. *(1 mark)*

2 A cell of e.m.f. of 1.4 V is connected across a resistor. The terminal p.d. is 1.0 V. Explain why the p.d. across the resistor is not 1.4 V. *(2 marks)*

3 In **Q2** the resistor has resistance 20 Ω. Calculate the internal resistance *r*. *(2 marks)*

4 In the circuit shown in Figure 4, the cell has an e.m.f. of 1.5 V, the variable resistor is set to 100 Ω, and the resistance of the thermistor is 68 Ω. Calculate the voltmeter reading. *(2 marks)*

5 In Figure 3, $R_2 = 4R_1$. Calculate V_{out} in terms of V_{in}. *(2 marks)*

6 Explain how the calculation in **Q4** would be affected if the voltmeter has a finite resistance. *(3 marks)*

Chapter 10 Practice questions

1 A resistor is connected across a chemical cell.

Which statement is correct about terminal p.d.?

The terminal p.d. ...

A is the 'lost volts'

B is the e.m.f. of the cell

C is the p.d across the external resistor

D is the p.d. across the internal resistance. *(1 mark)*

2 Figure 1 shows part of a circuit with three resistors. The resistance of each resistor is R.

What is the total resistance of the circuit in terms of R?

A $\frac{R}{3}$ **C** $\frac{2R}{3}$

B $\frac{R}{2}$ **D** $3R$ *(1 mark)*

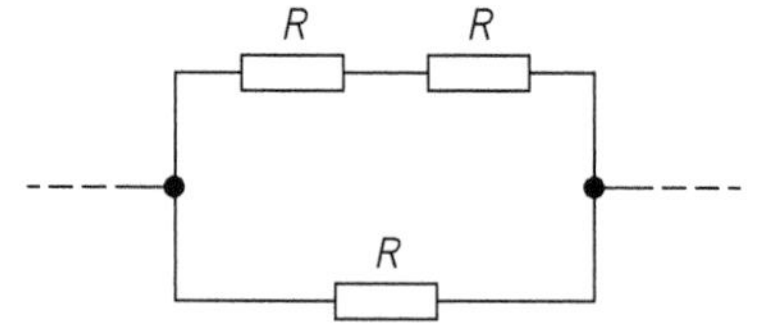

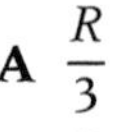

▲ **Figure 1**

3 A battery of negligible internal resistance is connected to a resistor of resistance R. The current in the resistor is 3.0 A. A resistor of resistance 20 Ω is connected in series with the resistor of resistance R to the same battery. The current in the circuit is now 1.0 A.

What is the value of R?

A 10 Ω **C** 30 Ω

B 20 Ω **D** 40 Ω *(1 mark)*

4 Figure 2 shows a circuit consisting of resistors.

The battery has negligible internal resistance.

The current in the 330 Ω resistor is 2.7 mA. Calculate:

a the total resistance of the circuit; *(3 marks)*

b the e.m.f. ε of the battery. *(4 marks)*

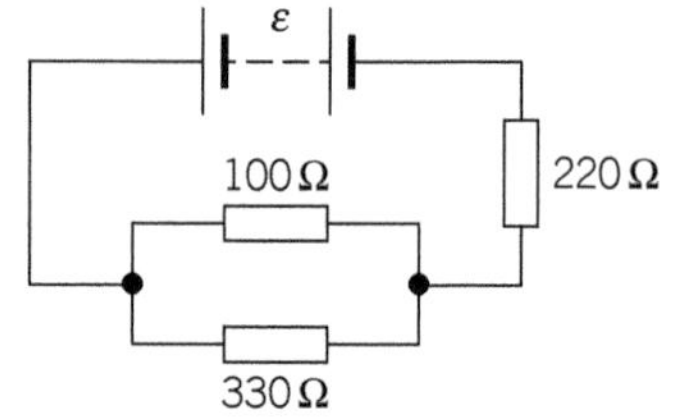

▲ **Figure 2**

5 **a** State Kirchhoff's second law. *(1 mark)*

b Figure 3 shows an electrical circuit.

The battery has e.m.f. 4.5 V and internal resistance r. The potential difference across the filament lamp is 3.0 V and it dissipates energy at a rate of 0.60 W.

Calculate:

i the resistance of the lamp; *(2 marks)*

ii the total resistance of the circuit connected across the terminals of the battery; *(2 marks)*

iii the internal resistance r. *(2 marks)*

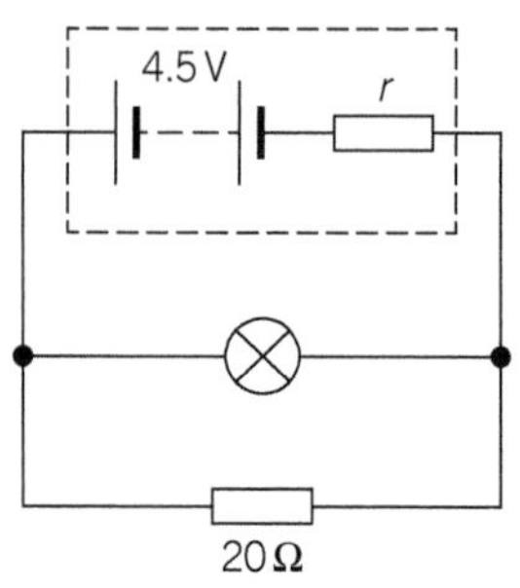

▲ **Figure 3**

6 Figure 4 shows a potential divider circuit designed by a student to monitor the temperature of a room.

The battery has e.m.f. 6.0 V and negligible internal resistance. The digital voltmeter has an infinite resistance. The resistance of the variable resistor is adjusted to 120 Ω. The current delivered by the battery is 15 mA when the thermistor is at room temperature.

a Calculate the resistance of the thermistor at room temperature. *(3 marks)*

b Describe the effect on the voltmeter reading when the resistance of the variable resistor is increased. The temperature of the thermistor is the same as before. *(3 marks)*

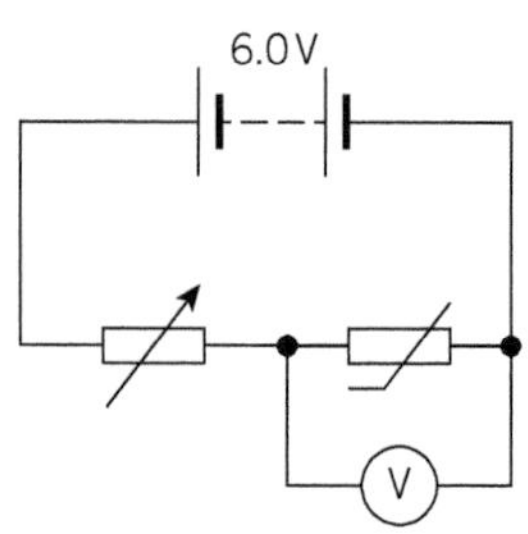

▲ **Figure 4**

Chapter 11 Waves 1

In this chapter you will learn about ...

- [ ] Transverse and longitudinal waves
- [ ] Wave properties
- [ ] Reflection and diffraction
- [ ] Polarisation
- [ ] Intensity
- [ ] Electromagnetic waves
- [ ] Refraction and refractive index
- [ ] Critical angle
- [ ] Total internal reflection

11 WAVES 1
11.1 Progressive waves
11.2 Wave properties

Specification reference: 4.4.1

11.1 Progressive waves

A progressive wave is an oscillation that travels through space. All progressive waves transfer energy from one place to another.

Two types of waves

In a **transverse wave** the oscillations or vibrations are perpendicular to the direction of energy transfer. (See Figure 1.)

In a **longitudinal wave** the oscillations are parallel to the direction of energy transfer. (See Figure 2.)

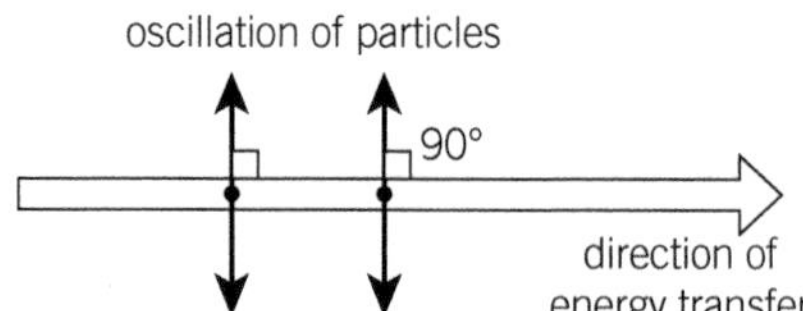

▲ **Figure 1** *Transverse*

11.2 Wave properties

Several quantities are needed to describe and analyse waves. Learn the quantities defined in Table 1 and take care not to confuse them.

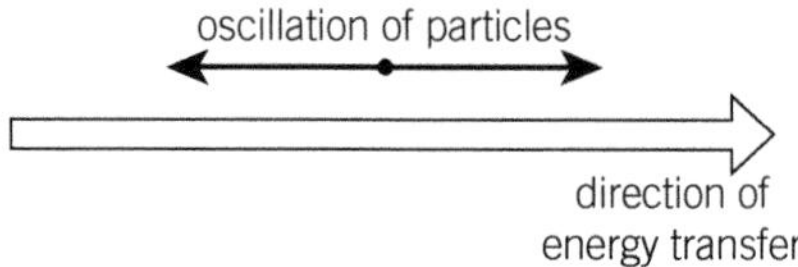

▲ **Figure 2** *Longitudinal*

Definitions

▼ **Table 1**

Quantity and definition	Comments
Displacement is the distance from the equilibrium position in a particular direction.	Symbol → s; unit → m; a vector quantity
Amplitude is the maximum displacement of a wave. (See Figure 3.)	Symbol → A; unit → m
Wavelength is the distance between two adjacent points on a wave that are oscillating in phase (in step). (See Figure 3.)	Symbol → λ; unit → m; it is also the distance between adjacent peaks or troughs
Period is the time taken for the wave to move a distance of one whole wavelength.	Symbol → T; unit → s; it is also the time taken for a medium particle to make one complete oscillation
Frequency is the number of wavelengths passing though a point per unit time.	Symbol → f; unit → Hz; it is also the number of oscillations of a medium particle per unit time
Wave speed is the distance travelled by the wave per unit time.	Symbol → v; unit → m s^{-1}

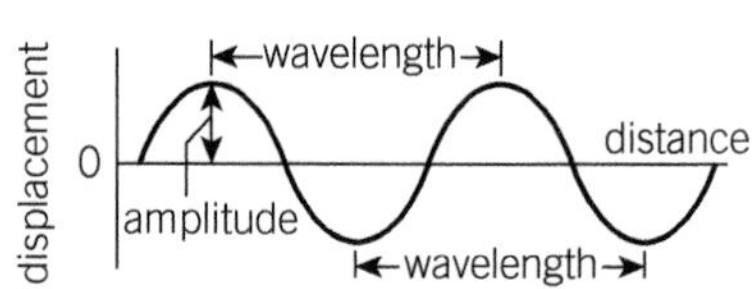

▲ **Figure 3** *A displacement–distance graph showing amplitude and wavelength*

Frequency and period

The relationship between period T and frequency f is $f = \frac{1}{T}$

For example, when $T = 0.010$ s then $f = \frac{1}{0.010} = 100$ Hz.

Wave equation

The relationship between wave speed v, wavelength λ, and frequency f is called the wave equation. The wave equation $v = f\lambda$ can be used for any periodic wave.

Proof:

- Distance travelled by wave in one oscillation = λ.
- There are f oscillations per unit time.
- Therefore, distance travelled per unit time = wave speed $v = f \times \lambda$.

Graphs

A displacement–distance graph of the wave is a snapshot of the wave, see Figure 4.

Imagine the wave travelling to the right. After a time of one period *T*, the peak A will be where peak E is shown.

A displacement–time graph of the wave shows the motion of a specific point in the medium, see Figure 5. The time for one complete oscillation is the period *T*.

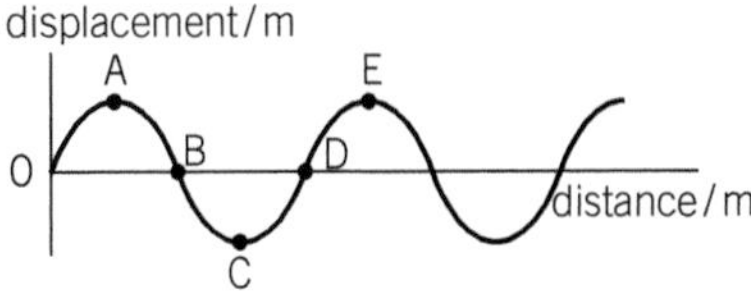

▲ **Figure 4** *A displacement–distance graph of a wave*

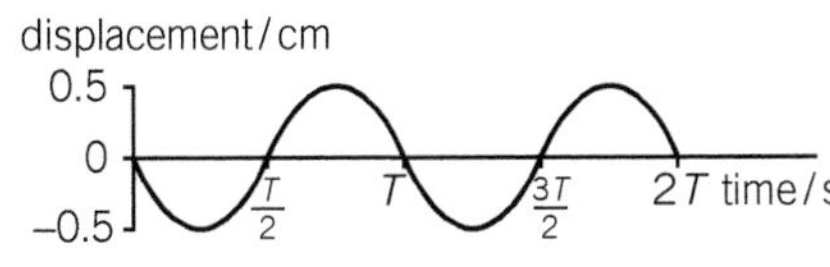

▲ **Figure 5** *A displacement–time graph shows the oscillations of a specific particle*

Phase difference

Phase difference is the fraction of an oscillation between two oscillating particles.

The term phase difference can be used for two points on two different waves or along the same wave.

Phase difference is measured in degrees or in radians, see Table 2.

▼ **Table 2**

separation between points	phase difference / °	phase difference / rad
λ	360	2π
$\frac{\lambda}{2}$	180	π
$\frac{\lambda}{4}$	90	$\frac{\pi}{2}$
x	$\frac{x}{\lambda} \times 360$	$\frac{x}{\lambda} \times 2\pi$

Maths: Angles

$2\pi = 360°$ therefore 1 rad = 57.3°

Worked example: What can we hear?

The audible frequency range for humans is 20 Hz to 20 kHz.

Calculate the longest and shortest wavelengths humans can hear.

speed of sound in air = 340 m s^{-1}

Step 1: Use the wave equation to calculate the longest wavelength.

$v = 340\ \text{m s}^{-1}$ $\quad f = 20\ \text{Hz}$ $\quad \lambda = ?$

$$\lambda = \frac{v}{f} = \frac{340}{20} = 17\text{m}$$

Step 2: Use the wave equation to calculate the shortest wavelength.

$v = 340\ \text{m s}^{-1}$ $\quad f = 20 \times 10^{3}\ \text{Hz}$ $\quad \lambda = ?$

$$\lambda = \frac{v}{f} = \frac{340}{20 \times 10^{3}} = 1.7 \times 10^{-2}\text{m}$$

The longest wavelength is 17 m and the shortest wavelength is 1.7 cm (2 s.f.).

Summary questions

1. Explain what is meant by a frequency of 10 Hz. *(1 mark)*
2. A wave has wavelength of 2.0 cm. State the distance travelled by the wave in one period. *(1 mark)*
3. Water waves have frequency 30 Hz and wavelength 4.0 cm. Calculate the speed of the water waves. *(1 mark)*
4. Bats can emit sound waves of frequency 150 kHz. The speed of sound is 340 m s^{-1}. Calculate the wavelength corresponding to this frequency. *(2 marks)*
5. Use Figure 4 to determine the phase difference between points:
   a B and D; *(2 marks)*
   b A and D. *(2 marks)*
6. A wave has a wavelength of 20 cm. Calculate the distance between two points with a phase difference of 300°. *(2 marks)*

11.3 Reflection and refraction
11.4 Diffraction and polarisation

Specification reference: 4.4.1

11.3 Reflection and refraction

All progressive waves can be reflected and refracted.

Reflection

Reflection occurs when a wave changes direction at a boundary between two different media, remaining in the original medium.

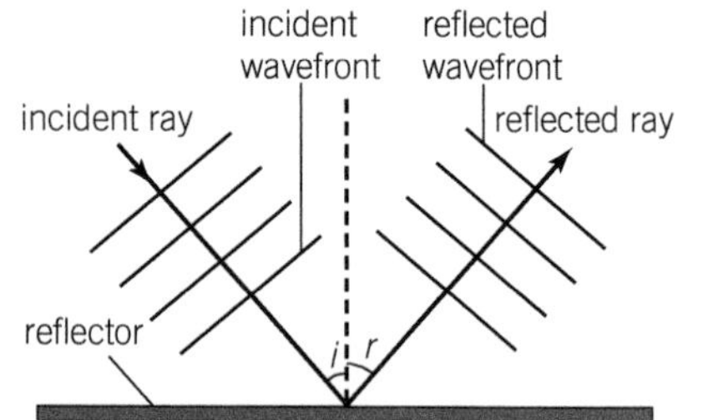

▲ **Figure 1** *Reflection*

Light is reflected when you look at yourself in the mirror. Figure 1 shows the reflection of a wave. The separation between adjacent wavefronts is equal to the wavelength. A wavefront is a line joining all neighbouring points oscillating in phase.

- The incident and reflected rays are in the same plane.
- The angle of incidence is equal to the angle of reflection. (The angles are measured relative to the normal – dotted line.)

Refraction

Refraction occurs when a wave changes direction as it changes speed when it passes from one medium to another.

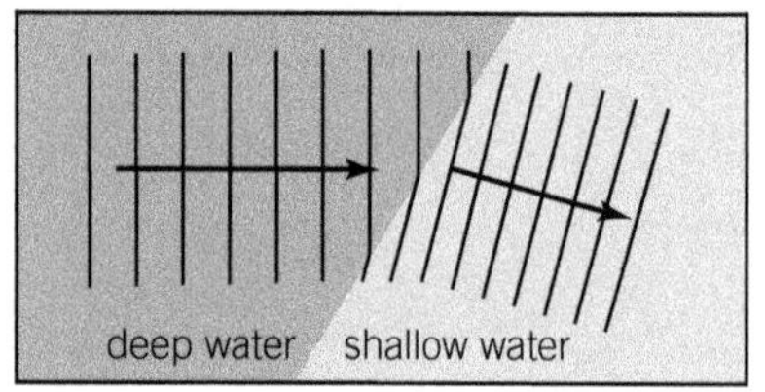

▲ **Figure 2** *Refraction*

Refraction will be covered more in Topic 8.3, Refractive index.

Figure 2 shows the refraction of water waves at the boundary between deep and shallow water.

- The frequency of the refracted waves remains constant.
- The speed, and therefore the wavelength, of the wave changes.

11.4 Diffraction and polarisation

All waves can be diffracted. Only transverse waves can be polarised.

Diffraction

Diffraction is the spreading of a wave when it passing through a narrow gap or around an obstacle.

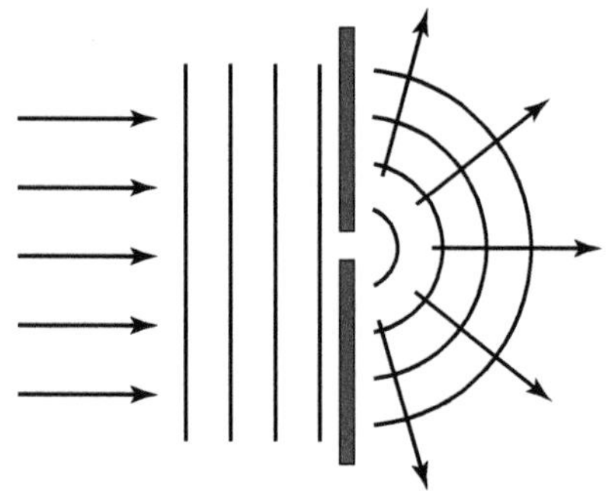

▲ **Figure 3** *Diffraction*

Figure 3 shows the diffraction of a wave at a gap.

- The speed, wavelength, and frequency of a wave do not change when diffraction occurs.
- There is significant diffraction when the width of the gap is similar to the wavelength of the incident wave. There is very little diffraction when the wavelength is much much smaller than the width of the gap.

Polarisation

A plane polarised wave has oscillations in one plane only.

Ordinary light is unpolarised with oscillations in many planes at right angles to the direction of travel. Unpolarised light passing through a single polarising filter (Polaroid) will be plane polarised. Figure 4 shows polarised and unpolarised oscillations.

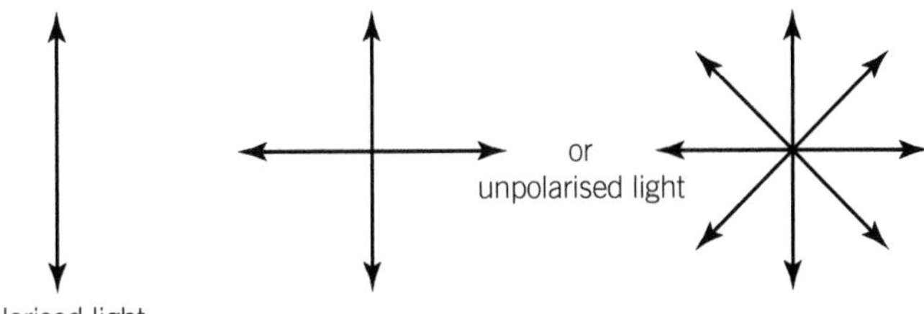

▲ **Figure 4** *Polarised and unpolarised*

You can show polarisation effects using two polarising filters. Figure 5 shows the effects of using two polarising filters. In (a), the transmission axes of the filters are aligned – light passing through the filters is plane polarised. In (c), the transmission axes of the filters are at right angles – light is blocked by the overlapping filter.

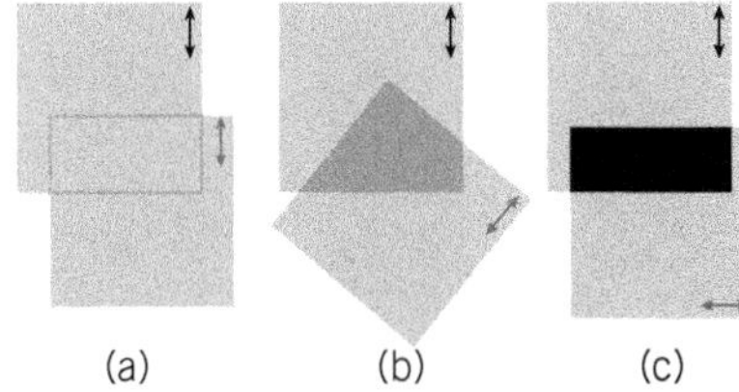

▲ **Figure 5** *Using polarising filters*

Worked example: What can we hear?

Estimate the frequency of sound that can be diffracted at a slit of width 3.0 cm. The speed of sound in air is 340 m s^{-1}.

Step 1: Estimate the wavelength of sound.

The wavelength λ must be similar to the width for diffraction to occur. Therefore, λ is about 3.0 cm.

Step 2: Write down the quantities given.

$v = 340\,\text{m s}^{-1}$ $\qquad \lambda \approx 3.0 \times 10^{-2}\,\text{m}$ $\qquad f = ?$

Step 3: Use the wave equation to estimate the frequency f of sound.

$$f = \frac{v}{\lambda} = \frac{340}{3.0 \times 10^{-2}} = 1.13 \times 10^{4}\,\text{Hz}$$

The frequency is about 11 kHz (2 s.f.).

Summary questions

1 State the quantities that remain the same when light is reflected off a mirror. *(1 mark)*

2 Explain why sound cannot be polarised. *(1 mark)*

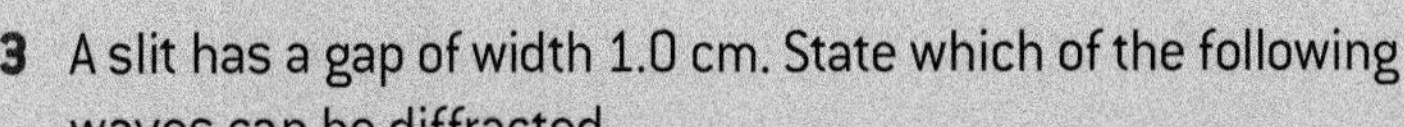

3 A slit has a gap of width 1.0 cm. State which of the following waves can be diffracted.

   a Light of wavelength 500 nm *(1 mark)*

   b Microwaves of wavelength 3.0 cm. *(1 mark)*

4 State one quantity that changes when a wave refracts. *(1 mark)*

5 Draw a diagram to show the diffraction of parallel wavefronts at the edge of an obstacle. *(1 mark)*

6 Use Figure 6 to state and explain the change of wavelength as light travels from glass into water. *(3 marks)*

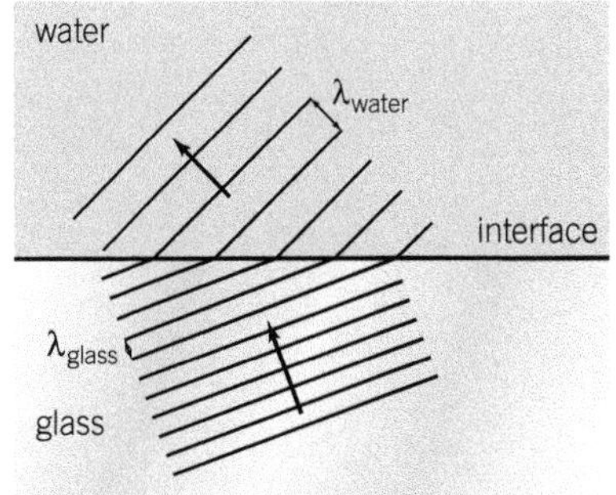

▲ **Figure 6**

11.5 Intensity
11.6 Electromagnetic waves
11.7 Polarisation of electromagnetic waves

Specification reference: 4.4.1, 4.4.2

11.5 Intensity

Sunlight incident at the top of the Earth's atmosphere delivers about 1 kW of power per m^2 or $1 kW\,m^{-2}$. This is the intensity of sunlight. Intensity must not be confused with power or energy.

Intensity

The **intensity** of a progressive wave is defined as the radiant power passing at right angles through a surface per unit area.

Intensity is measured in watts per square metre ($W\,m^{-2}$). The equation relating intensity I, area A, and incident radiant power P is

$$I = \frac{P}{A}$$

Inverse square law

The intensity from a point source of power P decreases with distance r from the source as the power spreads over a larger surface area. The intensity is given by

$$I = \frac{P}{4\pi r^2} \propto \frac{1}{r^2}$$

The intensity obeys an inverse square with distance.

You can use the expression above for the Sun, a filament lamp, and so on.

Intensity and amplitude

The intensity I for a wave is related to the amplitude A of the wave by the following expression

$I \propto A^2$

This means that doubling the amplitude of the wave will quadruple the intensity.

11.6 Electromagnetic waves

Light is just a small part of the spectrum of electromagnetic (EM) waves. Figure 1 shows the electromagnetic spectrum with typical wavelengths of the various regions or principal radiations.

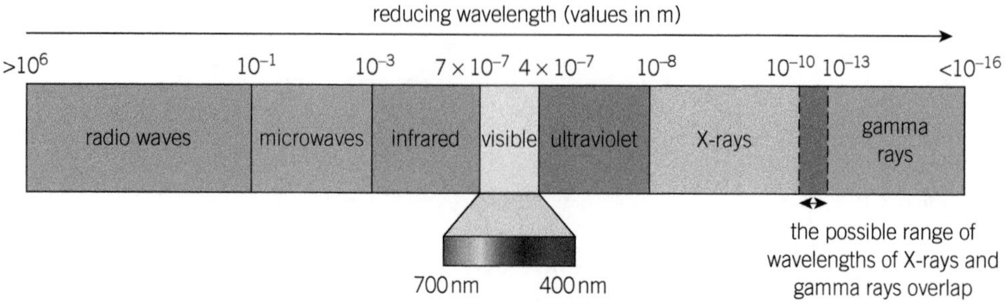

▲ **Figure 1** *The electromagnetic spectrum showing the principal radiations or regions*

Electromagnetic spectrum and properties of EM waves

All electromagnetic waves:

- can travel though a vacuum
- travel at the speed c of $3.0 \times 10^8\,\mathrm{m\,s^{-1}}$ in a vacuum
- are transverse waves
- consist of oscillating magnetic and electric fields at right angles to each other.

You can use the wave equation $c = f\lambda$, where f is the frequency of the electromagnetic wave of wavelength λ.

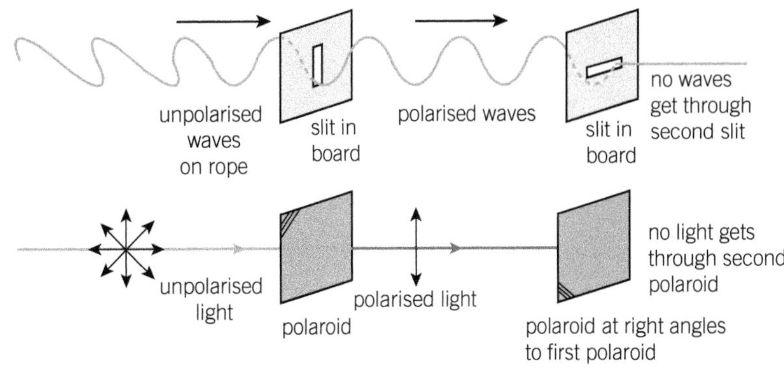

▲ **Figure 2** *Polarisation of light*

11.7 Polarisation of electromagnetic waves

You have already met the idea of plane polarised waves in Topic 11.4, Diffraction and polarisation. All electromagnetic waves can be plane polarised because they are all transverse waves.

Revision tip

You cannot use Polaroids to plane polarise microwaves – you have to use a metal grille.

Light

Visible light can be plane polarised using polarising filters (Polaroid). Figure 2 compares the polarisation of unpolarised waves on a rope and unpolarised light.

Microwaves

The microwave transmitters available in schools and colleges produce plane polarised waves with the electric field oscillating in the vertical plane. In order to show the transmission and absorption of these polarised waves you have to use a metal grille, see Figure 3.

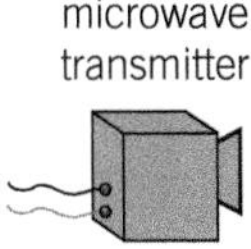

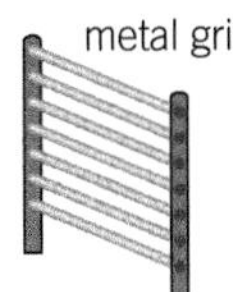

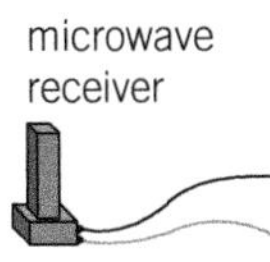

▲ **Figure 3** *A metal grille can be used to show polarisation effects*

- With the metal grille horizontal (as in Figure 3), the vertically plane polarised microwaves are transmitted through the grille.
- With the metal grille vertical, the vertically plane polarised microwaves are absorbed by the free electrons within the metal rods of the grille and therefore there is almost no transmission.

Worked example: Aerial

The length of an aerial used for detecting radio waves is about half the wavelength of the radio waves.

Calculate the length of an aerial for detecting radio waves of frequency 60 MHz.

Step 1: Use the wave equation to calculate the wavelength of the radio waves.

$c = 3.0 \times 10^8\,\mathrm{m\,s^{-1}}$ $\quad f = 60 \times 10^6\,\mathrm{Hz}$ $\quad \lambda = ?$

$$\lambda = \frac{c}{f} = \frac{3.0 \times 10^8}{60 \times 10^6} = 5.0\,\mathrm{m}$$

Step 2: Calculate the length of the aerial.

$$\text{length} = \frac{\lambda}{2} = \frac{5.0}{2}$$

length = 2.5 m (2 s.f.)

Summary questions

1. State the unit of power and intensity. *(1 mark)*
2. Explain why electromagnetic waves can be plane polarised. *(1 mark)*
3. State and explain the effect on the intensity of a wave when its amplitude is halved. *(2 marks)*
4. Identify an electromagnetic wave that has frequency of 1.5×10^{17} Hz. *(3 marks)*
5. The radiant power incident on the surface of the Earth from the Sun is about $1050\,\mathrm{W\,m^{-2}}$. The Earth is at a distance of 1.5×10^{11} m from the Sun. Estimate the total power emitted by the Sun. *(3 marks)*
6. A solar panel of $1.2\,\mathrm{m^2}$ has an efficiency of 20%. Use the information given in **Q5** to estimate the number of solar panels required to produce an electrical output of 1.0 kW. *(3 marks)*

11.8 Refractive index
11.9 Total internal reflection

Specification reference: 4.4.2

11.8 Refractive index

A ray of light incident at an angle to the boundary between two transparent materials will be reflected and refracted at this boundary. Refraction occurs because the speed of light changes as it enters a different material. The angle of refraction depends on the refractive index of each material.

Definition

The **refractive index** of a transparent material is defined as the speed of light in a vacuum divided by the speed of light in the material.

The refractive index n of a material is given by the equation

$$n = \frac{c}{v}$$

where c is the speed of light in a vacuum and v is the speed in the material. Refractive index has no unit. By definition, $n = 1$ for vacuum. For air $n = 1.0003 \approx 1.00$ (3 s.f.) and for glass about 1.5 (2 s.f.).

Revision tip
The refractive index of a material can never be less than 1.

Refraction law

Figure 1 shows the path of a ray of light travelling from a material of refractive index n_1 into another material of refractive index n_2. The ray of light is refracted at the boundary. The incidence angle θ_1 and refracted angle θ_2 are measured relative to the normal.

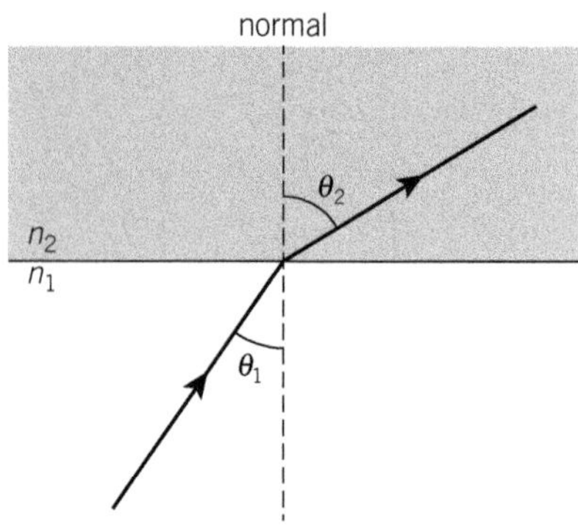

▲ **Figure 1** *Refraction at a boundary (the reflected ray is not shown)*

Experiments show that

$$n_1 \sin\theta_1 = n_2 \sin\theta_2$$

Figure 2 shows an arrangement used to investigate refraction in the laboratory.

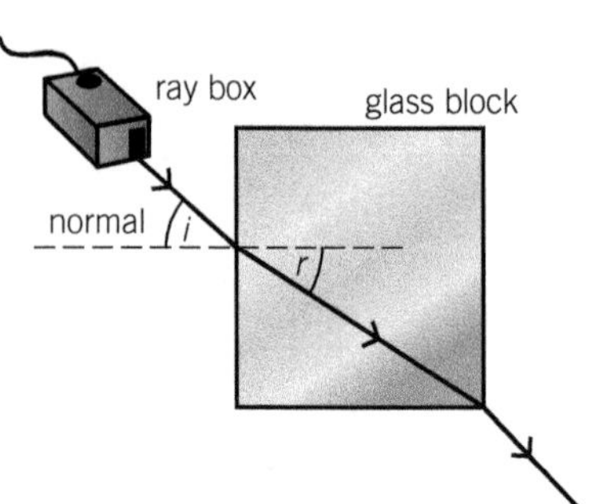

▲ **Figure 2** *An arrangement used to determine the refractive index of glass*

With $n_1 \approx 1.00$, $\theta_1 = i$ (angle of incidence), $\theta_2 = r$ (angle of refraction), and $n_2 = n$ (refractive index of glass), we can use the equation below to determine the refractive index of the glass.

$$1.00 \times \sin i = n \sin r$$

or

$$n = \frac{\sin i}{\sin r}$$

11.9 Total internal reflection

There is no refraction when **total internal reflection** (TIR) of light occurs at the boundary between two transparent materials. All the energy of the wave is returned back into the material. See Figure 3.

Total internal reflection only takes place when:

- light in a material travels towards a material of lower refractive index
- the angle of incidence is greater than the critical angle.

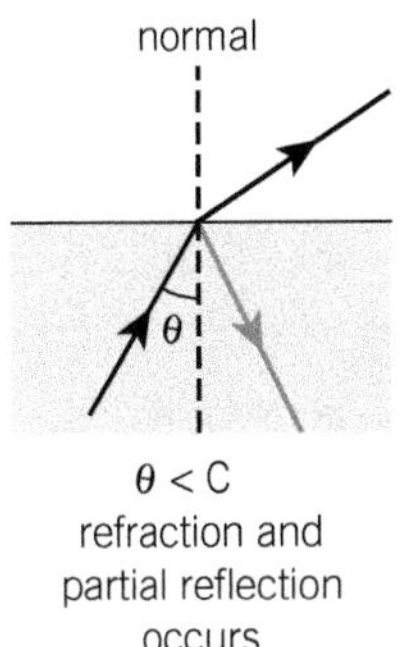

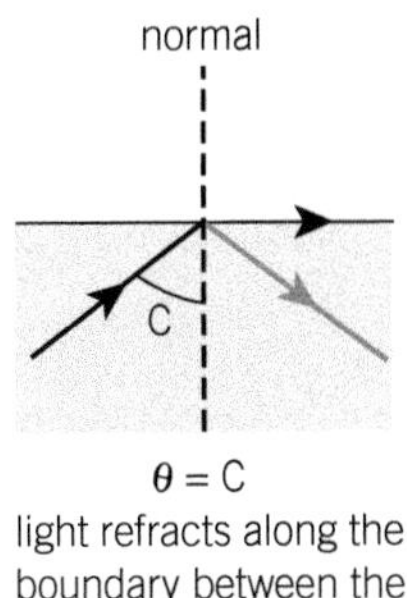

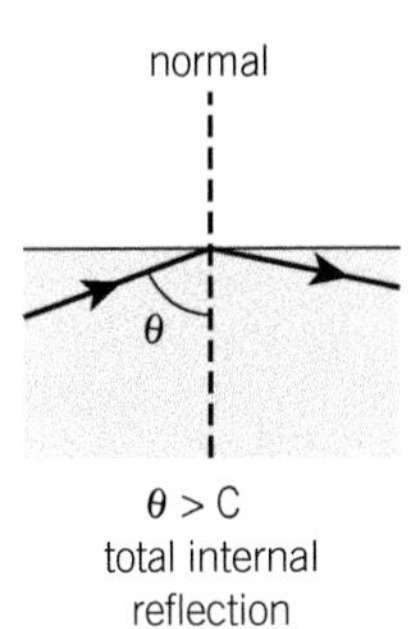

▲ **Figure 3** *Refraction and total internal reflection*

Critical angle

Consider light incident at the boundary between a material and air. The refractive index of the material is n. At the critical angle C the angle of refraction is 90°.

The refraction law $n_1 \sin\theta_1 = n_2 \sin\theta_2$ gives

$n \times \sin C = 1.00 \times \sin 90°$ (n_2 for air is 1.00)

or

$\sin C = \frac{1}{n}$

Revision tip

You can only use $\sin C = \frac{1}{n}$ for a material–air boundary. For the boundary between two materials such as glass and water, you have to go back to basics with $n_1 \sin\theta_1 = n_2 \sin\theta_2$

Worked example: Diamonds in water

Diamonds are submerged into water. The refractive index of diamond is 2.42 and the refractive index of water is 1.33. Calculate the critical angle for light in diamond.

Step 1: Write down the quantities given in the question.

$n_1 = 2.42$ $\theta_2 = C$ $n_2 = 1.33$ $\theta_2 = 90°$

Step 2: Use the refraction law to calculate the critical angle.

$n_1 \sin\theta_1 = n_2 \sin\theta_2$

$2.42 \times \sin C = 1.33 \times \sin 90°$ ($\sin 90° = 1$)

$\sin C = 0.550$

$C = 33°$ (2 s.f.)

Summary questions

1 State the refractive index of a vacuum. *(1 mark)*

2 The speed of light in a transparent material is 0.50 c, where c is the speed of light in a vacuum. Calculate the refractive index of the material. *(1 mark)*

3 According to a student there will be total internal reflection of light travelling from air into glass. Explain why this is incorrect. *(2 marks)*

4 The refractive index of glass is 1.50. Calculate the speed of light in a glass block. *(2 marks)*

5 Light travels from air into a transparent material. The angle of incidence is 70° and the angle of refraction is 35°. Calculate the refractive index of the material. *(2 marks)*

6 Use the information given in the worked example to calculate the critical angle for light in diamond when the diamond is in air. *(3 marks)*

Chapter 11 Practice questions

1 Which is the correct order for the increasing frequencies of the following electromagnetic waves?

A infrared	ultraviolet	X-rays	microwaves
B X-rays	infrared	ultraviolet	microwaves
C ultraviolet	microwaves	infrared	X-rays
D microwaves	infrared	ultraviolet	X-rays

(1 mark)

2 Which statement is correct about light from a laser travelling from air into glass?

A The speed of light increases.

B The intensity of light increases.

C The frequency remains the same.

D The wavelength of light decreases. *(1 mark)*

3 What is the frequency of an electromagnetic wave of wavelength 30 μm?

A 1.0×10^{10} Hz

B 1.0×10^{13} Hz

C 1.0×10^{16} Hz

D 1.0×10^{19} Hz *(1 mark)*

4 Light travels from a transparent material of refractive index 1.75 towards the boundary between the material and the air. The refractive index of air is 1.00.

What is the critical angle for light in the material?

A 30°

B 35°

C 55°

D 60° *(1 mark)*

5 **a** State two properties of electromagnetic waves. *(2 marks)*

b A space probe close to Jupiter emits a radio signal of frequency 8.0 GHz.

Jupiter is about 6.3×10^{11} m away from the Earth.

i Calculate the wavelength of the electromagnetic wave emitted by the probe. *(2 marks)*

ii Identify the part of the electromagnetic spectrum the waves in i belong to. *(1 mark)*

iii Estimate the time taken for the signal to travel from the probe to the Earth. *(2 marks)*

6 **a** Explain what is meant by the statement

The refractive index of glass is 1.54. *(1 mark)*

b Calculate the speed of light in the glass in (a). *(1 mark)*

c Figure 1 shows light from a laser incident at the boundary between water and the glass from (a).

The refractive index of water is 1.33. The angle of incidence of the light in the water is 40°.

i Calculate the angle of refraction. *(3 marks)*

ii Complete Figure 1 to show the refraction and reflection of light at the boundary. *(2 marks)*

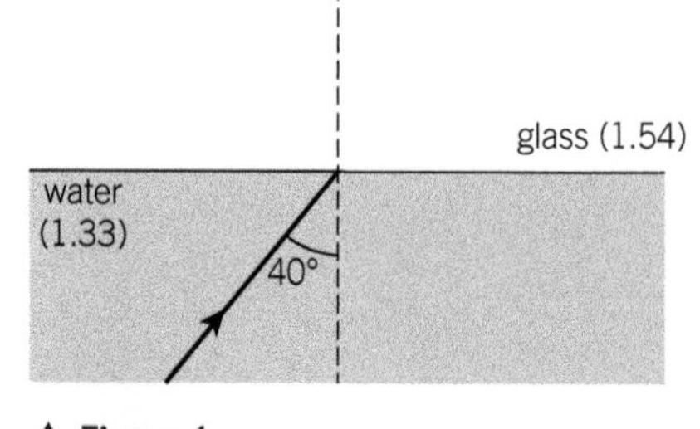

▲ **Figure 1**

Chapter 12 Waves 2

In this chapter you will learn about ...

- [ ] Superposition of waves
- [ ] Coherence
- [ ] Constructive and destructive Interference
- [ ] Path difference
- [ ] Phase difference
- [ ] The Young double-slit experiment
- [ ] Stationary waves
- [ ] Harmonics

12 WAVES 2
12.1 Superposition of waves
12.2 Interference
12.3 The Young double-slit experiment

Specification reference: 4.4.3

12.1 Superposition of waves

When two or more progressive waves meet, they combine and then pass through each other. The term used for waves of the same type combining is called superposition.

Principle of superposition of waves

The **principle of superposition** states that when two waves meet at a point the resultant displacement at that point is equal to the sum of the displacements of the individual waves.

For example, if the displacements at a point from two waves are −2.1 cm and +5.0 cm, then the resultant displacement is simply +2.9 cm.

12.2 Interference

Coherence refers to waves emitted from two sources having a constant phase difference.

All coherent sources emit waves of the same type, same frequency, and wavelength.

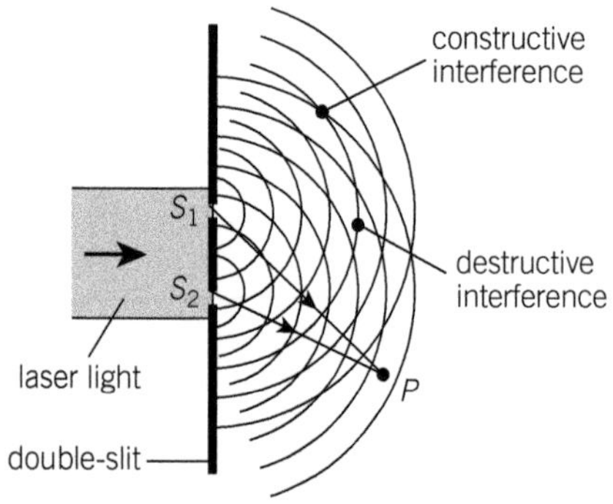

▲ **Figure 1** *Interference of waves diffracted at narrow slits*

Path difference and phase difference

Interference is the superposition of waves from two coherent sources.

Figure 1 shows laser light incident at two narrow slits S_1 and S_2. The light is diffracted at each slit and interference occurs where the diffracted waves overlap. The slits behave as coherent sources. The type of interference, constructive or destructive, at point P depends on the path difference or the phase difference, see Table 1. The path difference is the extra distance travelled by one of the waves from the slits, that is,

$$\text{path difference} = S_1P - S_2P$$

▼ **Table 1**

Type of interference	Observation at P	Path difference in terms of wavelength λ	Phase difference
Constructive 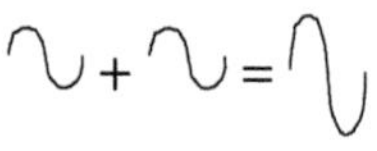	Light: bright 'fringe' Sound: loud sound Microwaves: large signal	$0, \lambda, 2\lambda, \ldots n\lambda$ (n is an integer)	0° or 2π rad
Destructive 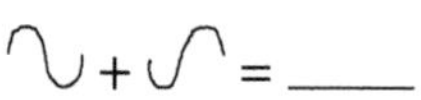	Light: dark 'fringe' Sound: quiet sound Microwaves: small signal	$\frac{\lambda}{2}, \frac{3\lambda}{2}, \ldots (n + \frac{\lambda}{2})\lambda$ (n is an integer)	180° or π rad

Common misconception

Path difference and phase difference are often confused. Path difference is a *length* and phase difference is an *angular measurement* in degrees or radians.

12.3 The Young double-slit experiment

The Young double-slit experiment uses two narrow slits to determine the wavelength of monochromatic light.

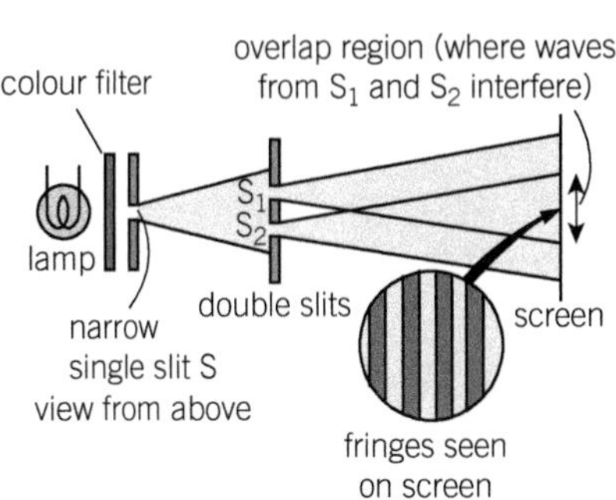

▲ **Figure 2** *Young double-slit experiment using a filament lamp and colour filter*

Light

Figures 2 and 3 show two arrangements for the Young double-slit experiment. The wavelength λ of the monochromatic light can be determined using the equation

$$\lambda = \frac{ax}{D} \qquad (a << D)$$

where a is the separation between the narrow slits, x is the separation between adjacent bright (or dark) fringes, and D is the separation between the slits and the screen.

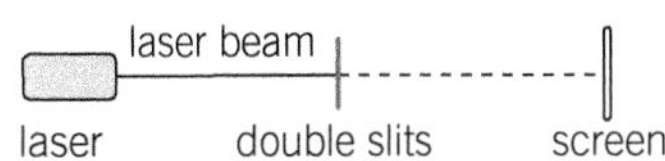

▲ **Figure 3** *Young double-slit experiment using a laser*

Microwaves and sound

The equation $\lambda = \frac{ax}{D}$ and the arrangement shown in Figure 4 can be used to determine the wavelength of microwaves. For determining the wavelength of sound, the slits are replaced by two loudspeakers connected to the same signal generator and the sound is detected using a microphone connected either to an intensity-meter or an oscilloscope.

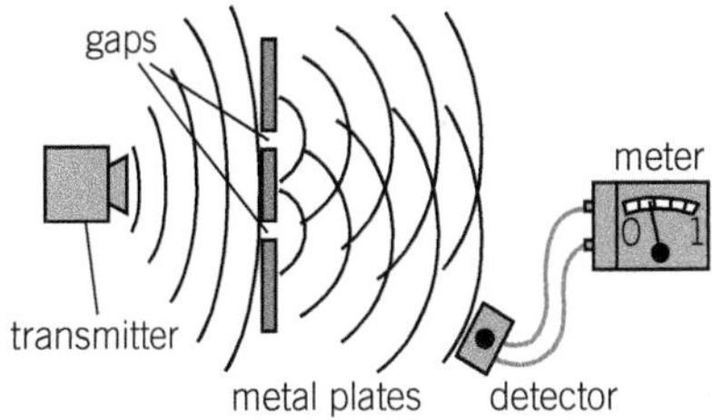

▲ **Figure 4** *Experiment to determine wavelength of microwaves*

Worked example: Fringes

Describe the effect on the fringes observed in a Young double-slit experiment when the separation between the slits is decreased.

Step 1: Write the equation for the fringe separation x.

$$x = \frac{\lambda D}{a}$$

Step 2: Carefully describe the effect on the fringes using the equation above.

The fringe separation x is inversely proportional to the separation a between the slits.

Therefore, the fringe pattern spreads out when the separation between the slits is decreased.

Summary questions

1. Explain what is meant by coherent sources. *(1 mark)*
2. Describe constructive interference in terms of path difference and phase difference. *(2 marks)*
3. The waves shown in Figure 1 have wavelength λ of 2.0 cm. State and explain the type of interference at point P when:
   - **a** $S_1P = 11.0$ cm and $S_2P = 7.0$ cm; *(2 marks)*
   - **b** $S_1P = 7.2$ cm and $S_2P = 6.2$ cm *(2 marks)*
4. Use the results below from a Young double-slit experiment to calculate the wavelength.
   $a = 0.10$ mm $\quad x = 3.1$ cm $\quad D = 6.20$ m *(2 marks)*
5. Red light of wavelength 650 nm incident at two slits produces bright fringes separated by 5.2 mm. The same arrangement produces bright fringes separated by 4.1 mm when green light is used. Calculate the wavelength of this green light. *(2 marks)*
6. In **Q4**, estimate the absolute uncertainty in the wavelength. *(3 marks)*

12.4 Stationary waves
12.5 Harmonics
12.6 Stationary waves in air columns

Specification reference: 4.4.4

12.4 Stationary waves

A stationary (or standing) wave is the result of superposition of two progressive waves with the same frequency (and ideally the same amplitude) travelling in opposite directions. All waves can produce stationary waves.

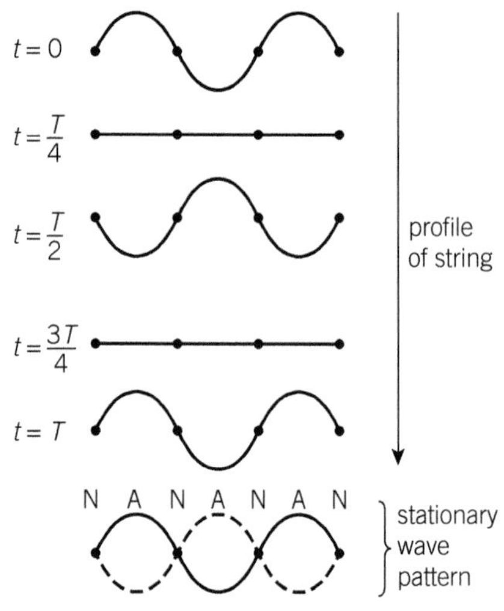

▲ **Figure 1** *A typical stationary wave pattern on a string*

Understanding stationary waves

Figure 1 shows a stationary wave on a stretched string with three 'loops'. The profile of the string is the result of superposition of progressive waves travelling to the left and to the right. The progressive and stationary waves have the same period T and frequency f.

At time $t = 0$, the two progressive waves combine constructively, at time $t = \frac{T}{4}$, they combine destructively, and so on.

Some important points:

- A stationary wave has nodes and antinodes. A **node** is a point that has zero amplitude and an **antinode** is a point of maximum amplitude.
- separation between adjacent nodes (or antinodes) $= \frac{\lambda}{2}$
- separation between adjacent node and antinode $= \frac{\lambda}{4}$
- You can determine the speed v of the **progressive wave** using the wave equation $v = f\lambda$.

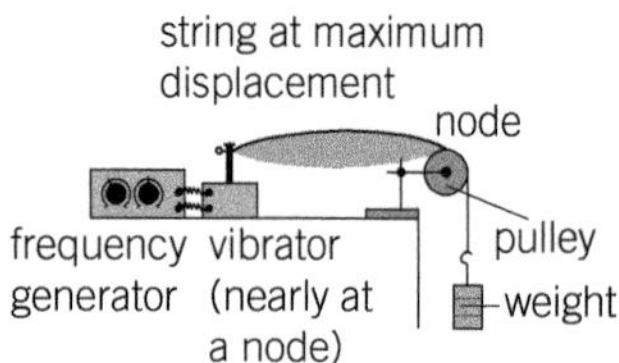

▲ **Figure 2** *Stationary waves on a stretched string*

Comparing progressive and stationary waves

Table 1 compares progressive and stationary waves.

▼ **Table 1**

	Progressive wave	Stationary wave
Energy transfer	Yes	No
Frequency	The same for all points.	The same for all points, except at the nodes where it is zero.
Amplitude	The same for all points.	Varies from zero at the nodes to maximum at the antinodes.
Phase difference ϕ between two points	$\phi = \frac{x}{\lambda} 360°$ where x is the separation between the points.	$\phi = 180°n$ where n is the number of nodes between the points.

N = node A = antinode
(dotted line shows string half a cycle earlier)

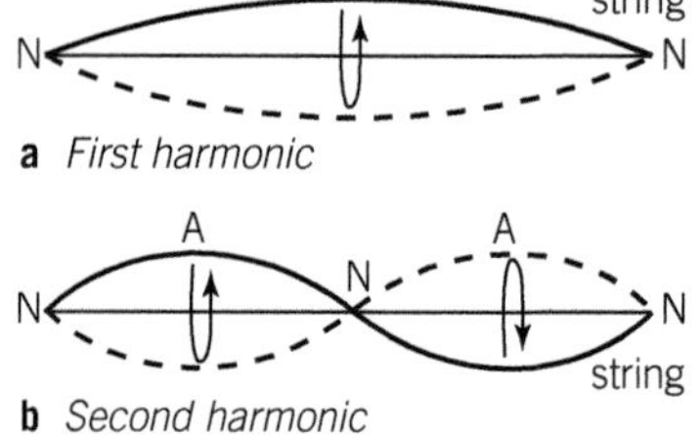

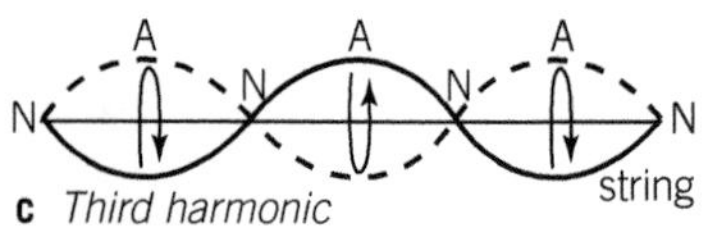

▲ **Figure 3** *Stationary wave patterns*

12.5 Harmonics

Figure 2 shows how stationary waves can be produced on a stretched string. The transverse waves on the string are reflected at the pulley-end. These reflected waves combine with those from the vibrator to produce the stationary waves.

Understanding harmonics

Figure 3 shows the first three stationary wave patterns produced on the stretched string – these are called harmonics.

The **fundamental frequency** of the stationary wave is the lowest frequency of vibration for a given arrangement.

The first harmonic (fundamental mode of vibration) is a stationary wave that has frequency equal to the fundamental frequency. All other harmonics have a frequency that is a multiple of the fundamental frequency.

Figure 4 shows how stationary waves can be investigated using microwaves.

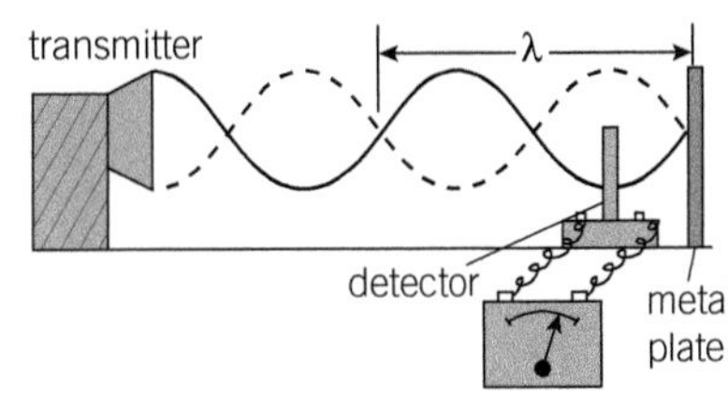

▲ **Figure 4** *Microwaves can also produce stationary waves*

12.6 Stationary waves in air columns

Sound resonates at certain frequencies in an air-filled tube. You can produce a loud sound by holding a vibrating tuning fork close to the open end of a tube and adjusting the length of the air column.

Closed and open tubes

Figures 5 and 6 show the stationary waves produced in a tube open at only one end and in a tube open at both ends.

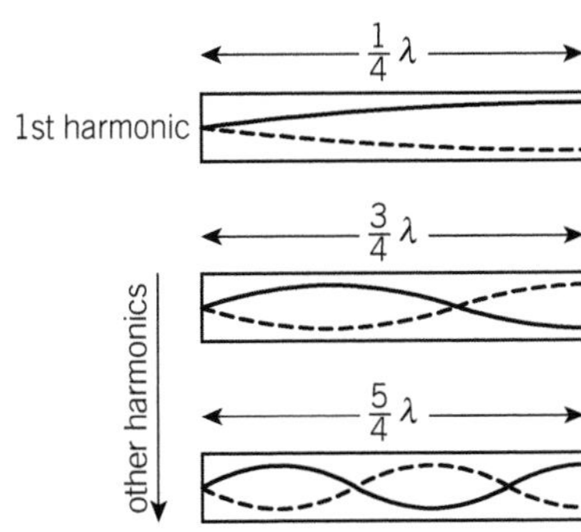

▲ **Figure 5** *Stationary waves in a tube open at only one end*

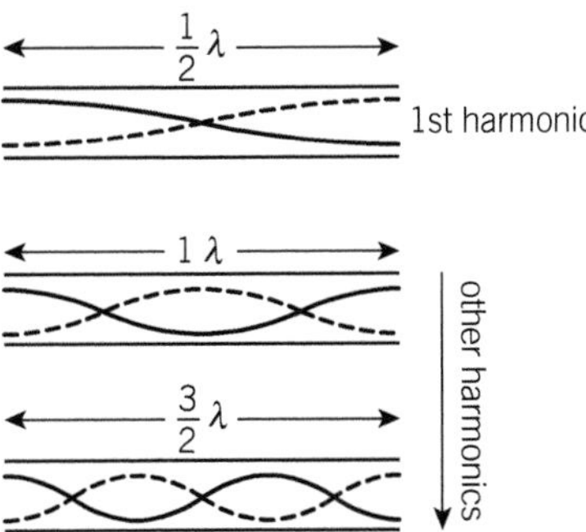

▲ **Figure 6** *Stationary waves in a tube open at both ends*

Worked example: Vibrating string

A string is stretched between two points separated by 15 cm. The string vibrates at a fundamental frequency of 120 Hz. Calculate the speed of the progressive waves on the string.

Step 1: Determine the wavelength of the progressive wave.

separation between successive nodes = 15 cm

Therefore, $\lambda = 2 \times 15 = 30$ cm

Step 2: Use the wave equation to determine the speed v of sound.

$v = ?$ $f = 120$ Hz $\lambda = 0.30$ m

$v = f\lambda = 120 \times 0.30$

$v = 36\ \text{m s}^{-1}$ (2 s.f.)

Summary questions

1 State the phase difference of all points between adjacent nodes of a stationary wave. *(1 mark)*

2 The separation between adjacent node and antinode is 10 cm. State and explain the wavelength of the progressive wave. *(2 marks)*

3 Explain how a stationary wave is produced in the arrangement shown in Figure 4. *(2 marks)*

4 The separation between adjacent antinodes of a string is 28 cm. The string vibrates at 60 Hz. Calculate the speed of the progressive waves on the string. *(3 marks)*

5 The length of the tube in Figure 5 is 30 cm. The speed of sound is $340\ \text{m s}^{-1}$. Calculate the fundamental frequency. *(3 marks)*

6 In Figure 6, the third harmonic has frequency 500 Hz. Calculate the length of the tube. *(3 marks)*

Chapter 12 Practice questions

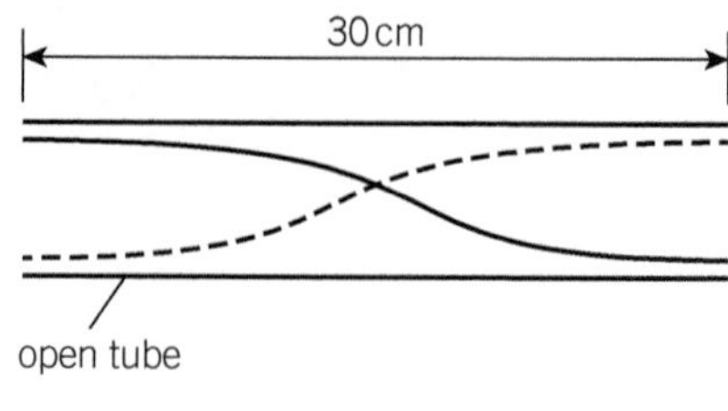

▲ **Figure 1**

1 Figure 1 shows a stationary wave drawn by a student for sound waves in an open tube.

What is the wavelength of the sound wave?

A 15 cm **B** 30 cm **C** 45 cm **D** 60 cm *(1 mark)*

2 Two progressive waves, of the same wavelength of 2.0 cm, combine together. The path difference between the two waves is 0.5 cm.

What is the phase difference between these two waves?

A 45° **B** 90° **C** 180° **D** 270° *(1 mark)*

3 In the Young double-slit experiment with monochromatic light, the distance between the screen and the slits is decreased.

Which statement is correct about adjacent bright fringes seen on the screen?

A There is no change.

B Their colour changes.

C They get further apart.

D They get closer together. *(1 mark)*

4 A microwave transmitter is placed in front of a metal sheet. A stationary wave is formed between the metal sheet and the transmitter with adjacent nodes separated by 3.0 cm.

What is the frequency of the microwaves?

A 5.0×10^{7} Hz **C** 5.0×10^{9} Hz

B 1.0×10^{8} Hz **D** 1.0×10^{10} Hz *(1 mark)*

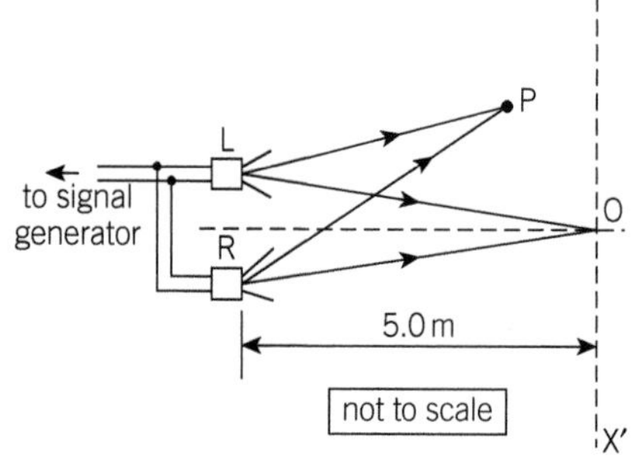

▲ **Figure 2**

5 **a** Explain what is meant by a progressive wave. *(1 mark)*

b Figure 2 shows two loudspeakers L and R connected to the same signal generator.

i Explain the common feature of the sound waves emitted from L and R. *(1 mark)*

ii At point P, a microphone picks up a strong signal. Explain this in terms of interference of the waves from L and R. *(2 marks)*

iii The separation between the loudspeakers is 1.4 m. The frequency of sound emitted from the loudspeakers is 4.0 kHz. The microphone is now moved along the straight line from X to X′. This line is 5.0 m from the loudspeakers. The separation between adjacent regions of loud sound along XX′ is 30 cm.

Calculate:

1 the wavelength of sound; *(3 marks)*

2 the speed of sound. *(2 marks)*

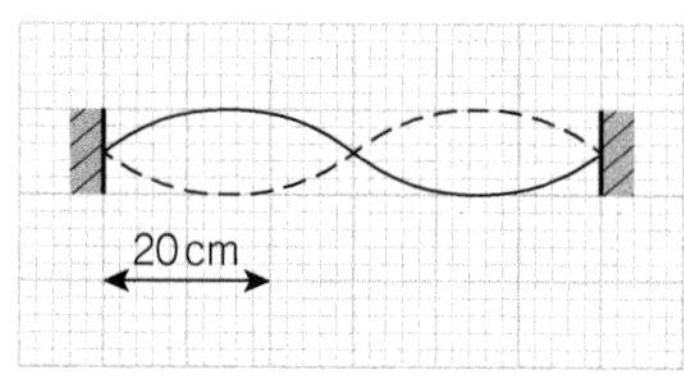

▲ **Figure 3**

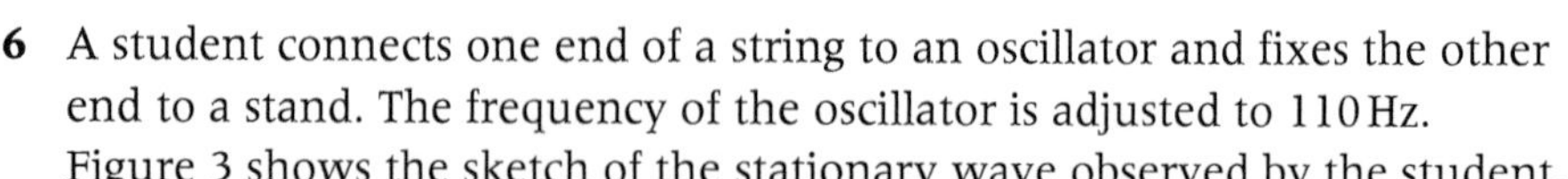

6 A student connects one end of a string to an oscillator and fixes the other end to a stand. The frequency of the oscillator is adjusted to 110 Hz. Figure 3 shows the sketch of the stationary wave observed by the student.

a Use Figure 3 to determine the speed of the progressive waves on the string. *(3 marks)*

b The frequency of the oscillator is increased slightly.

Suggest what happens to the profile of the string. *(1 mark)*

c The frequency of the oscillator is increased further until a stationary wave with three 'loops' is produced. Predict the frequency of the oscillator. *(3 marks)*

Chapter 13 Quantum physics

In this chapter you will learn about ...

- [ ] Photons
- [ ] Electronvolt
- [ ] Photoelectric effect
- [ ] Work function
- [ ] Threshold frequency
- [ ] Einstein's photoelectric equation
- [ ] Wave–particle duality
- [ ] De Broglie waves
- [ ] Electron diffraction

13 QUANTUM PHYSICS
13.1 The photon model
13.2 The photoelectric effect

Specification reference: 4.5.1, 4.5.2

13.1 The photon model

Electromagnetic waves travel through space as waves – the diffraction and interference of electromagnetic waves provides evidence for this. However, when electromagnetic waves interact with matter, they do so as packets of discrete energy called photons.

A **photon** is a quantum of electromagnetic energy.

Photons

The quantum of energy E of a photon is related to the frequency f and wavelength λ of electromagnetic radiation by the following equations

$$E = hf \qquad \text{and} \qquad E = \frac{hc}{\lambda}$$

where h is the Planck constant and c is the speed of light in a vacuum. The experimental value for h is 6.63×10^{-34} J s.

Electronvolts

Revision tip

Multiply by 1.60×10^{-19} when converting from eV to J, and divide by 1.60×10^{-19} when converting from J to eV.

The **electronvolt** (eV) is a convenient unit of energy when dealing with particles and photons.

The electronvolt is defined as the energy gained by an electron travelling through a potential difference of 1 volt.

1 eV = potential difference × charge on electron = $1.0\,\text{V} \times 1.60 \times 10^{-19}\,\text{C}$

1 eV = 1.60×10^{-19} J

LEDs and *h*

An LED emits light when the potential difference across it exceeds a threshold p.d.(voltage) V. Photons are emitted when electrons within the LED lose energy. A single photon is produced when a single electron transfers energy.

energy transferred by electron ≈ energy of photon

$$eV \approx \frac{hc}{\lambda}$$

Different colour emitting diodes can be used to determine an approximate value for the Planck constant.

- Plot a graph of threshold p.d. V against $\frac{1}{\lambda}$
- The gradient G of the straight line through the origin should be equal to $\frac{hc}{e}$
- Therefore, $h \approx \frac{Ge}{c}$

13.2 The photoelectric effect

Electromagnetic radiation can be used to remove electrons from the surface of metals. This effect is known as the **photoelectric effect**. A negatively charged gold-leaf electroscope and zinc plate can be used to demonstrate this effect, see Figure 1.

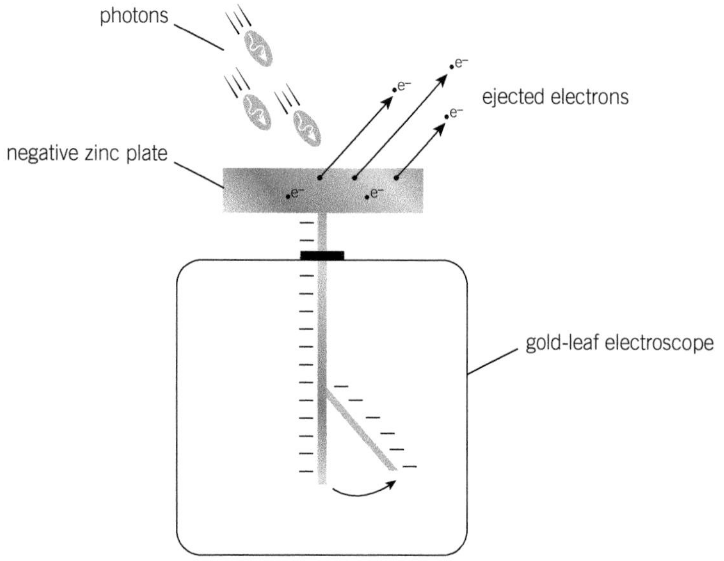

▲ **Figure 1** *A gold-leaf electroscope*

Observations:

- Visible light, no matter how intense, does not remove electrons from the zinc plate.
- Ultraviolet radiation, even of very low intensity, instantaneously ejects electrons from the zinc plate. The gold-leaf collapses as photoelectrons are removed.

Work function and threshold frequency

The wave model cannot explain the observations above because intensity, and not frequency, should be the factor that should govern the removal of electrons. You need to know about work function and threshold frequency to understand the photoelectric effect.

The **work function** of a metal is the minimum energy required to free an electron from the surface of the metal.

The **threshold frequency** is the minimum frequency of the incident electromagnetic radiation that will free electrons from the surface of the metal.

Quantum rules

- A single photon can only interact with a single electron on the surface of the metal.
- Energy is conserved in the photon–electron interaction.
- An electron is only released when the photon energy hf is equal, or greater than, the work function ϕ of the metal.
- At the threshold frequency f_0, energy of photon = work function.
- Intensity of the incident radiation is related to the number of photons per unit time – doubling the intensity doubles the number of photons per unit time.

Worked example: Threshold frequency

The work function of a metal is 4.90 eV. Calculate the threshold frequency for this metal.

Step 1: Calculate the work function in joules (J).

$$\phi = 4.90 \times 1.60 \times 10^{-19} = 7.84 \times 10^{-19}\ \text{J}$$

Step 2: Calculate the threshold frequency f_0.

energy of photon = work function

$$hf_0 = \phi$$

$$f_0 = \frac{7.84 \times 10^{-19}}{6.63 \times 10^{-34}} = 1.18 \times 10^{15}\ \text{Hz (3 s.f.)}$$

Summary questions

1 State the SI unit of work function. *(1 mark)*

2 Convert 100 eV into J. *(1 mark)*

3 State the energy in eV gained by an electron travelling through a p.d. of 1000 V. *(1 mark)*

4 Suggest why visible light incident on a zinc plate cannot eject any electrons. *(1 mark)*

5 Calculate the wavelength of a photon of energy 8.0×10^{-19} J. *(3 marks)*

6 An LED emits light of wavelength 560 nm. The radiant power of the LED is 40 mW. Calculate the number of photons emitted from the LED per second. *(4 marks)*

7 The work function of a metal is 4.30 eV. Electromagnetic radiation of wavelength 400 nm is incident on the metal. Explain if this radiation produces any photoelectrons. *(4 marks)*

13.3 Einstein's photoelectric equation
13.4 Wave–particle duality

Specification reference: 4.5.2, 4.5.3

13.3 Einstein's photoelectric equation

The photoelectric equation is based on two important ideas – the principle of conservation of energy and the quantum physics rule of one-to-one interaction between a photon and an electron.

Einstein's equation

When an electron is emitted from the surface of a metal, we have:

energy of photon = work function of the metal + maximum kinetic energy of emitted electron

or

$$hf = \phi + KE_{max}$$

The electron is loosely held by the attraction of the positive ions of the metal. The work function ϕ is the minimum energy it takes to free an electron. The emitted electrons can collide with other electrons and the positive ions and this is why the electrons emerge with a range of kinetic energies. The maximum KE_{max} is still given by the equation above.

Observations:

- Not a single electron can be emitted when $hf < \phi$. Increasing the intensity just increases the rate of the photons but the energy of individual photons is still less than the work function.
- At the threshold frequency f_0, the kinetic energy of the emitted electrons is zero. Therefore, $hf_0 = \phi$.

Revision tip

$1\text{ eV} = 1.60 \times 10^{-19}\text{ J}$

Use units consistently when using the equation $hf = \phi + KE_{max}$

KE_{max} against f graph

The photoelectric equation can be rearranged as

$$KE_{max} = hf - \phi$$

The work function ϕ and Planck constant h can be determined by plotting a graph of KE_{max} against f, see Figure 1.

- The gradient of the straight-line graph is h.
- The intercept with the vertical axis is $-\phi$.
- The intercept with the horizontal axis is the threshold frequency f_0.

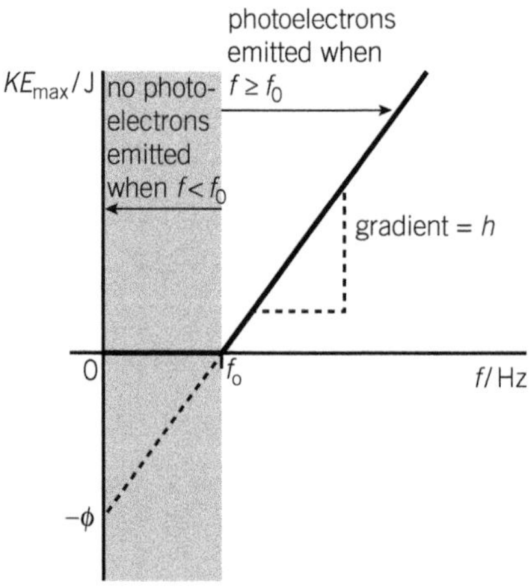

photoelectrons emitted when $f \geq f_0$
KE_{max} / J
no photo-electrons emitted when $f < f_0$
gradient = h
0
f_0
f / Hz
$-\phi$

▲ **Figure 1** *The gradient of this graph is always h and is independent of the metal used*

13.4 Wave–particle duality

Electromagnetic waves have a dual nature – they travel as waves and they interact as 'particles' (photons). Louis de Broglie, in 1923, proposed that matter, which includes particles such as electrons, must also have a dual nature.

De Broglie equation

A particle travelling through space with momentum p has an associated wavelength λ given by the de Broglie equation

$$\lambda = \frac{h}{p} \quad \text{or} \quad \lambda = \frac{h}{mv}$$

where h is the Planck constant, m is the mass of the particle, and v is its speed.

Revision tip

The waves associated with moving particles are also known as matter waves. These waves are not electromagnetic waves.

Evidence

The wave-like behaviour of electrons can be demonstrated using an electron diffraction tube. Accelerated electrons pass through a thin sheet of graphite. The separation between the atomic layers of graphite is very similar to the de Broglie wavelength of the electrons. Electrons are diffracted by the 'gaps' between the atoms of graphite and therefore demonstrate wave-like behaviour. Figure 2 shows the diffraction rings produced by electrons having travelled through graphite (or a thin metal foil).

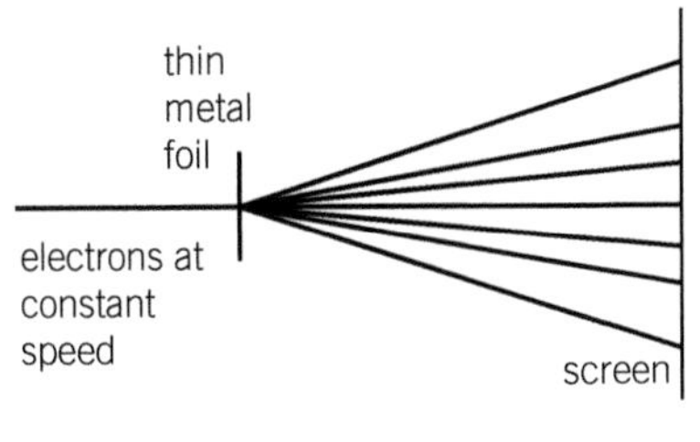

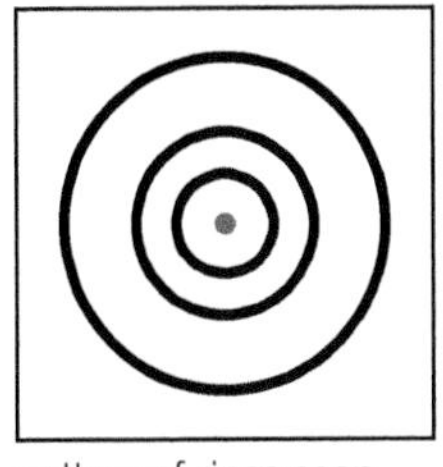

pattern of rings seen on the screen

▲ **Figure 2** *Electrons are diffracted by the gaps between atoms*

Worked example: Diffracting people?

Make suitable estimations to explain why a person running through a metre-wide gap in a wall will not be diffracted.

Step 1: Make suitable estimates for quantities required for the de Broglie equation.

mass $m = 70\,\text{kg}$ speed $v = 8.0\,\text{m s}^{-1}$

Step 2: Calculate the de Broglie wavelength.

$$\lambda = \frac{h}{p} = \frac{h}{mv} = \frac{6.63 \times 10^{-34}}{70 \times 8.0}$$

wavelength $= 1.2 \times 10^{-36}\,\text{m} \approx 10^{-36}\,\text{m}$ (2 s.f.)

Step 3: Provide a suitable explanation.

For noticeable diffraction, the wavelength must be similar to the width of the gap.

The person's de Broglie wavelength is 10^{36} times smaller than the width of the gap therefore there will be no diffraction.

Summary questions

1. State one quantity conserved when a photon interacts with an electron. *(1 mark)*
2. A photon has energy 7.2×10^{-19} J. It removes an electron from a metal of work function 6.9×10^{-19} J. Calculate the maximum kinetic energy of the electron. *(2 marks)*
3. Calculate the maximum speed of the electron in **Q2**. *(2 marks)*
4. The maximum wavelength that produces photoemission from a metal is 2.1×10^{-7} m. Calculate the work function of the metal. *(3 marks)*
5. Electromagnetic radiation of wavelength 320 nm is incident on a metal of work function 2.3 eV. Calculate the maximum kinetic energy of the emitted photoelectrons. *(4 marks)*
6. Calculate the de Broglie wavelength of an electron accelerated through a p.d. of 1.5 V. *(4 marks)*

Chapter 13 Practice questions

1 Electromagnetic radiation incident on metal ejects electrons from the surface of the metal.

Which statement about the kinetic energy (KE) of the electrons is correct?

A The electrons have the same KE.

B The maximum KE of the electrons depends on the intensity.

C The KE of electrons is always greater than the work function.

D The maximum KE depends on the incident wavelength. *(1 mark)*

2 What is the typical energy of a photon of infrared radiation?

A 4×10^{-24} J **C** 4×10^{-19} J

B 4×10^{-20} J **D** 4×10^{-17} J *(1 mark)*

3 The energy of a photon is related to the wavelength λ or the frequency f of the electromagnetic radiation.

Which graph will produce a straight line?

A E against f **B** E against λ **C** $E\lambda$ against f **D** Ef against λ *(1 mark)*

4 Electrons accelerated through a potential difference V have a de Broglie wavelength λ.

What is the wavelength of electrons accelerated through a potential difference of $4V$?

A $\frac{\lambda}{4}$ **B** $\frac{\lambda}{2}$ **C** 2λ **D** 4λ *(1 mark)*

5 **a** State what is meant by the photoelectric effect. *(1 mark)*

b Electromagnetic radiation incident on a metal ejects electrons from its surface.

Explain this observation in terms of photons and electrons. *(3 marks)*

c A researcher is investigating photoelectric effect using a new material.

Figure 1 shows a graph of maximum kinetic energy KE_{max} of the photoelectrons against frequency f of the incident electromagnetic radiation.

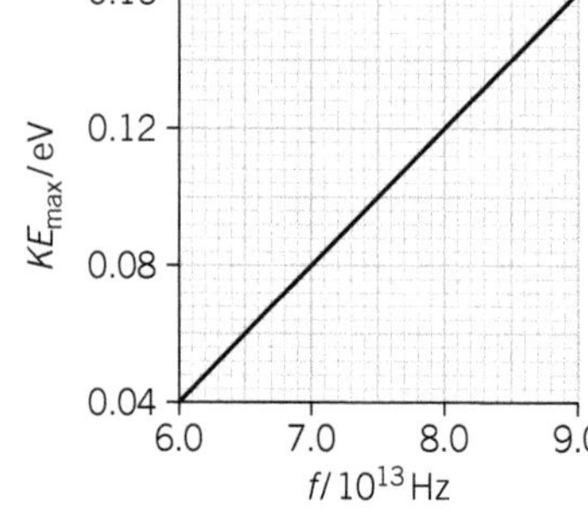

▲ **Figure 1**

i Explain why the graph is linear. *(2 marks)*

ii Use Figure 1 to determine the work function of the material in eV. *(4 marks)*

iii Use your answer to (ii) to determine the threshold frequency. *(2 marks)*

6 The work function of a metal is 3.7 eV. The metal is charged negatively and then it is illuminated with electromagnetic waves of wavelength 300 nm. Photoelectrons are emitted from the surface of the metal.

a Explain the term work function. *(1 mark)*

b Calculate the maximum kinetic energy, in eV, of the photoelectrons. *(4 marks)*

c Calculate the longest wavelength of electromagnetic radiation that will produce photoelectric effect. *(2 marks)*

7 **a** Explain what is meant by the de Broglie wavelength of an electron. *(2 marks)*

b Fast-electrons can be used to probe into the structure of nuclei. Such electrons have a de Broglie wavelength of about 2.0×10^{-15} m.

Calculate the momentum of the electrons. *(2 marks)*

c Briefly described an experiment that confirms that electrons behave as waves. *(3 marks)*

Module 5 Newtonian world and astrophysics

Chapter 14 Thermal physics

In this chapter you will learn about ...

- ☐ Temperature
- ☐ Celsius scale
- ☐ Kelvin scale
- ☐ Thermal equilibrium
- ☐ Kinetic model of matter
- ☐ Brownian motion
- ☐ Internal energy
- ☐ Specific heat capacity
- ☐ Specific latent heat

14 THERMAL PHYSICS
14.1 Temperature
14.2 Solids, liquids, and gases

Specification reference: 5.1.1, 5.1.2

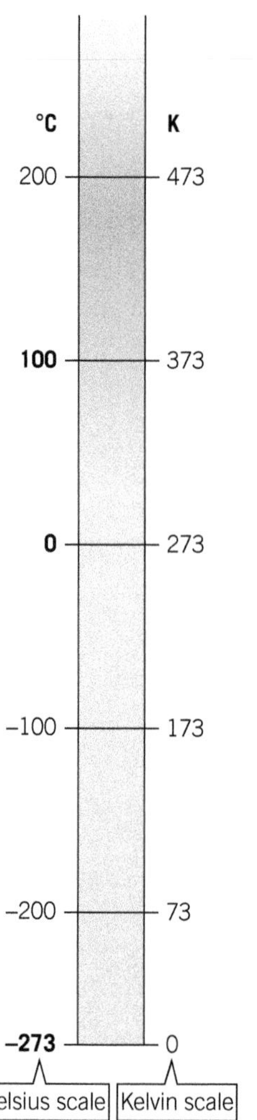

▲ **Figure 1** *Temperatures in °C and K*

Common misconception

Do not confuse heat (thermal) energy and temperature. Energy is measured in joules and temperature in °C or K.

Revision tip: Temperature in K

Thermodynamic temperature in K is always a positive number.

Synoptic link

You will find more information on electrostatic potential energy in Topic 22.5, Electrical potential and energy.

14.1 Temperature

The temperature of a substance is a number which indicates its level of hotness on some chosen scale.

The **Celsius** temperature scale is defined in terms of two accurately reproducible fixed points. The lower fixed point of 0 °C is the temperature of pure melting ice. The upper fixed point of 100 °C is the temperature of steam from boiling water under atmospheric pressure of 1.01×10^5 Pa. The interval between the upper and lower fixed points is divided into 100 equal degrees.

Kelvin scale

The two fixed points for the **thermodynamic temperature scale** are **absolute zero**, 0 K, which is the lowest possible temperature, and the triple point of water, 273.16 K.

0 °C is equivalent to 273.15 K. A change of 1 **kelvin** is equal to a change of 1 degree Celsius, see Figure 1. You can use the equations below to convert temperature between the Celsius and thermodynamic scales.

$$T(\text{K}) \approx \theta(°\text{C}) + 273$$

$$\theta(°\text{C}) \approx T(\text{K}) - 273$$

Thermal equilibrium

When a hot object is in contact with a cooler object, then there is a net flow of thermal energy from the hot object to the cooler object. The temperature of the hot object will decrease and the temperature of the cooler object will increase. Eventually both objects will reach the same temperature.

Two or more objects are in **thermal equilibrium** when there is no net transfer of thermal energy between them.

14.2 Solids, liquids, and gases

The three states or **phases** of matter are solid, liquid, and gas. The **kinetic model** describes how all substances are made up of atoms or molecules.

Kinetic model

In solid and liquid phases, the molecules experience attractive electrostatic force and have negative electrostatic potential energy. The negative simply means that external energy is required to pull apart the molecules. The potential energy is lowest in solids, higher in liquids, and at its highest (0 J) in gases.

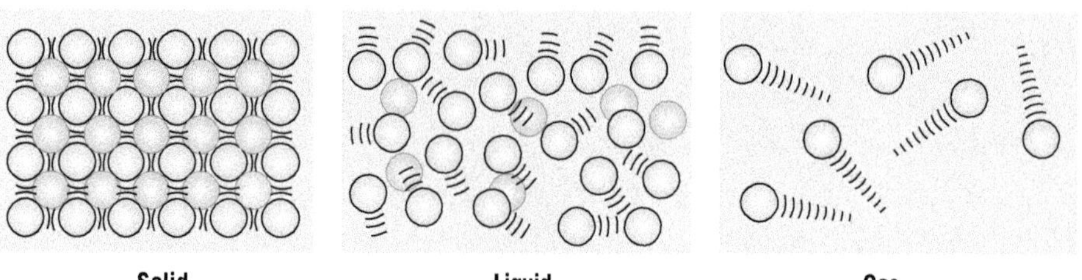

▲ **Figure 2** *Solid, liquid, and gas*

Solid phase

The molecules:

- vibrate about fixed positions and are closely packed together
- have a mixture of kinetic energy (KE) and electrostatic potential energy (PE).

Liquid phase

The molecules:

- can move around each other
- when compared with the solid state:
  - have greater KE
  - have greater mean separation
  - experience smaller attractive electrostatic force
  - have greater PE.

Gas phase

The molecules:

- move freely and rapidly
- collide elastically with each other and move randomly with different speeds in different directions
- have greater mean separation than that in the liquid phase
- experience negligible electrostatic force
- have maximum PE of 0 J.

Brownian motion

Smoke particles in air exhibit random motion because of collisions with air molecules. At a given temperature, the mean kinetic energy of the smoke particles is the same as that of the air molecules. The more massive smoke particles have much smaller speeds than the air molecules.

Synoptic link

The mean kinetic energy of molecules of a gas is directly proportional to the thermodynamic temperature. You will learn more about this in Topic 15.4, The Boltzmann constant.

Summary questions

1 State one difference between temperature and heat (thermal) energy. *(1 mark)*

2 Describe how you could determine 0 °C on an unmarked mercury-in-glass thermometer. *(1 mark)*

3 Convert the following temperatures into K:
a −270 °C **b** −52 °C **c** 20 °C **d** 10 °C **e** 2000 °C *(5 marks)*

4 Convert the following temperatures into °C:
a 3 K **b** 273 K **c** 380 K **d** 500 K **e** 5500 K *(5 marks)*

5 An ice cube at −10 °C is placed in a warm room. It eventually melts and becomes water at 20 °C.
Compare the electrostatic potential energy and kinetic energy of the molecules of water at 20 °C with ice at −10 °C. *(4 marks)*

6 Smoke particles in air are observed to have very small mean speed. Explain how you can deduce that the mean speed of the air molecules is much greater. *(3 marks)*

14.3 Internal energy
14.4 Specific heat capacity
14.5 Specific latent heat

Specification reference: 5.1.2, 5.1.3

14.3 Internal energy

The internal energy of a substance is defined as the sum of the randomly distributed kinetic and potential energies of atoms or molecules within the substance.

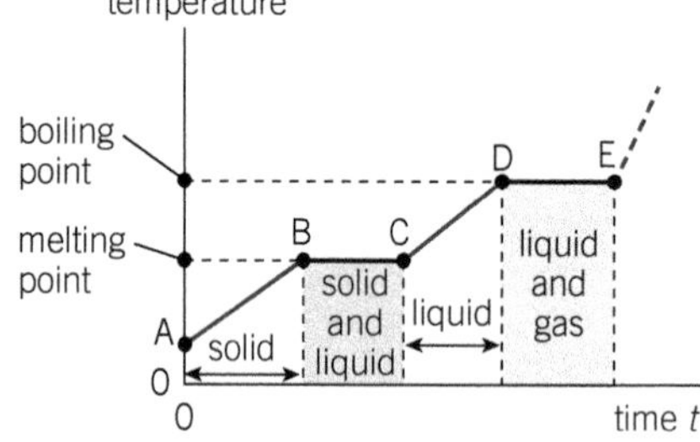

▲ **Figure 1** *Variation of temperature of a substance showing phase changes*

Figure 1 shows the variation of temperature with time for a solid heated at a constant rate. Table 1 summaries what happens to the total energy of the molecules.

▼ **Table 1**

Section	Kinetic energy	Potential energy	Internal energy
A to B (solid)	Increases	No change	Increases
B to C (phase change from solid to liquid)	No change	Increases	Increases
C to D (liquid)	Increases	No change	Increases
D to E (phase change from liquid to gas)	No change	Increases	Increases

Absolute zero

The **internal energy** of a substance depends on its temperature. At absolute zero, 0 K, all molecules stop moving. The internal energy of the substance is a *minimum* and is entirely due to electrostatic potential energy of the molecules.

14.4 Specific heat capacity

The **specific heat capacity** of a substance is defined as the energy required per unit mass to change the temperature by 1 K (or 1 °C).

The specific heat capacity c of a substance is given by the equation

$$c = \frac{E}{m\Delta\theta} \qquad \text{or} \qquad E = mc\Delta\theta$$

where E is the energy supplied to the substance, m is the mass of the substance, and $\Delta\theta$ is the change in temperature of the substance. Specific heat capacity has units $\text{J kg}^{-1}\text{K}^{-1}$.

Determining specific heat capacity *c*

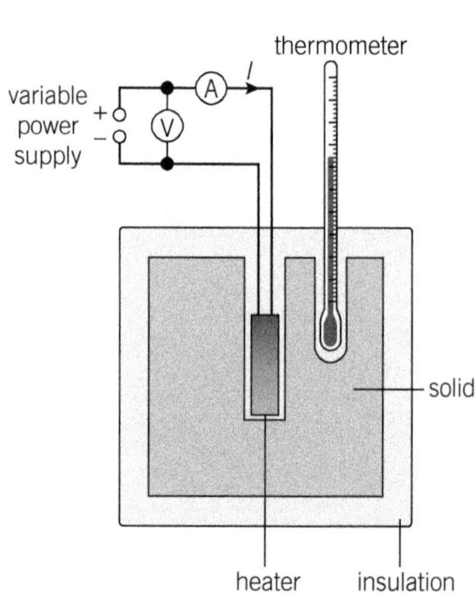

▲ **Figure 2** *Arrangement for determining specific heat capacity of a metal.*

The electrical arrangement shown in Figure 2 can be used to determine c of a metal – it can easily be adapted for a liquid. The method of mixtures is illustrated in the worked example.

The energy supplied by the heater in a time t is equal to VIt, where V is the p.d. across the heater and I is the current in the heater. The specific heat capacity of the metal is given by the equation

$$c = \frac{VIt}{m \times [\theta_f - \theta_i]}$$

where m is the mass of the metal block and θ_i and θ_f are the initial and final temperatures of the block.

 **Worked example: Specific heat capacity of a liquid**

100 g of water at 80 °C is poured into 200 g of liquid at 10 °C. The final temperature of the mixture is 50 °C. The specific heat capacity of water is 4200 J kg^{-1} K^{-1}. Calculate the specific heat capacity c of the liquid. Assume there are no heat losses.

Step 1: Calculate the energy transferred to the liquid.

$$E = mc\Delta\theta = 0.100 \times 4200 \times [80 - 50] = 12\,600\ \text{J}$$

Step 2: Calculate c.

$$12\,600 = 0.200 \times c \times [50 - 10]$$

$$c = 1575\ \text{J kg}^{-1}\text{K}^{-1} \approx 1600\ \text{J kg}^{-1}\text{K}^{-1}\ (2\ \text{s.f.})$$

14.5 Specific latent heat

The term latent means 'hidden'.

The **specific latent heat** of a substance L is defined as the energy required to change the phase per unit mass while at constant temperature.

You can use the equation below to calculate the specific latent heat L of a substance.

$$L = \frac{E}{m} \qquad \text{or} \qquad E = mL$$

where E is the energy supplied to change phase of a mass m of the substance. Specific latent heat has units J kg^{-1}.

Specific latent heat of fusion L_f is used when a substance changes from solid to liquid and **specific latent heat of vaporisation** L_v is used when it changes from liquid to gas.

Determining latent heats

The heater arrangement in Figure 2 can be adapted to determine either L_f or L_v by changing the phase of a substance of mass m in a time t. The energy E supplied by the heater is given by $E = VIt$.

Therefore

$$VIt = mL.$$

Summary questions

For water: $c = 4200\ \text{J kg}^{-1}\text{K}^{-1}$ $L_f = 3.3 \times 10^4\ \text{J kg}^{-1}$ $L_v = 2.3 \times 10^6\ \text{J kg}^{-1}$

1 Explain what is meant by absolute zero. *(1 mark)*

2 Calculate the energy needed to change 10 g of ice at 0 °C to water at 0 °C. *(2 marks)*

3 A 200 W heater is used to heat water. How long would it take 300 g of water at 20 °C to reach boiling point? *(3 marks)*

4 How much longer would it take to completely boil the water in **Q3**? *(3 marks)*

5 A metal block of mass 200 g and temperature 20 °C is placed in 120 g of water initially at 100 °C. The final temperature of the water is 87 °C. Calculate the specific heat capacity c of the metal. *(4 marks)*

6 Calculate how long it would take for a 120 W heater to change 500 g of ice at 0 °C into water at 20 °C. *(4 marks)*

Chapter 14 Practice questions

1 Two objects X and Y are in thermal equilibrium. Which statement is correct?

A There is a net transfer of energy between X and Y.

B X and Y have the same amount of thermal energy.

C X and Y have the same temperature in kelvin.

D X and Y have the same specific heat capacity. *(1 mark)*

2 What is specific latent heat of fusion measured in base units?

A $\mathrm{m\,s^{-2}}$

B $\mathrm{m^2\,s^{-2}}$

C $\mathrm{m^2\,s^{-2}\,K^{-1}}$

D $\mathrm{kg\,m^2\,s^{-2}}$ *(1 mark)*

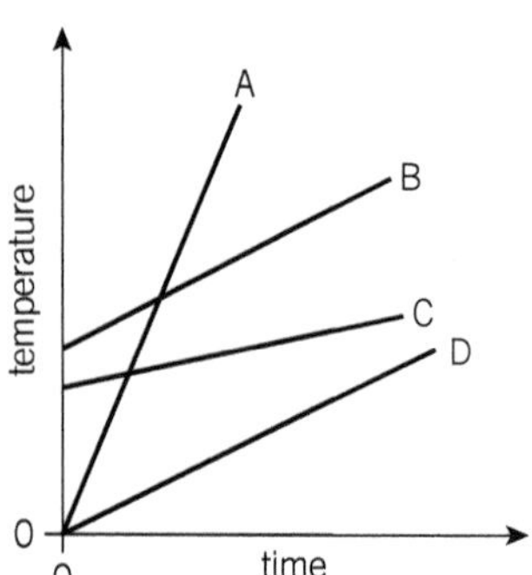

▲ **Figure 1**

3 Four solids A, B, C, and D are heated using the same heater. The solids have the same mass.

A temperature–time graph is plotted for each solid on the same axes (see Figure 1).

Which solid has the smallest value of specific heat capacity? *(1 mark)*

4 A beaker with 120 g of water at 15 °C is placed inside a freezer. It takes 6.0 minutes for the temperature of the water to drop to 0 °C.

specific heat capacity of water = $4200\,\mathrm{J\,kg^{-1}\,K^{-1}}$

specific latent heat of fusion of ice = $3.3 \times 10^5\,\mathrm{J\,kg^{-1}}$

a Calculate the average rate of energy loss from the water. *(3 marks)*

b Calculate the time it would take for the water at 0 °C to turn into ice at 0 °C. Assume that the rate of energy transfer is the same as your answer to **a**. *(3 marks)*

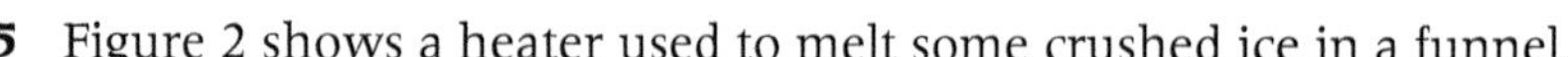

5 Figure 2 shows a heater used to melt some crushed ice in a funnel.

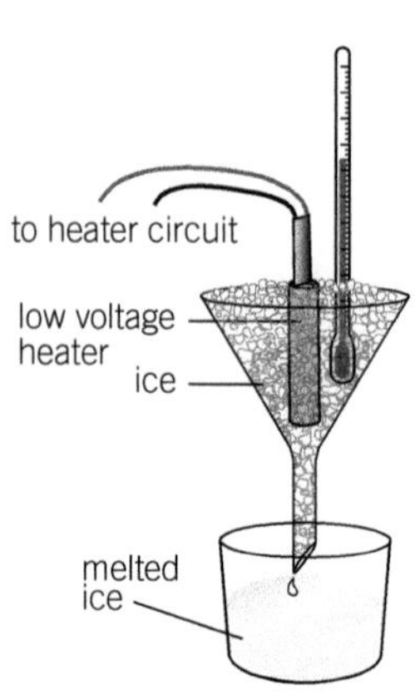

▲ **Figure 2**

The power of the heater is 25 W. The temperature of the ice remains at 0 °C. The table below shows the variation with time t of the total mass m of the water from the melting ice.

t / s	0	30	60	90	120	150
m / g	0	2.7	5.5	8.2	10.9	14.0

a Plot a graph of m against t and draw a line of best fit through the data points. *(2 marks)*

b Use the graph to determine the gradient of the straight line. *(1 mark)*

c Use your answer to **b** to calculate the specific latent heat of fusion of ice. *(3 marks)*

d Explain why your value the specific latent heat of fusion of ice is smaller than the accepted value of $3.3 \times 10^5\,\mathrm{J\,kg^{-1}}$. *(1 mark)*

6 A block of iron of mass 210 g is quickly transferred from a very hot oven into 500 g of water in a beaker. The temperature of water increases from 20 °C to 37 °C.

specific heat capacity of water = $4200\,\mathrm{J\,kg^{-1}\,K^{-1}}$

specific heat capacity of iron = $450\,\mathrm{J\,kg^{-1}\,K^{-1}}$

a State and explain the transfer of energy between the block and water. *(1 mark)*

b Calculate the energy gained by the water. *(2 marks)*

c Estimate the initial temperature of the iron block. Assume there is no transfer of energy to the surroundings. *(3 marks)*

Chapter 15 Ideal gases

In this chapter you will learn about ...

- [ ] Kinetic theory of gases
- [ ] The mole and Avogadro constant
- [ ] Boyle's law
- [ ] Equation of state for ideal gases
- [ ] Absolute zero
- [ ] Root mean square (r.m.s.) speed
- [ ] Boltzmann constant
- [ ] Mean kinetic energy of gas molecules
- [ ] Internal energy of ideal gas

15 IDEAL GASES
15.1 The kinetic theory of gases
15.2 Gas laws

Specification reference: 5.1.4

15.1 The kinetic theory of gases

The kinetic theory of gases is a model used to describe the behaviour of the particles in an **ideal gas**. The assumptions made in the kinetic model for an ideal gas are:

- The gas contains a very large number of atoms (or molecules) moving in random directions with random speeds.
- The volume of the gas atoms is negligible compared with the volume of the gas in the container.
- The atoms collide elastically with each other and with the container walls.
- The time of collision of atoms is negligible compared with the time between the collisions.
- The electrostatic forces between atoms are negligible except during collisions.

The mole and the Avogadro constant

The SI base unit for the **amount of substance** is the **mole**. One mole of *any* substance has 6.02×10^{23} particles (atoms or molecules). The number 6.02×10^{23} is known as the **Avogadro constant**, N_A.

One mole is defined as the amount of substance that contains as many elementary entities as there are atoms in 0.012 kg (12 g) of carbon-12.

You can calculate the total number N of atoms or molecules in a substance using the equation

$$N = n \times N_A$$

where n is the number of moles of the substance.

Explaining pressure

A gas exerts pressure because of repeated elastic collisions of the gas atoms with the container walls.

A gas atom of mass m travelling at a speed u collides with the container wall and bounces back at the same speed u. The change of momentum is $2mu$ (magnitude only). It then travels to the opposite wall and returns back after a time t and makes another collision. The average force f exerted *on* the atom by the wall is given by Newton's second law, $f = \frac{2mu}{t}$. According to Newton's third law, the atom exerts an equal but opposite force *on* the wall. The total force F on the wall is from the random collisions of all the atoms inside the container and the pressure p exerted on the wall is $\frac{F}{A}$, where A is the area of the wall.

15.2 Gas laws

The relationships between the thermodynamic temperature T, pressure p, and volume V of an ideal gas can be described by a few simple **gas laws**.

Boyle's law

The pressure exerted by a fixed amount of gas is inversely proportional to its volume, provided its temperature remains constant.

This can be expressed mathematically as

$$p \propto \frac{1}{V} \qquad \text{or} \qquad pV = \text{constant}$$

Pressure, volume, and temperature

Experiments on ideal gases also show that

$$p \propto T \text{ at constant volume}$$

and

$$V \propto T \text{ at constant pressure}$$

Figure 1 shows an arrangement you can use to determine the value of absolute zero in °C. A linear graph is produced by plotting pressure p against temperature θ in °C. The extrapolated graph intersects the temperature axis at absolute zero (−273 °C).

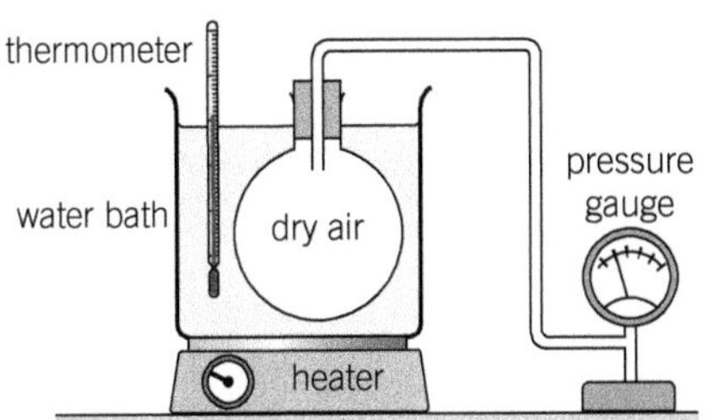

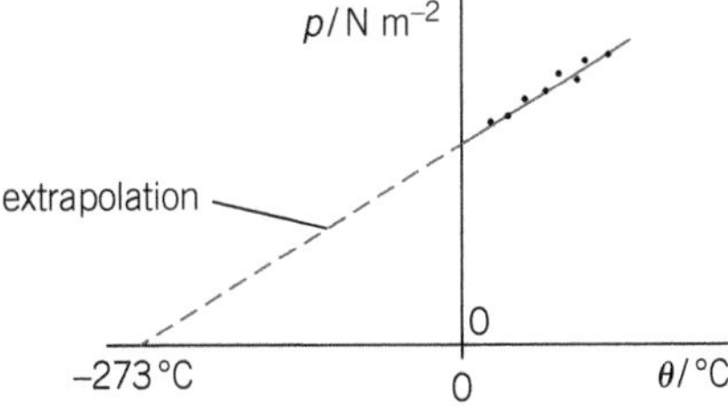

▲ **Figure 1** *Determining absolute zero*

Equation of state for ideal gases

The three relationships above can be combined together to give $\frac{pV}{T}$ = constant.

The 'constant' is equal to nR, where n is the number of moles and R is the **molar gas constant**, and is equal to $8.31\,\text{J mol}^{-1}\,\text{K}^{-1}$. The **equation of state for an ideal gas** is $pV = nRT$.

Revision tip

The temperature T is always in kelvin K and not °C.

Worked example: Balloon

Air trapped in a balloon has volume $3.4 \times 10^{-3}\,\text{m}^3$, temperature 20 °C, and pressure 1.2×10^5 Pa. Calculate the number of air molecules inside the balloon.

Step 1: Rearrange the equation of state with n as the subject and calculate n.

$$n = \frac{pV}{RT} = \frac{1.2 \times 10^5 \times 3.4 \times 10^{-3}}{8.31 \times (273 + 20)} = 0.1676$$

Step 2: Calculate the number N of air molecules.

$$N = n \times N_A = 0.1676 \times 6.02 \times 10^{23}$$

$$N = 1.0 \times 10^{23} \text{ (2 s.f.)}$$

Summary questions

1. Calculate the number of molecules in 3.0 moles of air. *(1 mark)*
2. The molar mass of oxygen is $0.032\,\text{kg mol}^{-1}$. Calculate the mass of a single molecule. *(2 marks)*
3. The temperature of a gas changes from 100 °C to 200 °C at constant volume. Calculate the factor by which the pressure increases. *(3 marks)*
4. Describe and explain what happens to the volume of a bubble of air rising up through water. *(3 marks)*
5. Calculate the density of 1 mole of air at temperature of −50 °C and pressure of 1.0×10^5 Pa. molar mass of air = $29\,\text{g mol}^{-1}$ *(4 marks)*
6. Estimate the mass of air in a small room. *(4 marks)*

15.3 Root mean square speed
15.4 The Boltzmann constant

Specification reference: 5.1.4

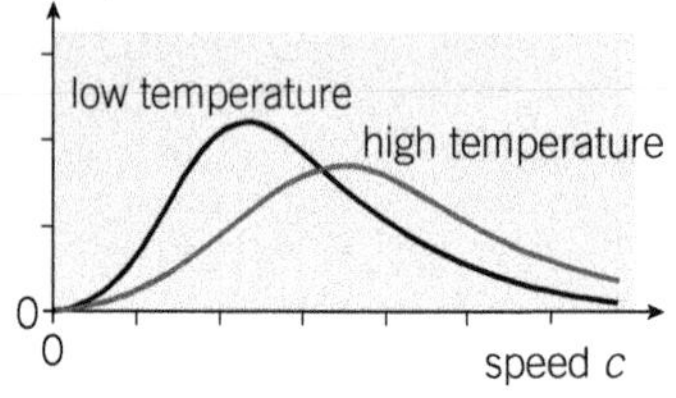

▲ **Figure 1** *Maxwell-Boltzmann distribution showing the spread of speed of the molecules of a gas*

15.3 Root mean square speed

The pressure exerted by a gas and the mean kinetic energy of molecules of a gas are related not to the mean speed of the molecules, but to the root mean square (r.m.s.) speed of the molecules.

r.m.s. speed

The atoms or molecules of a gas move in random directions and have a range of speeds. Figure 1 shows the spread in the speed of molecules – the graph is known as **Maxwell-Boltzmann distribution**.

Consider a container with N atoms. The velocity of each atom is c. The random motion implies that the mean velocity of the atoms is zero.

Here are some important terms:

- mean speed = sum of speed of all the atoms divided by N or simply
$$\bar{c} = \frac{\Sigma\, speed}{N}$$
- mean square speed = sum of velocity² of all the atoms divided by N or
$$\overline{c^2} = \frac{\Sigma c^2}{N}$$
- root mean square speed or r.m.s. speed = $c_{\text{r.m.s.}} = \sqrt{\text{mean square speed}}$

For 4 atoms with velocities $-200\,\text{m}\,\text{s}^{-1}$, $+300\,\text{m}\,\text{s}^{-1}$, $-430\,\text{m}\,\text{s}^{-1}$, and $+330\,\text{m}\,\text{s}^{-1}$, you can show that the mean velocity is $0\,\text{m}\,\text{s}^{-1}$, the mean speed is $315\,\text{m}\,\text{s}^{-1}$, and the r.m.s. speed is $325\,\text{m}\,\text{s}^{-1}$.

Maths: Mathematical symbols

The sigma Σ symbol stands for 'sum of ...' and the 'bar' on top of quantities stands for 'mean of ...'.

Pressure at the microscopic level

The kinetic theory model of gases, based on Newtonian mechanics, shows that

$$pV = \tfrac{1}{3}Nm\overline{c^2}$$

where p is the pressure exerted by the gas, V is the volume of the gas, N is the number of atoms, or molecules, in the gas, m is the mass of each atom, and $\overline{c^2}$ is the mean square speed of the atoms. As you see later, this is an important equation because it gives us a better understanding of the mean kinetic energy of the gas atoms and the thermodynamic temperature of the gas.

15.4 The Boltzmann constant

The **Boltzmann constant** k is equal to the molar gas constant R divided by Avogadro constant N_A.

Therefore

$$k = \frac{R}{N_A} = \frac{8.31}{6.02 \times 10^{23}} = 1.38 \times 10^{-23}\,\text{J}\,\text{K}^{-1}$$

The equation of state $pV = nRT$, can also be written as

$$pV = n(kN_A)T$$

or

$$pV = NkT$$

where N is the number of atoms or molecules in the gas. Be careful – the lower case n stands for the 'number of moles of gas' and the capital N represents the 'number of atoms or molecules'.

Mean kinetic energy and temperature

The proof below shows the derivation of an important equation ($\frac{1}{2}m\overline{c^2} = \frac{3}{2}kT$) – it relates the thermodynamic temperature of the gas to the mean kinetic energy of the atoms or molecules.

Main equations: $pV = nRT$, $pV = \frac{1}{3}Nm\overline{c^2}$, $N = nN_A$, and $R = kN_A$

$$pV = \frac{1}{3}Nm\overline{c^2} = nRT$$

Therefore

$$\frac{1}{3}m\overline{c^2} = \frac{nRT}{N} = \frac{RT}{N_A} = kT$$

This can be written as

$$\frac{1}{2}m\overline{c^2} = \frac{3}{2}kT$$

In the equation above, $\frac{1}{2}m\overline{c^2}$ is the mean kinetic energy of the gas molecules.

Note:

At a particular temperature T:

- all gas atoms have the same mean kinetic energy
- the greater mass atoms have smaller r.m.s. speed.

Internal energy

The internal energy of an ideal gas is entirely in the form of kinetic energy. The potential energy is zero because of the negligible electrostatic forces between atoms. Therefore

internal energy of a gas = $N \times \frac{3}{2}kT$

or

internal energy $\propto T$

Worked example: Air molecules

Calculate the r.m.s. speed of oxygen molecules in a room at a temperature of 20 °C.
mass of oxygen molecule = 5.3×10^{-26} kg

Step 1: Derive the equation for the r.m.s. speed.

$$\frac{1}{2}m\overline{c^2} = \frac{3}{2}kT$$

mean square speed = $\overline{c^2} = \frac{3kT}{m}$

r.m.s. speed = $c_{r.m.s.} = \sqrt{\frac{3kT}{m}}$

Step 2: Substitute values to determine the r.m.s. speed.

$$c_{r.m.s.} = \sqrt{\frac{3 \times 1.38 \times 10^{-23} \times (273+20)}{5.3 \times 10^{-26}}} = 480\ \text{m s}^{-1}\ \text{(2 s.f.)}$$

Revision tip

Mean kinetic energy of molecules is proportional to the thermodynamic temperature T.

Summary questions

1 The velocities of 5 atoms in m s^{-1} are −100, −200, 150, 200, and 300. Calculate the mean velocity and the mean speed of these atoms. *(2 marks)*

2 Calculate the mean square speed and the r.m.s. speed of the atoms in **Q1**. *(2 marks)*

3 Calculate the mean kinetic energy of gas molecules at a temperature of 200 °C. *(2 marks)*

4 In the demonstration of Brownian motion with smoke particles in air, explain why the smoke particles have a much smaller speed than the air molecules. *(2 marks)*

5 Calculate the r.m.s. speed of the molecules in **Q3** given the mass of each molecule is 4.8×10^{-26} kg. *(3 marks)*

6 Calculate the internal energy of 2.0 moles of gas atoms at a temperature of 0 °C. *(3 marks)*

Chapter 15 Practice questions

1 Which statement is correct about absolute zero?
The atoms of a substance:

A have no total energy

B have no kinetic energy

C have no internal energy

D have no potential energy *(1 mark)*

2 The molar mass of carbon is $12\,\text{g}\,\text{mol}^{-1}$.
How many atoms of carbon are there in 1.0 kg?

A 5.0×10^{22}

B 7.2×10^{22}

C 7.2×10^{24}

D 5.0×10^{25} *(1 mark)*

3 The actual velocity of four atoms in $\text{m}\,\text{s}^{-1}$ is +200, +300, −500, and +600.
What is the r.m.s. speed of the atoms?

A $180\,\text{m}\,\text{s}^{-1}$

B $280\,\text{m}\,\text{s}^{-1}$

C $430\,\text{m}\,\text{s}^{-1}$

D $450\,\text{m}\,\text{s}^{-1}$ *(1 mark)*

4 The r.m.s. speed of gas atoms is $500\,\text{m}\,\text{s}^{-1}$ at 50 °C. The temperature of the gas is doubled to 100 °C.
What is the r.m.s. speed of the atoms at this higher temperature?

A $540\,\text{m}\,\text{s}^{-1}$

B $580\,\text{m}\,\text{s}^{-1}$

C $710\,\text{m}\,\text{s}^{-1}$

D $1000\,\text{m}\,\text{s}^{-1}$ *(1 mark)*

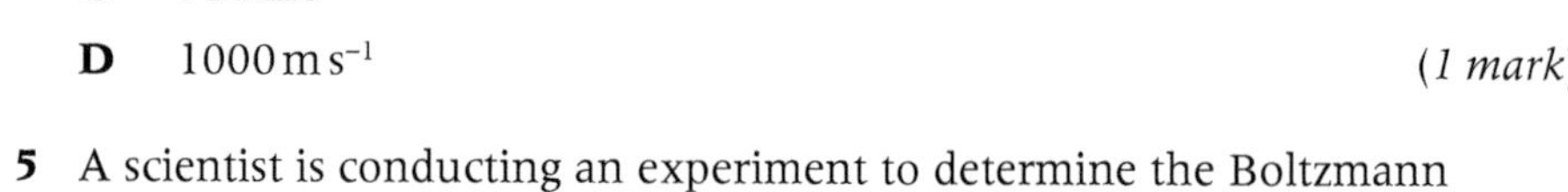

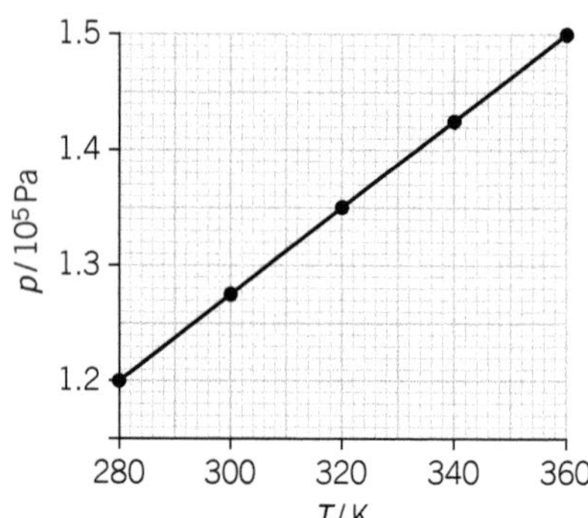

▲ **Figure 1**

5 A scientist is conducting an experiment to determine the Boltzmann constant k using 1.0 mole of gas trapped in a container of volume $2.2 \times 10^{-2}\,\text{m}^2$. Figure 1 shows a graph of pressure p exerted by the gas against temperature T in kelvin.

a Write an equation relating pressure p exerted by the gas, its volume V, absolute temperature T, and the molar gas constant R. *(1 mark)*

b Explain why the graph shown in Figure 1 is a straight-line graph. *(1 mark)*

c Use Figure 1 to determine a value for k. *(3 marks)*

6 A container of volume $5.2 \times 10^{-4}\,\text{m}^3$ has 0.65 g of trapped air. The air pressure inside the container is 1.2×10^5 Pa. The molar mass of air is about $30\,\text{g}\,\text{mol}^{-1}$.
Calculate:

a the density of the air in the container; *(1 mark)*

b the number of moles of air in the container; *(1 mark)*

c the temperature in °C of the air in the container; *(3 marks)*

d the mean kinetic energy of the molecules in the container. *(2 marks)*

Chapter 16 Circular motion

In this chapter you will learn about ...

- ☐ Angular velocity
- ☐ Radians
- ☐ Circular motion
- ☐ Centripetal force
- ☐ Centripetal acceleration

16 CIRCULAR MOTION
16.1 Angular velocity and the radian
16.2 Centripetal acceleration

Specification reference: 5.2.1, 5.2.2

16.1 Angular velocity and the radian

There are many examples of objects moving in circular paths – a moon orbiting a planet, a car on a roundabout, an electron in a uniform magnetic field, and so on. It is easier to analyse the motion of such objects in terms of their angular velocity.

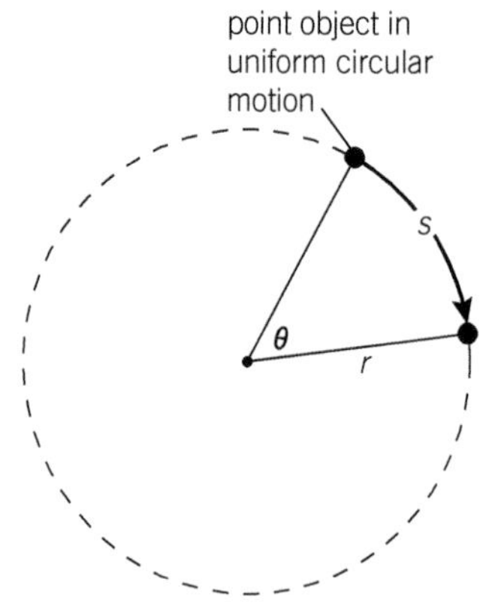

▲ **Figure 1** *The angle θ in radians is given by the equation* $\theta = \frac{s}{r}$

The radian

Angles can be measured in degrees. There are 360° in a full circle. Angles can also be measured in radians.

A **radian** is the angle subtended by a circular arc with a length equal to the radius of the circle.

An angle of 1 radian, or 1 rad, is about 57°. In general, the angle θ in radians is defined as follows:

$$\text{angle (in radians)} = \frac{\text{arc length}}{\text{radius of circle}} = \frac{s}{r} \text{ (see Figure 1)}$$

For a complete circle, s = circumference = $2\pi r$. Therefore, 360° is the same as 2π rad. You can also show that 180° = π, 90° = $\frac{\pi}{2}$, and so on.

Angular velocity ω

The **angular velocity** of an object moving in a circle is defined as the rate of change of angle.

The equation for angular velocity ω is

$$\omega = \frac{\theta}{t}$$

where θ is the angle through which an object moves in a time t. Angular velocity is measured in radians per second, rad s^{-1}. In a time t equal to one period T, the object will move through an angle θ equal to 2π radians. Therefore

$$\omega = \frac{2\pi}{T}$$

The frequency f of the rotating object is given by $f = \frac{1}{T}$. Therefore, we can also use the following equation for angular velocity:

$$\omega = 2\pi f$$

Revision tip

To avoid confusing the terms **velocity** and **angular velocity**, just remember that velocity has units **m s^{-1}** and angular velocity is measured in **rad s^{-1}**.

16.2 Centripetal acceleration

Figure 2 shows an object moving in a circle at a constant speed v. The velocity of the object is continually changing because its direction of travel changes. The object must be accelerating because acceleration is defined as the rate of change of velocity.

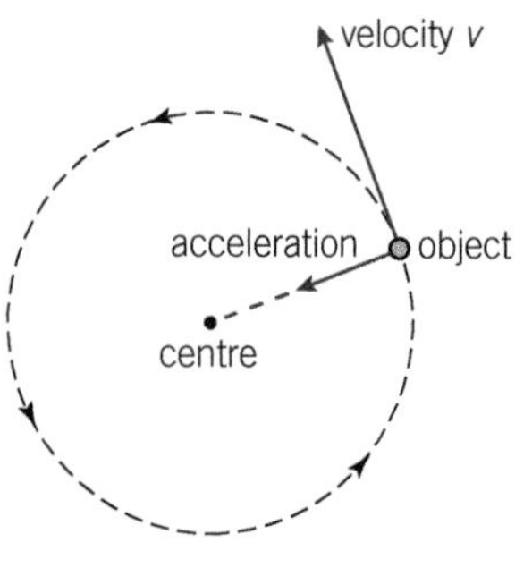

▲ **Figure 2** *An object moving in a circle has acceleration*

Speed in a circle

You can determine the speed v of an object moving in a circle of radius r at a constant angular velocity ω as follows:

$$v = \frac{\text{circumference}}{\text{period}}$$

$$v = \frac{2\pi r}{T}$$

However, $\omega = \frac{2\pi}{T}$, therefore

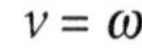

$$v = \omega r$$

Centripetal force and acceleration

For an object travelling in a circle of radius r at a constant speed v, it has a constant acceleration a towards the centre of the circle. This acceleration is known as **centripetal acceleration**. Centripetal simple means *'towards the centre'*. The equation for the centripetal acceleration is

$$a = \frac{v^2}{r}$$

Since $v = \omega r$, we can also use the equation

$$a = \omega^2 r$$

According to $F = ma$, the resultant force on the object and the acceleration must be in the same direction – towards the centre of the circle. This resultant force is referred to as **centripetal force**.

Model answer: Mercury

The planet Mercury orbits the Sun in 88 days in a circle of radius 5.8×10^{10} m. Calculate its centripetal acceleration.

Answer

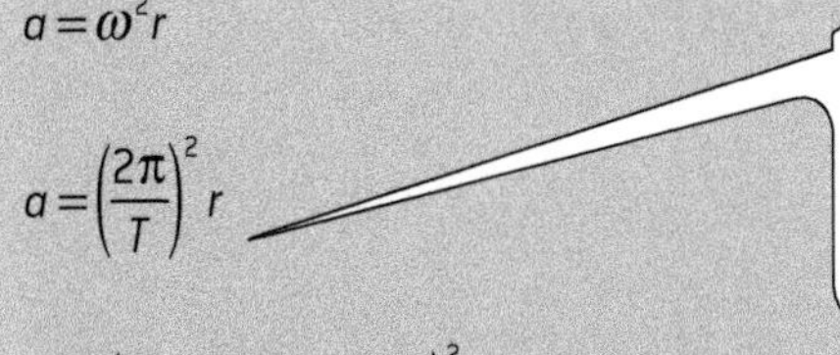

$a = \omega^2 r$

$a = \left(\frac{2\pi}{T}\right)^2 r$

You can calculate ω separately, but this technique is good – the substitution of values is later, so there is less chance of making errors.

$a = \left(\frac{2\pi}{88 \times 24 \times 3600}\right)^2 \times 5.8 \times 10^{10}$

The period T must be in seconds.

$a = 4.0 \times 10^{-2}\ \text{m s}^{-2}$

The data is given to 2 s.f., so is the final answer.

Summary questions

1 Change the following angles into radians:
a 45° b 5.0° c 420° *(3 marks)*

2 Calculate the speed of an object travelling in a circle of radius 50 cm with an angular velocity of 15 rad s^{-1}. *(2 marks)*

3 An object travels through an angle of 45° in 2.5 s. Calculate its angular velocity in rad s^{-1}. *(2 marks)*

4 An aircraft moves in a circular path of radius 12 km at a constant speed of 150 m s^{-1}. Calculate the angular velocity and centripetal acceleration of the aircraft. *(4 marks)*

5 A car travelling at 20 m s^{-1} over a humpback bridge has acceleration equal to the acceleration of free fall. Calculate the radius of curvature of the bridge. *(2 marks)*

6 Explain why the speed of an object moving in a circle is not affected by the centripetal force. *(2 marks)*

16.3 Exploring centripetal forces

For an object (planet, car, aeroplane, etc.) moving in a circle of radius r at a constant speed v, the centripetal acceleration a can be calculated using either

$$a = \frac{v^2}{r}$$

or

$$a = \omega^2 r$$

where ω is the angular velocity. The resultant force on the object, which is the **centripetal force** F, can be calculated using Newton's second law $F = ma$. Therefore

$$F = \frac{mv^2}{r}$$

and

$$F = m\omega^2 r$$

Revision tip

In the equations $F = \frac{mv^2}{r}$ and $F = m\omega^2 r$, F is the *total* or *resultant* force acting on the object which must *point towards the centre of the circle*.

Investigating centripetal forces

Figure 1 shows an arrangement you can use to demonstrate and analyse circular motion. A rubber bung of mass m is whirled in a horizontal circle at a constant speed. The speed v can be measured directly using a motion sensor. Alternatively, the time t for N revolutions could be timed using a stopwatch and v calculated using $v = \frac{2prN}{t}$.

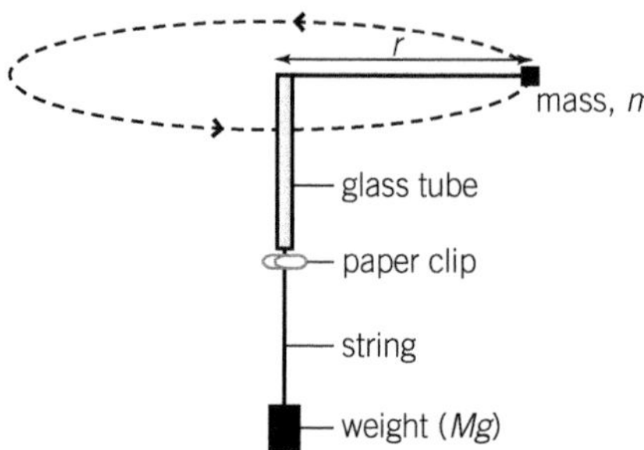

▲ **Figure 1** *Whirling bung experiment*

For this arrangement, the tension in the string provides the centripetal force. Therefore

$$F = \frac{mv^2}{r} = Mg$$

The radius r of the circle could be kept constant and different values obtained for v as m is changed.

A graph of v^2 against m should be a straight line passing through the origin.

Car on a roundabout

Figure 2 shows a car moving round a roundabout.

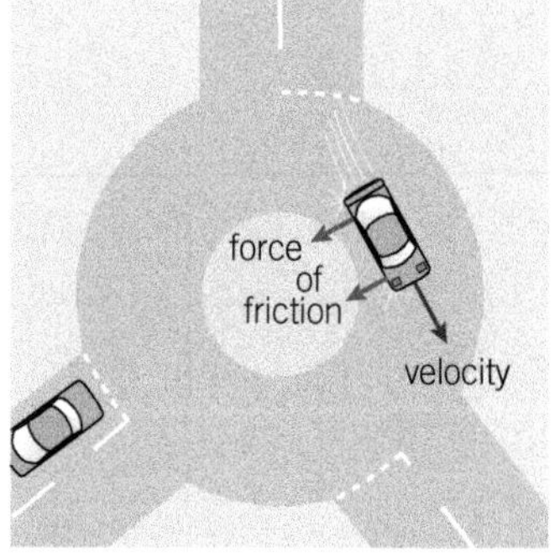

▲ **Figure 2** *Car on a roundabout*

The centripetal force is provided by the frictional force F_R between the tyres and the road. Therefore

$$F_R = \frac{mv^2}{r}$$

Vertical circle

Figure 3 shows an object tied to a string whirled in a vertical circle of constant radius r and a constant speed v. The centripetal force on the object must be constant and equal to $\frac{mv^2}{r}$. This can only happen if the tension in the string changes.

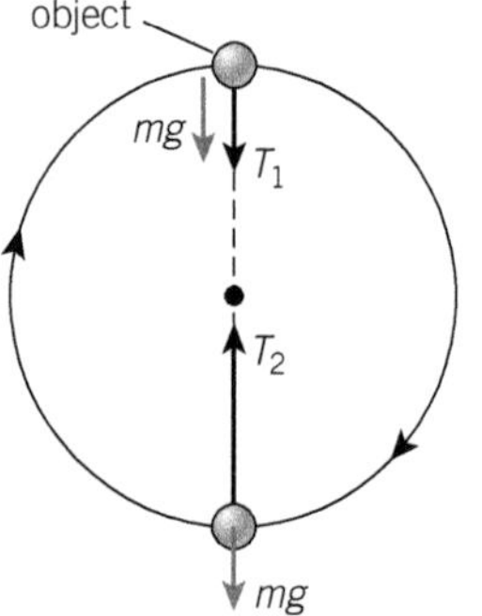

▲ **Figure 3** *Motion in a vertical circle*

At the top of the motion: centripetal force = weight + tension = $mg + T_1$

Therefore

$$mg + T_1 = \frac{mv^2}{r}$$

$$T_1 = \frac{mv^2}{r} - mg$$

At the bottom of the motion: centripetal force = tension − weight = $T_2 - mg$

Therefore

$$T_2 - mg = \frac{mv^2}{r}$$

$$T_2 = \frac{mv^2}{r} + mg$$

The tension in the string is greatest at the bottom.

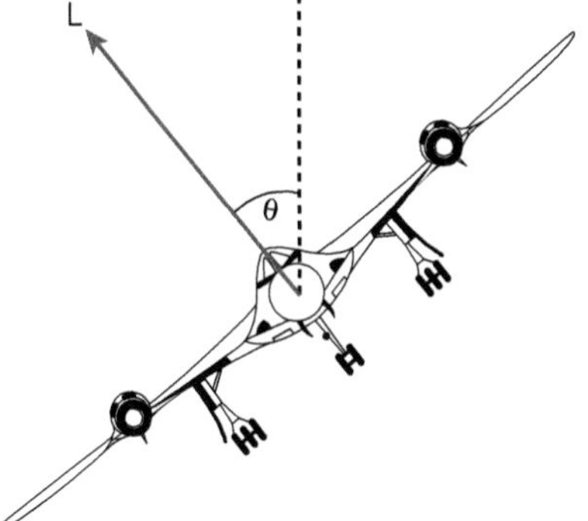

▲ **Figure 4** *An aeroplane turning*

Banked turn

Figure 4 shows an aeroplane turning or banking while flying horizontally. The lift force L is at an angle to the vertical. This force has both vertical and horizontal components. The weight W of the aeroplane acts vertically and has no horizontal component.

Vertically: Resultant force = 0

vertical component of L = weight

$$L\cos\theta = W$$

Horizontally: The horizontal component of L provides the centripetal force.

$$L\sin\theta = \frac{mv^2}{r}$$

Worked example: Whirling bung

An object of mass 50 g is attached to a light rod of length 10 cm and spun at a constant speed of 3.0 m s^{-1} in a *vertical* circle. Calculate the maximum tension in the rod.

Step 1: Calculate the centripetal force on the object.

$$F = \frac{mv^2}{r} = \frac{0.050 \times 3.0^2}{0.10} = 4.5\ \text{N}$$

Step 2: The maximum tension T is at the bottom of the motion. Write an equation for the tension and then substitute the values.

T − weight = 4.5

$T - (0.050 \times 9.81) = 4.5$

$T = 5.0$ N (2 s.f.)

Summary questions

1. Suggest what provides the centripetal force on a planet orbiting the Sun. *(1 mark)*
2. Calculate the centripetal force on a stone of mass 0.20 kg that is whirled in a horizontal circle of radius 0.50 m at a speed of 3.2 m s^{-1}. *(2 marks)*
3. A car of mass 900 kg is travelling at a constant speed round a roundabout of radius 30 m. The frictional force provided by the tyres is 4.3 kN. Calculate the speed of the car. *(2 marks)*
4. In the whirling bung experiment, what is the gradient of the v^2 against m graph equal to? *(2 marks)*
5. The Moon has mass 7.3×10^{22} kg and is at a distance of 3.8×10^8 m from the Earth. It takes 27 days to orbit the Earth. Calculate the gravitational force acting on the Moon. *(3 marks)*
6. Show that $\tan\theta = \frac{v^2}{rg}$ for an aeroplane turning (see Figure 4). *(3 marks)*

Chapter 16 Practice questions

1 What is the correct unit for angular velocity?

A $m\,s^{-1}$

B Hz

C rad

D $rad\,s^{-1}$ *(1 mark)*

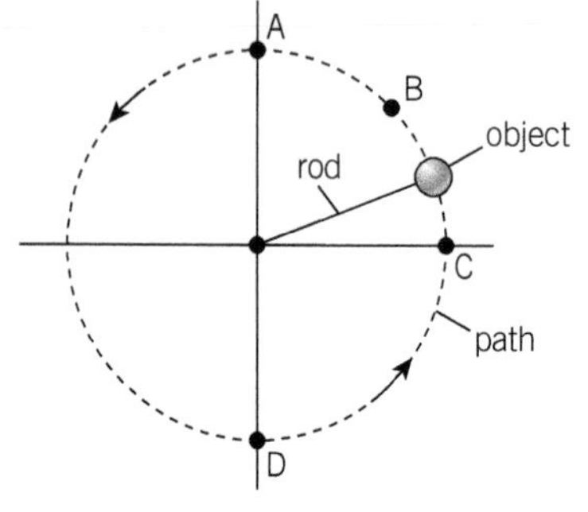

▲ **Figure 1**

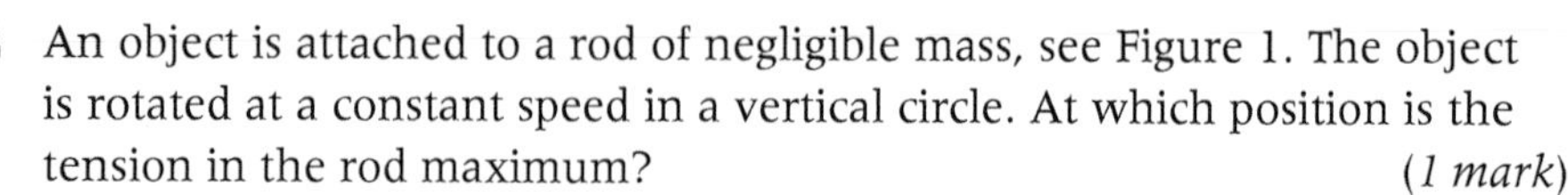

2 An object is attached to a rod of negligible mass, see Figure 1. The object is rotated at a constant speed in a vertical circle. At which position is the tension in the rod maximum? *(1 mark)*

3 An object moves in a circular path of radius 10 cm with a constant speed of $2.6\,m\,s^{-1}$.
What is the frequency of the object moving in the circle?

A 0.24 Hz

B 4.1 Hz

C 26 Hz

D 160 Hz *(1 mark)*

4 An object is moving in a circle. The centripetal force on the object is *F*. The speed of the object is now doubled and its radius halved.
What is the new value for the centripetal force in terms of *F*?

A *F*

B 2*F*

C 4*F*

D 8*F* *(1 mark)*

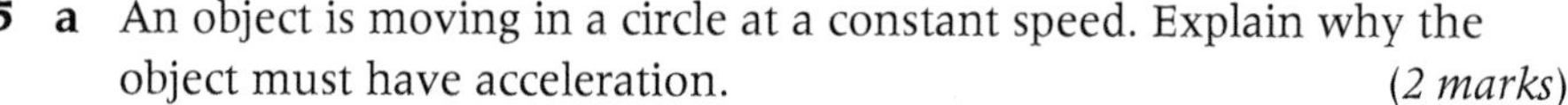

5 **a** An object is moving in a circle at a constant speed. Explain why the object must have acceleration. *(2 marks)*

b Figure 2 shows an object of mass 120 g on a revolving metal disc. The object moves in a circle of radius 10 cm.

i On Figure 2, show the direction of the frictional force acting on the object as it travels round in a circle. *(1 mark)*

ii The maximum frictional force acting on the object is equal to half the weight of the object. The speed of the disc is slowly increased. Calculate the maximum speed of the object for it to remain on the disc. *(4 marks)*

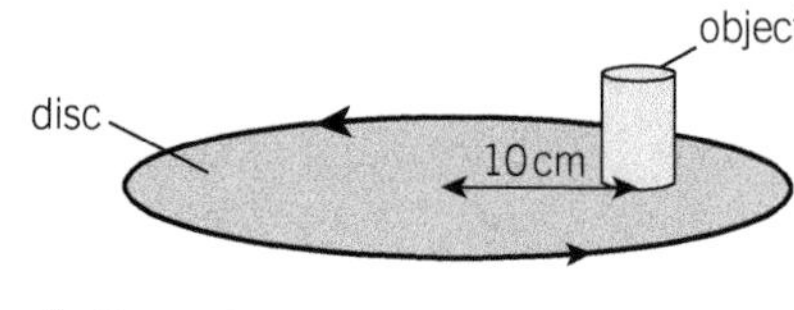

▲ **Figure 2**

6 An electron moves in a circle of radius 1.2 cm at a constant speed of $4.2 \times 10^{7}\,m\,s^{-1}$.

Calculate:

a the centripetal acceleration of the electron; *(2 marks)*

b the centripetal force on the electron; *(2 marks)*

c the period of the electron as it moves round in a circle. *(2 marks)*

7 Figure 3 shows a conical pendulum where an object of weight 1.2 N attached to the end of a string describes a horizontal circle.

The radius of the circular path is 15 cm and the string makes an angle of 45° to the vertical.

a Show that the tension in the string is 1.7 N. *(1 mark)*

b Show that the centripetal force acting on the object is 1.2 N. *(1 mark)*

c Calculate the speed *v* of the object. *(3 marks)*

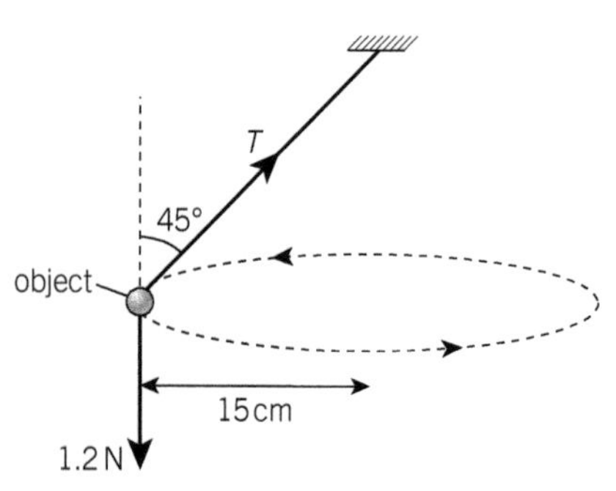

▲ **Figure 3**

Chapter 17 Oscillations

In this chapter you will learn about ...

- ☐ Simple harmonic motion
- ☐ Phase difference
- ☐ Period and frequency of oscillations
- ☐ Graphs for simple harmonic motion
- ☐ Energy of a simple harmonic oscillator
- ☐ Damping
- ☐ Resonance

17 OSCILLATIONS
17.1 Oscillations and simple harmonic motion

Specification reference: 5.3.1

17.1 Oscillations and simple harmonic motion

A swinging pendulum, an oscillating mass at the end of a vertical spring, and a vibrating ruler fixed at one end are all examples of free oscillations. Simple harmonic motion (SHM) is the simplest type of motion of oscillators. One important characteristic of SHM is that the period of the oscillations is independent of the amplitude of the oscillator. The oscillator keeps 'steady time' and is known as an **isochronous oscillator**. (In Greek, *iso* = *same* and *chronous* = *time*.)

Oscillatory motion

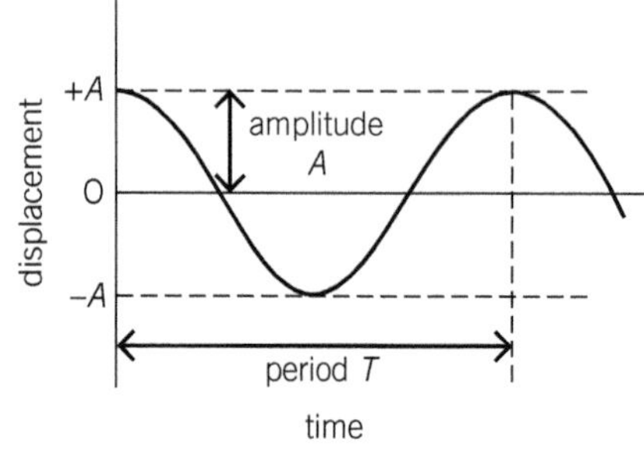

▲ **Figure 1** *The displacement–time graph for an oscillator*

Figure 1 shows the displacement against time graph for a simple harmonic oscillator.

Here are some important definitions:

- The **displacement** x is the distance from the equilibrium position in a particular direction.

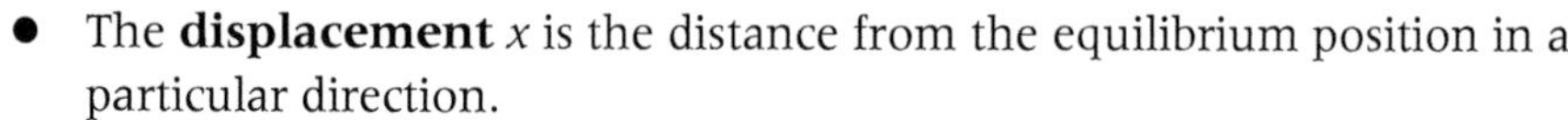

- The **amplitude** A is the maximum displacement of the oscillator from its equilibrium position.
- The **period** T is the time taken for the oscillator to complete one complete oscillation.
- The **frequency** f is the number of oscillations per unit time. Frequency is measured in hertz, Hz.

Simple harmonic motion

An oscillator executes **simple harmonic motion** when its acceleration is directly proportional to its displacement from its **equilibrium position**, and is directed towards the equilibrium position.

For such an oscillator

$$\text{acceleration} \propto -\text{displacement}$$

or

$$a \propto -x$$

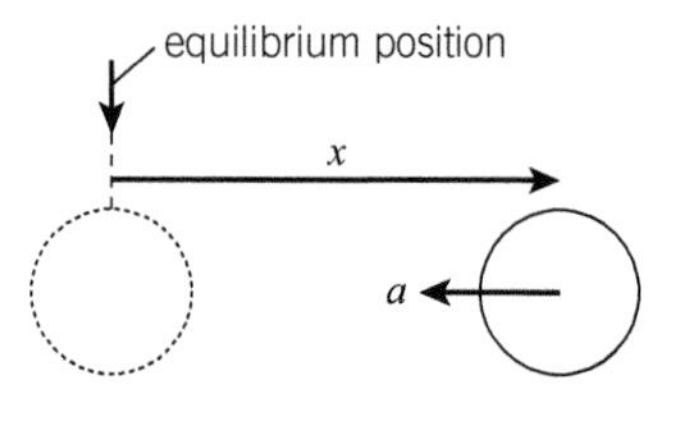

▲ **Figure 2** *Acceleration and displacement are in opposite directions*

The minus sign implies that the direction of the acceleration is always towards a fixed point (equilibrium position), see Figure 2. The constant of proportionality is ω^2, where ω is the angular frequency of the oscillator.

The angular frequency ω is defined by either the equation $\omega = 2\pi f$ or $\omega = \frac{2\pi}{T}$.

Common misconception: Omega?

ω is the **angular** *velocity* of the object moving in a *circle*. In oscillatory motion, ω is the **angular** *frequency* of the oscillator. The equations and unit for ω are the same for both. Use the terms carefully.

Phase difference

Phase difference is a useful quantity when comparing the motions of two simple harmonic oscillators with identical period of oscillations.

The **phase difference** is the fraction of an oscillation one oscillator leads or lags behind another.

Phase difference, as for waves, can be measured either in degrees or radians. Figure 3 illustrates the motions of two oscillators A and B. The oscillator A *lags* behind oscillator B by a time Δt.

The phase difference ϕ in radians between the motions is given by the equation

$$\phi = 2\pi \times \left(\frac{\Delta t}{T}\right) \text{ radians}$$

where T is the common period of the two oscillators.

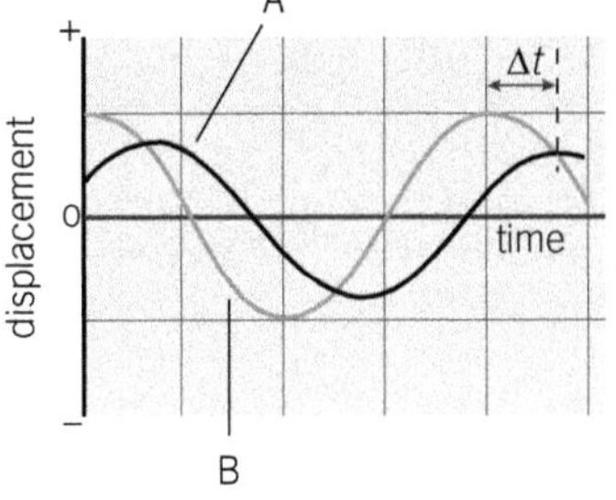

▲ **Figure 3** *There is a phase difference between the motions of oscillators A and B*

Investigationing period and frequency

The period T of a simple harmonic oscillator can be determined by measuring the time t taken for N oscillations; $T = \frac{t}{N}$. Having a large number of oscillations reduces the percentage uncertainty due to reaction time.

You can also use a motion sensor connected to a laptop or a data-logger to show that an oscillator keeps steady time (isochronous).

Worked example: Oscillating pendulum

A pendulum bob of mass 120 g oscillates with a period of 1.40 s and has amplitude of 50.0 cm.

Calculate the maximum resultant force acting on the pendulum bob.

Step 1: Derive an equation for the force F in terms of the quantities given.

$F = ma$, and $a_{max} = \omega^2 A$ with $\omega = \frac{2\pi}{T}$

Therefore, the maximum force $F_{max} = m\omega^2 A = \frac{4\pi^2 mA}{T^2}$

Step 2: Calculate the value for the maximum force.

$m = 0.120$ kg $\quad A = 0.500$ m $\quad T = 1.40$ s

$$F_{max} = \frac{4\pi^2 \times 0.120 \times 0.500}{1.40^2} = 1.21 \text{ N (3 s.f.)}$$

Summary questions

1 Calculate the angular frequency of an oscillator with period 2.00 s. *(1 mark)*

2 Describe how the acceleration of an oscillator changes as it travels from its equilibrium position to its maximum displacement. *(2 marks)*

3 An oscillator with frequency 1.5 kHz has amplitude 0.60 mm. Calculate the angular frequency and maximum acceleration of the oscillator. *(4 marks)*

4 Calculate the phase difference between the oscillators A and B shown in Figure 3 when the lag time is one-eighth of the period. *(1 mark)*

5 The acceleration of an oscillator is given by the equation $a = -400x$. Calculate its period. *(3 marks)*

6 The period T of a pendulum of length L is given by $T = 2\pi\left(\frac{L}{g}\right)^{\frac{1}{2}}$, where g is the acceleration of free fall.

A pendulum has length 5.0 m. The mass of the bob is 1.2 kg and the amplitude is 1.6 m. Calculate the maximum resultant force on the oscillating pendulum bob. *(4 marks)*

17.2 Analysing SHM
17.3 SHM and energy

Specification reference: 5.3.1, 5.3.2

17.2 Analysing SHM

The variation of the displacement x of a simple harmonic with time t is sinusoidal. This simply means that x against t is a sine or cosine shaped graph. You have already seen this in Figure 1 in Topic 17.1, Oscillations and simple harmonic motion.

Displacement equations

You need to be familiar with the following two equations for displacement x:

$$x = A\sin\omega t \text{ and } x = A\cos\omega t$$

Revision tip

Make sure your calculator is in *radian mode* when using the equations $x = A\cos\omega t$ and $x = A\sin\omega t$.

The amplitude of the oscillations is A and ω is the angular frequency. These equations are 'solutions' of the SHM equation $a = -\omega^2 x$.

- Use the sine version when $x = 0$ at $t = 0$.
- Use the cosine version when $x = A$ at $t = 0$.

Graphs

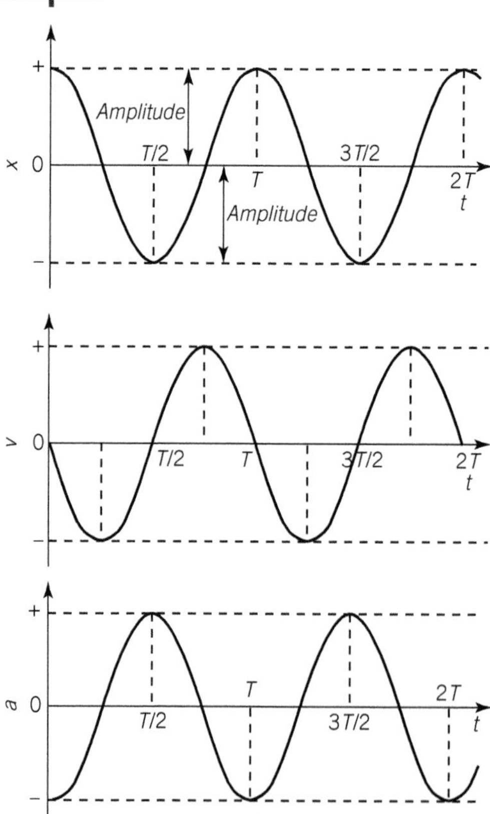

▲ **Figure 1** *Simple harmonic graphs*

Synoptic link

You learnt about displacement–time graphs and velocity–time graphs in Chapter 3, Motion.

The velocity v of the oscillator can be determined from the *gradient* of the displacement–time graph and the acceleration a from the *gradient* of the velocity–time graph. Figure 1 show x–t, v–t, and a–t graphs for a simple harmonic oscillator.

Velocity equations

Imagine a swinging pendulum. When displacement $x = 0$ (equilibrium position), the pendulum bob has maximum speed and the speed decreases as its displacement increases. The speed is momentarily zero when $x = A$. You can determine the velocity v of any simple harmonic oscillator using the equation $v = \pm\,\omega\sqrt{A^2 - x^2}$

The maximum speed is when $x = 0$ and is given by the equation $v_{max} = \omega A$.

17.3 SHM and energy

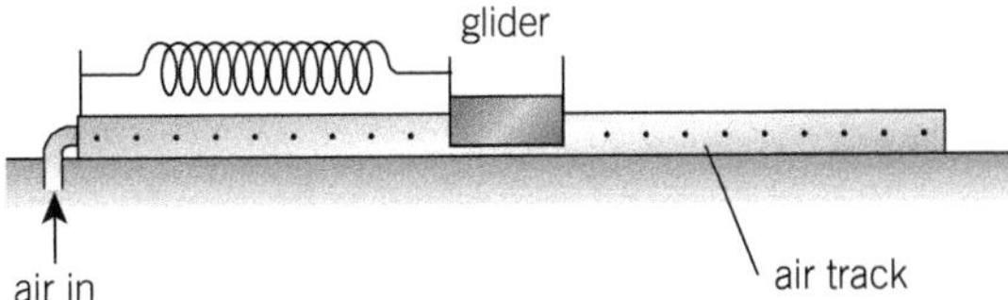

▲ **Figure 2** *A spring–mass oscillator has kinetic energy and potential energy*

The energy of an oscillator is made up of kinetic energy and potential energy. For example, a vibrating atom in a solid has kinetic energy because of its motion and electrostatic potential energy due to the electrical forces on the atom. For a spring–mass system (see Figure 2) oscillating horizontally, the stored energy is elastic potential energy in the spring.

Energy–displacement graph

The kinetic energy E_k of a simple harmonic oscillator of mass m is given by

$$E_k = \frac{1}{2}mv^2 = \frac{1}{2}m\omega^2(A^2 - x^2)$$

- E_k has a maximum value of $\frac{1}{2}m\omega^2A^2$ when $x = 0$.
- The total energy of the oscillator must also be $\frac{1}{2}m\omega^2A^2$.
- The potential energy E_p of the oscillator must be $\frac{1}{2}m\omega^2x^2$.

Figure 3 shows the variation of E_k and E_p with displacement x.

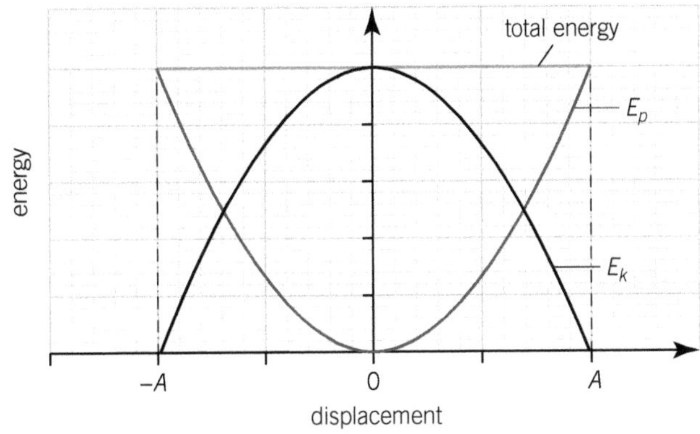

▲ **Figure 3** *The variation of potential and kinetic energies with displacement*

Worked example: Energy of an oscillating mass

A 700 g mass is hung from a spring. The mass is pulled vertically below its equilibrium and then released. It oscillates with a period of 0.75 s and amplitude of 32 cm.

Calculate the maximum kinetic energy E_{max} of the mass.

Step 1: Calculate the angular frequency ω.

$$\omega = \frac{2\pi}{T} = \frac{2\pi}{0.75} = 8.379\ \text{rad s}^{-1}$$

Step 2: The maximum speed v_{max} is when the mass is at its equilibrium. Calculate v_{max}.

$$v_{max} = \omega A = 8.379 \times 0.32 = 2.68\ \text{m s}^{-1}$$

Step 3: Calculate E_{max}.

$$E_{max} = \frac{1}{2}mv^2 = \frac{1}{2} \times 0.700 \times 2.68^2$$

$$E_{max} = 2.5\ \text{J (2 s.f.)}$$

Summary questions

1 State how the maximum speed of an oscillator is related to the amplitude of its motion. *(1 mark)*

2 Explain how the first and last graphs in Figure 1 illustrate simple harmonic motion. *(2 marks)*

3 The oscillations of a vibrating plate have amplitude 3.0 mm and period 40 ms. Calculate the maximum speed of the vibrating plate. *(2 marks)*

4 For the plate in **Q3**, calculate its speed when the displacement of the plate is 1.0 mm. *(2 marks)*

5 An oscillator of mass 180 g has period of 0.80 s and amplitude 20 cm. Calculate the potential energy of the oscillator when the displacement is 10 cm. *(3 marks)*

6 The oscillator in **Q5** has maximum displacement at time $t = 0$. Calculate the time t when its displacement is 7.5 cm for the first instant. *(2 marks)*

17.4 Damping and driving
17.5 Resonance

Specification reference: 5.3.3

17.4 Damping and driving

The motion of a mechanical system oscillating without any external forces is known as **free oscillations**. The frequency of the free oscillations is known as the **natural frequency** of the oscillator. In **forced oscillations**, an external periodic driving force is applied to an oscillator. The frequency of the driving force is known as the **driving frequency**.

Oscillations of a mechanical system are damped when an external force acts on the system which reduces the amplitude of the oscillations.

Damping

Damping is caused by frictional forces, for example air resistance and viscous drag in oil. For a mechanical system such as a car, damping is essential because without it, the car would keep bouncing up and down every time it went over a bump.

Figure 1 shows how the displacement of a damped oscillator varies with time. For **light damping**, the amplitude of the oscillator decreases by the same fraction every oscillation – the amplitude is said to decay exponentially.

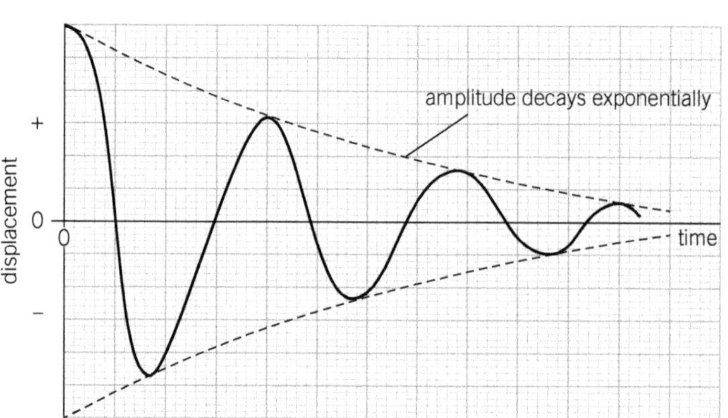

▲ **Figure 1** *Damped oscillations of an oscillator – the period remains the same*

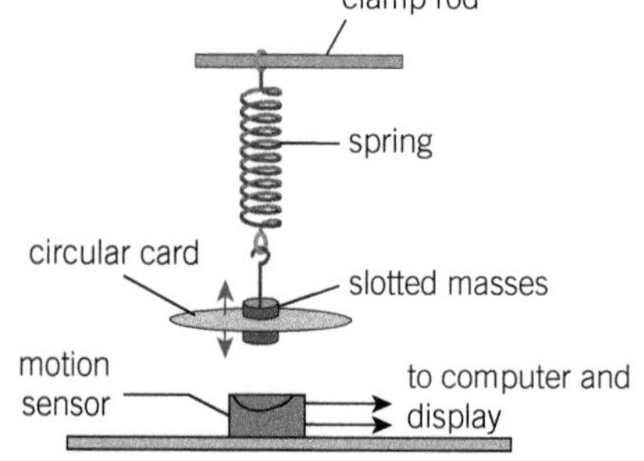

▲ **Figure 2** *Investigating damped oscillations using a motion sensor*

The damped oscillations of a spring–mass system can be investigated using the arrangement shown in Figure 2.

17.5 Resonance

Figure 3 shows an arrangement that may be used to investigate the forced oscillations of a mechanical oscillator. As the driver frequency is slowly increased from zero, the amplitude of oscillations of the forced spring–mass system increases until it reaches maximum amplitude at a particular frequency, and then the amplitude decreases again.

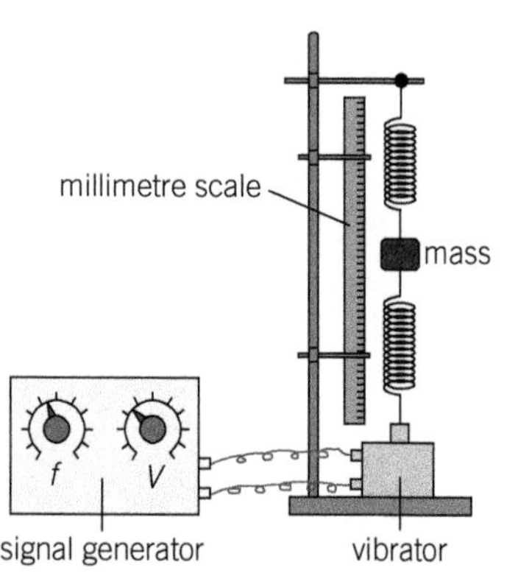

▲ **Figure 3** *Investigating forced oscillations and resonance*

Characteristic of resonance

The oscillator is in resonance when it has maximum amplitude when being forced to oscillate.

At resonance:

- the driving frequency is equal to the natural frequency f_0 of the forced oscillator (for light damped system)
- the forced oscillator absorbs maximum energy from the driver.

The frequency at the maximum amplitude is also known as the resonant frequency.

Figure 4 shows the variation of amplitude of the forced oscillator with the driving frequency for different degrees of damping. As the amount of damping is increased:

- the amplitude of vibration at any frequency decreases
- the peak on the graph becomes flatter and broader
- the maximum amplitude occurs at a lower frequency than f_0.

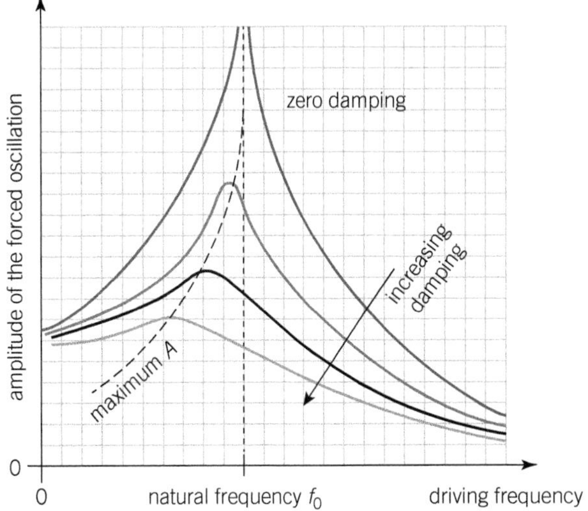

▲ **Figure 4** *Resonance curves*

Some examples of resonance: tuning circuit of a radio, MRI scanner, and wind instrument (e.g., organ pipe).

Worked example: Resonance

Figure 5 shows a resonance curve drawn by a student. The values indicated on the axes are correct, but the labelling has been omitted.

The mass of the oscillator is 100 g.

State the labelling for the axes and determine the total energy of the oscillator at resonance.

Step 1: Correctly identify the labels for the axes.

The *x*-axis should be 'driving frequency/Hz' and the *y*-axis should be 'amplitude/cm'.

Step 2: Determine the angular frequency ω.

$\omega = 2\pi f = 2\pi \times 4.0 = 25.1 \text{ rad s}^{-1}$

Step 3: Calculate the total energy.

The total energy of the oscillator is the maximum kinetic energy.

total energy $= \frac{1}{2} m\omega^2 A^2 = \frac{1}{2} \times 0.100 \times 25.1^2 \times 0.020^2$

total energy = 0.013 (2 s.f.)

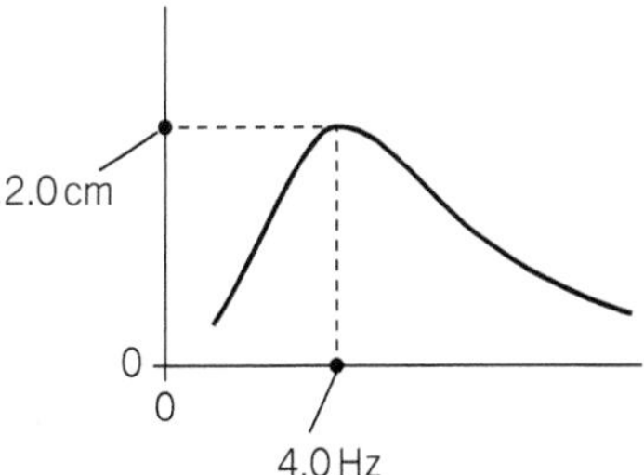

▲ **Figure 5**

Summary questions

1 State what is meant by resonance. *(1 mark)*

2 State what is meant by damped oscillations. *(1 mark)*

3 Describe how the 'sharpness' of the resonance curve is affected by the degree of damping. *(1 mark)*

4 Describe how the resonant frequency of a specific oscillator is affected by the degree of damping. *(1 mark)*

5 The amplitude of an oscillator decays exponentially. The amplitude decreases by half after every 10 oscillations. How many oscillations does it take for the amplitude to become an eighth of its initial amplitude? *(2 marks)*

6 The oscillator illustrated in the worked example is set up with less damping.
Resonance occurs at the same frequency as before but the amplitude increases to 20 cm.
Calculate the total energy of the oscillator under these conditions. *(3 marks)*

Chapter 17 Practice questions

1 What is the correct unit for angular frequency?

A $m\,s^{-1}$ | C rad

B Hz | D $rad\,s^{-1}$ *(1 mark)*

2 The maximum speed of a simple harmonic oscillator is v_{max} when it oscillates with amplitude A and period T.

What is the correct equation for v_{max}?

A $v_{max} = \frac{A}{T}$

B $v_{max} = \frac{2A}{T}$

C $v_{max} = \frac{\pi A}{T}$

D $v_{max} = \frac{2\pi A}{T}$ *(1 mark)*

3 The period of oscillations of a simple harmonic oscillator is T. The amplitude of the oscillator is now doubled.

What is the new value of its period in terms of T?

A $\frac{T}{2}$

B T

C $\sqrt{2}T$

D $2T$ *(1 mark)*

4 The acceleration a of a simple harmonic oscillator is related to its displacement x by the equation $a = -100x$.

What is the frequency of the oscillations?

A 0.63 Hz | C 63 Hz

B 1.6 Hz | D 100 Hz *(1 mark)*

▲ **Figure 1**

5 Figure 1 shows the end of a vibrating metal strip.

The metal strip oscillates with a simple harmonic motion with a frequency of 840 Hz. The distance between the two extreme positions of the end of the strip is 3.2 mm.

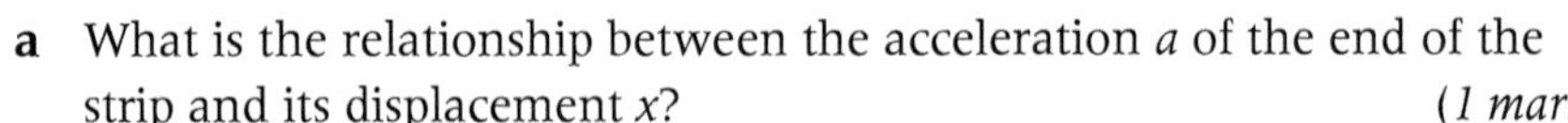

a What is the relationship between the acceleration a of the end of the strip and its displacement x? *(1 mark)*

b State the amplitude of the oscillating end of the strip. *(1 mark)*

c Calculate the maximum:

i velocity of the end of the strip; *(2 marks)*

ii acceleration of the end of the strip. *(2 marks)*

d Describe how the potential energy of the strip varies as the end of the strip travels from the equilibrium position to its amplitude. *(2 marks)*

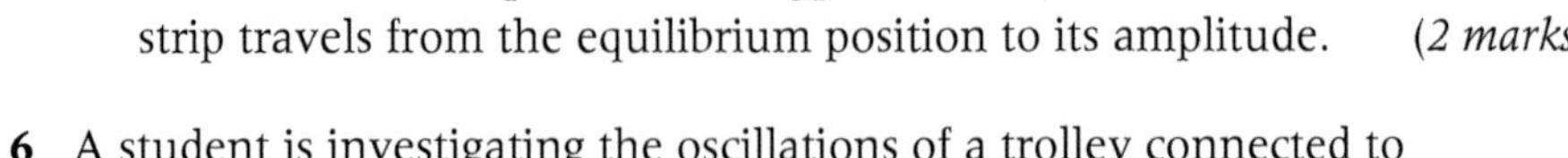

6 A student is investigating the oscillations of a trolley connected to springs. The maximum speed v_{max} of the trolley as it passes through the equilibrium position is measured and recorded using a data-logger.

Figure 2 shows the variation of v_{max} with the amplitude A of the oscillations.

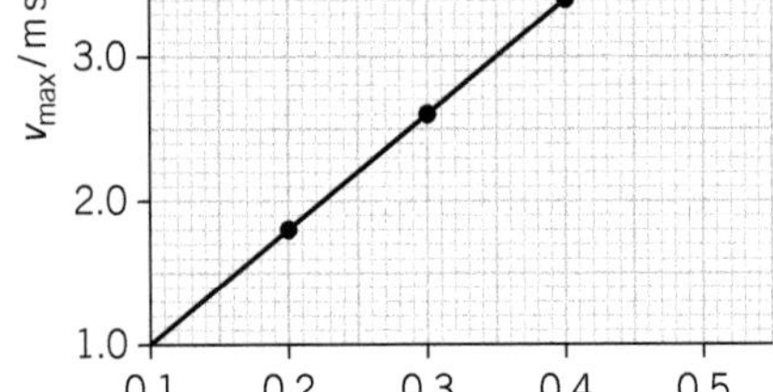

▲ **Figure 2**

a Write an equation for v_{max} in terms of A. *(1 mark)*

b Explain why the graph shown in Figure 2 is a straight line. *(1 mark)*

c i Determine the gradient of the line shown in Figure 2. *(1 mark)*

ii Calculate the frequency f of the oscillations. *(3 marks)*

Chapter 18 Gravitational fields

In this chapter you will learn about ...

- ☐ Gravitational field lines
- ☐ Gravitational field strength
- ☐ Newton's law of gravitation
- ☐ Radial gravitational field
- ☐ Kepler's laws
- ☐ Motion of planets and satellites
- ☐ Geostationary orbits
- ☐ Gravitational potential
- ☐ Gravitational potential energy
- ☐ Escape velocity

GRAVITATIONAL FIELDS
18.1 Gravitational fields
18.2 Newton's law of gravitation
18.3 Gravitational field strength for a point mass

Specification reference: 5.4.1, 5.4.2

18.1 Gravitational fields

All objects have mass and they all create a **gravitational field** in the space around them. A mass placed in the gravitational field of another object will experience *attractive* gravitational force. For example, the Earth creates a gravitational field and a person in this field will experience a gravitational force – we call this the weight of the person.

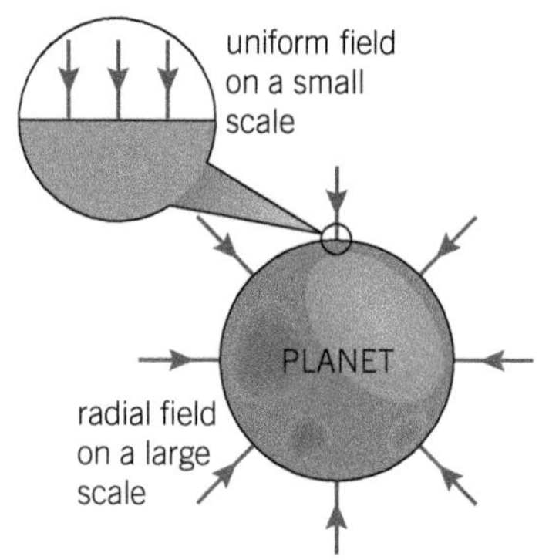

▲ **Figure 1** *Radial and uniform gravitational fields*

Field patterns

Gravitational field patterns can be mapped using **gravitational field lines** (or lines of force). The direction of the gravitational field is indicated by an arrow on the field lines. The **gravitational field strength** (see later) is indicated by the separation between the field lines (see Figure 1).

- A **uniform field** has equally spaced field lines.
- A **radial field** has straight field lines converging to a point at the centre of the object. The field strength gets smaller with increased distance from the object.

Gravitational field strength g

The gravitational field strength is the gravitational force experienced per unit mass on a small object at that point.

This can be written as

$$g = \frac{F}{m}$$

where F is the gravitational force on the object of mass m. The unit for g is $\mathrm{N\,kg^{-1}}$. According to Newton's second law, $\frac{F}{m}$ is the acceleration of free fall g therefore $a = g$. On the surface of the Earth $a = g = 9.81\,\mathrm{N\,kg^{-1}}$.

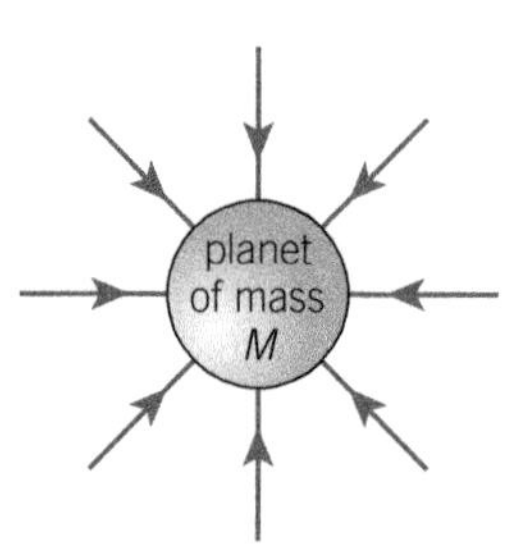

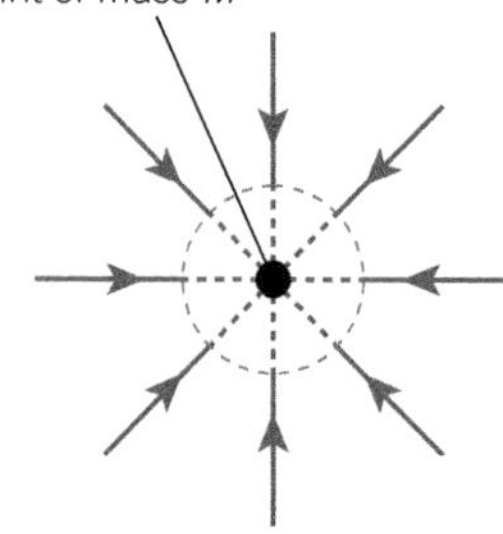

▲ **Figure 2** *A spherical mass M can be modelled as a point mass M*

18.2 Newton's law of gravitation

Newton's law of gravitation is a universal law that can be applied to all objects.

Newton's law of gravitation: Two point masses attract each other with a force that is directly proportional to the product of their masses and inversely proportional to the square of the separation.

Equation for Newton's law

According to Newton's law of gravitation

$$F \propto -\frac{Mm}{r^2}$$

where F is the gravitational force, M and m are the masses, and r is the separation. The minus sign means an attractive force F. The law can be written as an equation using the **gravitational constant** G ($6.67 \times 10^{-11}\,\mathrm{N\,kg^{-1}}$) as follows:

$$F = -\frac{GMm}{r^2}$$

The gravitational field of a spherical object (e.g., planet) can be modelled as a **point mass** at its centre, see Figure 2. So just remember that r is the centre-to-centre separation, see Figure 3.

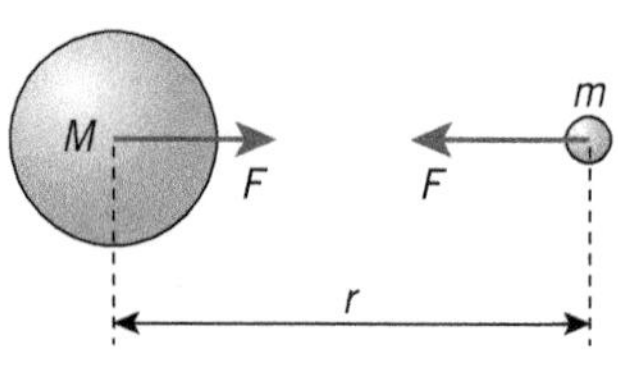

▲ **Figure 3** *The separation r is from centre-to-centre*

18.3 Gravitational field strength for a point mass

The **gravitational field strength** g depends on the distance r from the centre of a spherical mass. For the Earth, g is only equal to $9.81\,\mathrm{N\,kg^{-1}}$ at the surface. Beyond its surface, g obeys an inverse square law with distance.

Radial field

A spherical object of mass m produces a radial field. The gravitational field strength g at a distance r from the centre of the mass can be determined as follows:

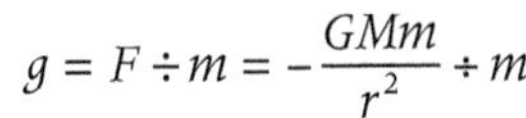

$$g = F \div m = -\frac{GMm}{r^2} \div m$$

Therefore

$$g = -\frac{GM}{r^2}$$

Figure 4 shows the variation of g with r for a point or spherical mass.

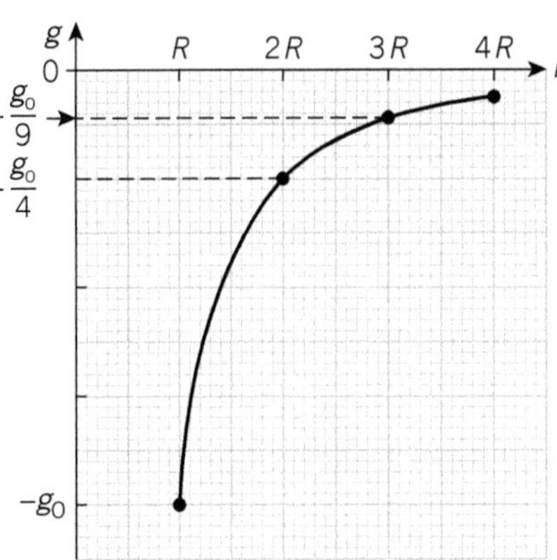

▲ **Figure 4** *Gravitational field strength g obeys an inverse square law*

Worked example: Mass of the Moon

The Moon has diameter 3480 km and surface gravitational field strength of $1.62\,\mathrm{N\,kg^{-1}}$.

Calculate the mass of the Moon.

Step 1: Write down the quantities needed to calculate the mass.

$r = 1740 \times 10^3\,\mathrm{m}$ $\quad g = 1.62\,\mathrm{N\,kg^{-1}}$ $\quad G = 6.67 \times 10^{-11}\,\mathrm{N\,m^2\,kg^{-2}}$

Step 2: Rearrange the equation for g and then substitute the values.

$$g = \frac{GM}{r^2} \text{ (magnitude only)}$$

$$M = \frac{gr^2}{G} = \frac{1.62 \times (1740 \times 10^3)^2}{6.67 \times 10^{-11}} = 7.35 \times 10^{22}\,\mathrm{kg} \text{ (3 s.f.)}$$

Revision tip: Inverse square law

$g \propto \frac{1}{r^2}$ which is an inverse square law. Doubling the distance r from the centre of the object will decrease g by a factor of $2^2 = 4$.

Summary questions

1 The gravitational force on an object of mass 50 kg is 190 N. Calculate the gravitational field strength. *(1 mark)*

2 The gravitational field strength at a point is $3.0\,\mathrm{N\,kg^{-1}}$. Calculate the gravitational force experienced by a space probe of mass 600 kg at this point. *(1 mark)*

3 Calculate the gravitational force between Mercury and the Sun. *(2 marks)*
mass of Sun = 2.0×10^{30} kg
mass of Mercury = 3.3×10^{23} kg
separation = 5.8×10^{10} m

4 The gravitational force experienced by a rocket is F on the surface of the Earth. Calculate the force (in terms of F) on the rocket at a *height* of 3 Earth radii. *(3 marks)*

5 The mass of Uranus is 8.7×10^{25} kg and its surface gravitational field strength is $10\,\mathrm{N\,kg^{-1}}$. Calculate its radius. *(3 marks)*

6 Compared with the Earth, Jupiter has 320 times greater mass and is 11 times larger. Estimate the gravitational field strength on the surface of Jupiter. *(3 marks)*

18.4 Kepler's laws
18.5 Satellites

Specification reference: 5.4.3

18.4 Kepler's laws

The German mathematician and astronomer Johannes Kepler (1571–1630) published his three laws of planetary motion in the 17th century based on the naked-eye planetary observations of the Danish astronomer Tycho Brahe.

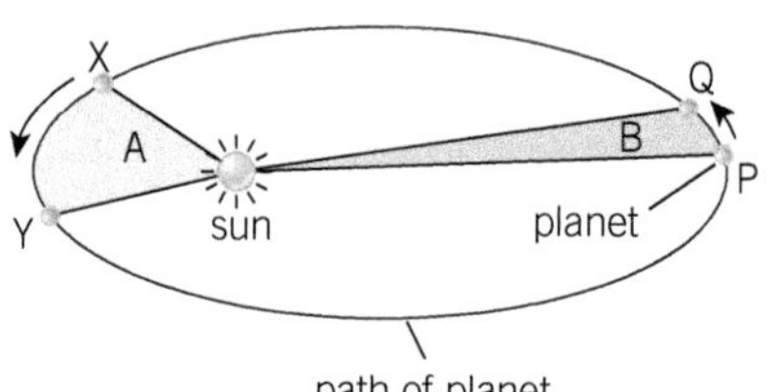

▲ **Figure 1** *A planet travels from X to Y and from P to Q in the same time. According to the second law, area A = area B*

The three laws

First law: All planets move in elliptical orbits, with the Sun at one focus.

Second law: A line that connects a planet to the Sun sweeps out equal areas in equal times. (See Figure 1.)

Third law: The square of the orbital period of any planet is directly proportional to the cube of the mean distance from the Sun.

The third law may be written as

$$T^2 \propto r^3 \text{ or } \frac{T^2}{r^3} = K$$

where K is a constant for the planets orbiting the Sun.

Although the third law originated for the planets in our Solar System, it can be applied to any system where objects orbit round a central mass. For example, you can apply this law to the satellites orbiting the Earth, to the moons of Jupiter, to planets orbiting other stars in our galaxy, and so on.

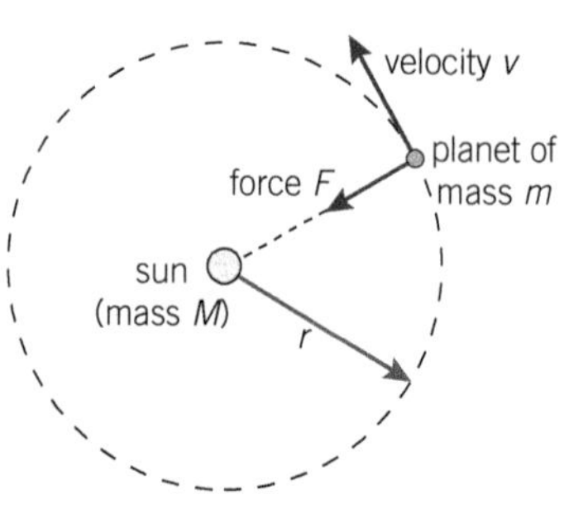

▲ **Figure 2** *A planet orbiting the Sun*

Modelling planetary motion

Figure 2 shows a planet of mass m moving at a speed v in a circular orbit of radius r round the Sun. The orbital period of the planet is T. The mass of the Sun is M. You can use Newton's law of gravitation and ideas from circular motion to show the validity of Kepler's third law.

The centripetal force on the planet is provided by the gravitational force.

Therefore

$$F = \frac{mv^2}{r}$$

$$\frac{GMm}{r^2} = \frac{mv^2}{r}$$

$$v^2 = \frac{GM}{r}$$

$$\left(\frac{2\pi r}{T}\right)^2 = \frac{GM}{r}$$

$$T^2 = \left(\frac{4\pi^2}{GM}\right)r^3$$

For the planets in the Solar System, GM is constant therefore

$$T^2 \propto r^3$$

which is Kepler's third law.

> **Revision tip: Graphs**
>
> 1 A graph of T^2 against r^3 will be a straight line of gradient $\frac{4\pi^2}{GM}$ and passing through the origin.
>
> 2 A graph of $\lg T$ against $\lg r$ should be linear, with a gradient of 1.5.

> **Synoptic link**
>
> Centripetal forces were covered in Topic 16.3, Exploring centripetal forces.

18.5 Satellites

Artificial **satellites** are used for space exploration, exploration of the Earth, communications, and so on. The physics of any satellite can be analysed using Newton's law of gravitation and equations for circular motion.

Geostationary

A satellite in a geostationary orbit:

- has an orbital period of 1 day round the Earth
- travels in the same direction as the rotation of the Earth
- has its orbit in the equatorial plane of the Earth.

The distance r of a satellite in a geostationary orbit can be calculated using the equation for Kepler's third law.

$$T^2 = \left(\frac{4\pi^2}{GM}\right)r^3$$

The radius of a geostationary orbit is about 6.6 Earth radii from the centre of the Earth.

Worked example: Moons of Jupiter

Jupiter's moon Io is 420 Mm from the centre of Jupiter and has a period of 1.8 days. Observations of Europa, another moon of Jupiter, shows it to have a period of 3.6 days. Calculate the radius of orbit of Europa.

Step 1: Write down the known and unknown quantities for Kepler's third law.

Io: $T = 1.8$ days and $r = 420$ Mm   Europa: $T = 3.6$ days and $r = ?$

Step 2: Use Kepler's third law to calculate r.

$$\frac{T^2}{r^3} = \text{constant}$$

$$\frac{1.8^2}{420^3} = \frac{3.6^2}{r^3}$$

You can work in any consistent units

$$r = \sqrt[3]{\frac{420^3 \times 3.6^2}{1.8^2}} = 670 \text{ Mm (2 s.f.)}$$

Summary questions

Data for the Earth: mass = 6.0×10^{24} kg   radius = 6400 km

1 State Kepler's third law. *(1 mark)*

2 State the orbital period of a satellite in geostationary orbit. *(1 mark)*

3 Calculate the speed of a satellite orbiting at a distance of 3000 km *above* the surface of the Earth. *(4 marks)*

4 Calculate the orbital period of the satellite in **Q3**. *(3 marks)*

5 For planets in the Solar System, explain why a graph of lg$T$ against lg$r$ has gradient 1.5. *(3 marks)*

6 Show that the radius of a geostationary orbit for the Earth is about 6.6 Earth radii. *(3 marks)*

18.6 Gravitational potential
18.7 Gravitational potential energy

Specification reference: 5.4.4

18.6 Gravitational potential

You need the idea of gravitational potential in order to understand the gain and loss in the gravitational potential energy of objects such as rockets and planets.

Gravitational potential V_g

The **gravitational potential** V_g at a point in a gravitational field is defined as the work done per unit mass to move an object to that point from infinity.

For a spherical object, such as a planet or star, the gravitational potential V_g is directly proportional to the mass m of the object and inversely proportional to the distance r from the centre of the object. The equation for V_g is

$$V_g = -\frac{GM}{r}$$

where G is the gravitational constant $6.67 \times 10^{-11}\,\mathrm{N\,m^2\,kg^{-2}}$.

- V_g has unit $\mathrm{J\,kg^{-1}}$.
- V_g is defined to be zero at infinity.
- The negative sign signifies that the gravitational force is attractive.

At the surface of the Earth, V_g is about $-63\,\mathrm{MJ\,kg^{-1}}$. It will take a minimum energy of 63 MJ to remove 1 kg mass all the way to infinity. The 1 kg mass will completely escape the gravitational influence of the Earth if given an energy slightly greater than 63 MJ.

Figure 1 shows the variation of V_g with distance r.

Revision tip: Graphs
Gravitational potential V_g is always negative.

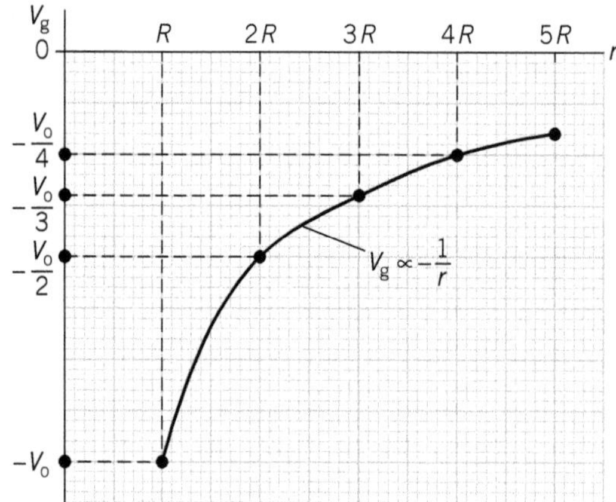

▲ **Figure 1** *V_g obeys an inverse law with distance r*

Equipotentials: Extension work

At a specific distance r from the centre of a spherical mass, V_g has a specific negative value. An equipotential is a line of equal gravitational potential. The equipotentials for a radial field are concentric circles. Figure 2 shows the equipotentials of an imaginary planet.

field lines
MJ kg^{-1}
−40
−60
−80
−100
planet
equipotentials
(with potential values shown)

▲ **Figure 2** *Equipotentials for a planet. The field lines are at right angles to the equipotentials*

18.7 Gravitational potential energy

The **gravitational potential energy** E of any object with mass m within a gravitational field is defined as the work done to move the mass from infinity to a point in a gravitational field.

From the definition of gravitational potential energy, we have

$$E = mV_g$$

$$E = -\frac{GMm}{r}$$

A comet, in an elliptical orbit, has both kinetic energy and gravitational potential energy. Its total energy in its orbit remains constant. Its kinetic energy increases as it gets closer to the Sun and its potential energy decreases.

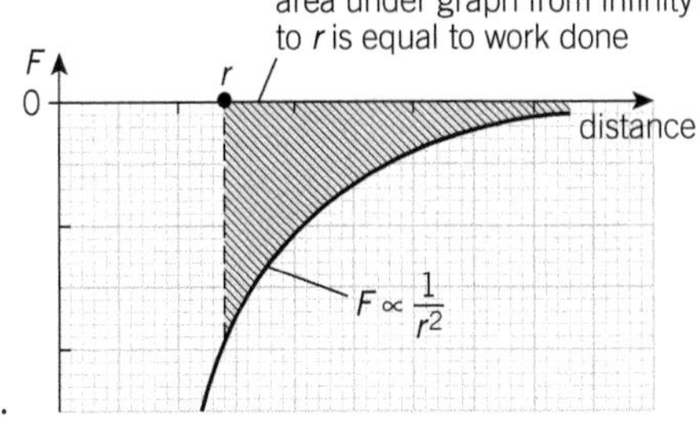

▲ **Figure 3** *The area under the force–distance graph is work done*

Understanding gravitational potential energy

Figure 3 shows the force F against distance graph for a spherical mass (planet). The area under the force–distance graph is equal to work done. The area under the graph from infinity to a distance r is equal to the gravitational potential energy E.

Quite often, we are interested in changes in potential and potential energy.

- change in gravitational potential ΔV_g = final V_g – initial V_g
- change in gravitational potential energy ΔE = final E – initial E

Escape velocity

Escape velocity of a projectile is the minimum velocity it must have to completely escape from the gravitational influence of a planet.

You can determine the escape velocity v from a planet of mass M and radius R as follows:

initial total energy = KE + PE = $\frac{1}{2}mv^2 - \frac{GMm}{R}$

final total energy of projectile = 0

Therefore

$$\frac{1}{2}mv^2 = \frac{GMm}{R}$$

$$v = \sqrt{\frac{2GM}{R}}$$

Worked example: Escaping from the Moon

The mass of the Moon is 7.4×10^{22} kg and it has radius 1700 km.

Calculate the escape velocity from the Moon.

Step 1: Write down the information given.

$M = 7.4 \times 10^{22}$ kg $R = 1700 \times 10^3$ m $G = 6.67 \times 10^{-11}$ N m^2 kg^{-2}

Step 2: Calculate the escape velocity v.

$$v = \sqrt{\frac{2GM}{R}} = \sqrt{\frac{2 \times 6.67 \times 10^{11} \times 7.4 \times 10^{22}}{1700 \times 10^3}}$$

$v = 2.4$ km s^{-1} (2 s.f.)

Summary questions

Data for the Earth: mass = 6.0×10^{24} kg, radius = 6400 km

1. State the maximum value for gravitational potential. *(1 mark)*
2. The gravitational potential on the surface of Venus is -54 MJ kg^{-1}. What is the energy required to remove a 1 kg mass from the surface of Venus to infinity? *(1 mark)*
3. The radius of Venus is 6.1 Mm. Use the information in **Q2** to determine its mass. *(2 marks)*
4. Calculate the gravitational potential at the surface of the Earth. *(2 marks)*
5. A rocket of mass 1200 kg is moved from the surface of the Earth to an orbit at a distance of 9400 km from the centre of Earth. Calculate the change in gravitational potential. *(3 marks)*
6. Calculate the change in gravitational potential energy of the rocket in **Q5**. *(2 marks)*

Chapter 18 Practice questions

1 A satellite is in a geostationary orbit.

What is the period of rotation of this satellite around the Earth?

A 3.60×10^3 s

B 4.32×10^4 s

C 8.64×10^4 s

D 3.16×10^7 s *(1 mark)*

2 A planet has a radius R. The gravitational field strength on the surface of the planet is g_0.

What is the distance from the *surface* of the planet where the gravitational field strength is $\frac{g_0}{9}$?

A $2R$

B $3R$

C $8R$

D $9R$ *(1 mark)*

3 Astronomers have recently discovered a distant star to have orbiting planets. The period of a planet is T and its distance from the star is r.

Which graph will *not* produce a straight line graph?

A T^2 against r^3

B T against $r^{1.5}$

C $\lg(T)$ against r

D $\lg(T)$ against $\lg(r)$ *(1 mark)*

4 **a** A small object is falling towards the centre of a planet. Explain why its acceleration of free fall at a point is the same as the gravitational field strength at that point. *(2 marks)*

b The Earth has a mean radius of 6400 km. The table below shows the magnitude of the gravitational field strength g and distance r from the centre of the Earth.

r / km	6400	8000	9600
g / N kg^{-1}	9.8	6.3	4.4

i Use the table to confirm the relationship between g and r. *(2 marks)*

ii Calculate the mass of the Earth. *(3 marks)*

iii Calculate the mean density of the Earth. *(2 marks)*

5 A planet of mass m is orbiting in a circular path at a distance r from the centre of the star. The mass of the star is M.

a Derive an equation for the speed v of the planet in its orbit. *(3 marks)*

b Neptune is about 30 times further from the Sun than the Earth.

Calculate the ratio $\frac{\text{speed of Earth}}{\text{speed of Neptune}}$. *(2 marks)*

6 Figure 1 shows the Moon. The value of the gravitational potential at its surface is -2.8 MJ kg^{-1}. The radius of the Moon is R.

a How much energy would it take to completely remove a 1.0 kg mass from the surface of the Moon? *(1 mark)*

b Determine the gravitational potentials at points A and B. *(3 marks)*

c Calculate the energy required to move a rocket of mass 3500 kg from A to B. *(2 marks)*

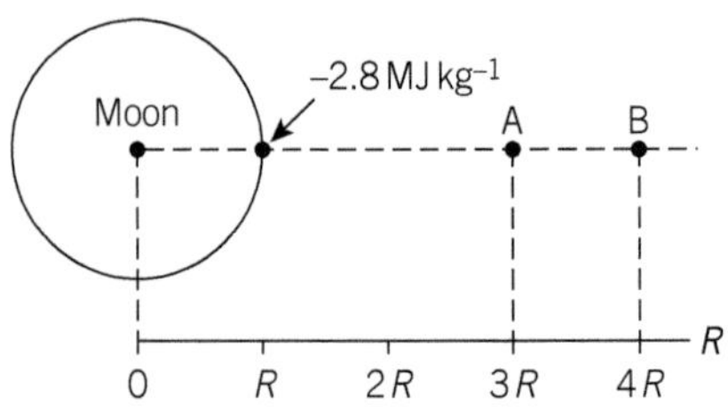

▲ **Figure 1**

Chapter 19 Stars

In this chapter you will learn about ...

- ☐ Stellar formation
- ☐ Life cycles of stars
- ☐ Hertzsprung–Russell (HR) diagrams
- ☐ Energy levels
- ☐ Line spectra
- ☐ Emission, absorption, and continuous spectra
- ☐ Diffraction grating
- ☐ Stellar luminosity
- ☐ Wien's displacement law
- ☐ Stefan's law

19 STARS

19.1 Objects in the Universe
19.2 The life cycle of stars
19.3 Hertzsprung–Russell diagram

Specification reference: 5.5.1

19.1 Objects in the Universe

The **Universe** is immense in size. It has about 10^{11} galaxies – each galaxy has about 10^{11} stars. Our local star is the Sun – it has orbiting planets, asteroids, comets, and dust.

Stellar formation

Stars are formed from large clouds of dust and gas (mainly hydrogen) called **nebulae**.

- The tiny gravitational force between dust and gas slowly brings some parts of the cloud together (gravitational collapse).
- The gravitational potential energy of the cloud decreases as its kinetic energy and temperature increases.
- *Fusion* of hydrogen nuclei into helium nuclei occurs when the temperature of the cloud is about 10^7 K. The energy released from the fusion reactions further increases the temperature. A hot ball of gas (star) is formed.
- The gravitational forces compress the star. The **radiation pressure** from the photons emitted during fusion and the **gas pressure** of the star push outwards. A stable star is formed when the force from the radiation and gas pressure is balanced by the gravitational force.

Synoptic link

There is more detail on nuclear fusion in Topic 26.4, Nuclear fusion.

19.2 The life cycle of stars

The rate of fusion subsides when most of the hydrogen is depleted in a star. The core of a star is made of rings of elements with iron at the centre. The outermost layers of the star has helium and a very small amount of hydrogen. The fusion of helium makes outermost layers of the star expand into a **red giant**.

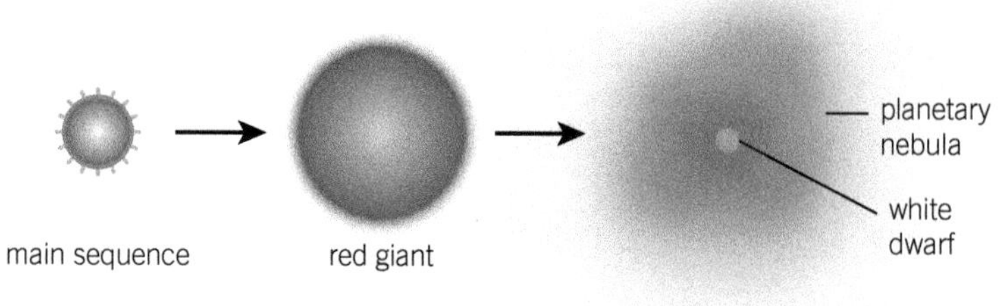

▲ **Figure 1** *Evolution of a low-mass star, for example the Sun*

Low-mass stars

Figure 1 shows the evolution of a star with mass between 0.5 $M_\odot$ and 10 $M_\odot$.

$M_\odot$ is the solar mass 2.0×10^{30} kg.

- Eventually most of the layers of the red giant around the core drift away into space as a **planetary nebula.**
- The hot core (30 000 K) left behind is called a **white dwarf.**
- The white dwarf is very dense. There is no fusion but it slowly leaks away photons created in its earlier evolution. It is prevented from gravitational collapse by **electron degeneracy pressure.** The maximum mass of a white dwarf is $1.44M_\odot$. This is called the **Chandrasekhar limit**.

Massive stars

Figure 2 shows the evolution of a star of mass greater than about $10M_\odot$.

- When no further fusion reactions occur, the layers in the core suddenly implode under gravitational forces and bounce off the solid core, leading to a shockwave that ejects core material into space. This event is called a (type II) **supernova**.
- The remnant core is a **neutron star** when its mass is greater than the Chandrasekhar limit. The core has tightly packed neutrons and extremely dense (~ 10^{17} kg m^{-3}).
- The remnant core is a **black hole** when its mass is greater than about $3M_\odot$. Black holes could be a singularity and have infinite density. Light cannot escape from a black hole.

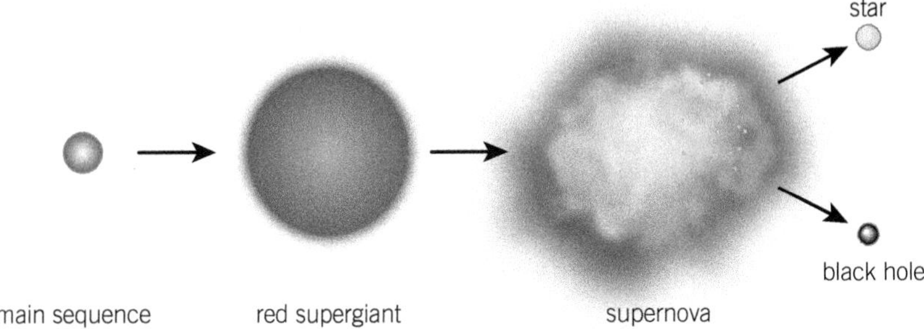

▲ **Figure 2** *Evolution of a massive star*

19.3 Hertzsprung–Russell diagram

The **Hertzsprung–Russell (HR) diagram** is a graph of stars in a galaxy, or a star cluster. **Luminosity** is plotted on the y-axis and the surface temperature of the stars on the x-axis. The temperature axis has temperature increasing from right to left (which is very odd).

The luminosity L of a star is the total radiant power of the star.

HR diagram

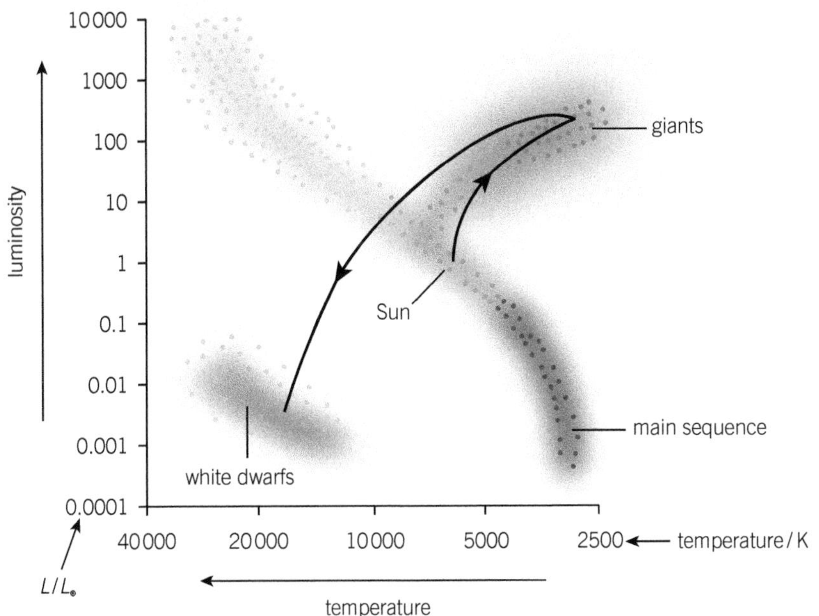

▲ **Figure 3** *HR diagram showing the likely evolution of the Sun. $L_\odot$ is the luminosity of the Sun*

Figure 3 shows the HR diagram for the stars in a galaxy. The bulk of the stars lie on the **main sequence** band. The white dwarfs and red giants form their own distinct regions.

Summary questions

1. State the Chandrasekhar limit. *(1 mark)*
2. Calculate the mass of a 10 $M_\odot$ star. *(1 mark)*
3. State the quantity that governs the fate of the remnant core of a massive star. *(1 mark)*
4. Compare the white dwarfs and the red giants on the HR diagram. *(2 marks)*
5. A particular neutron star has mass 4.0×10^{30} kg and a radius of only 11 km. Calculate the mass of a 1 mm^3 'grain' of neutron star matter. *(3 marks)*
6. Use you knowledge of escape velocity to estimate the maximum radius of a black hole of mass 5 $M_\odot$ *(3 marks)*

19.4 Energy levels in atoms
19.5 Spectra

Specification reference: 5.5.2

19.4 Energy levels in atoms

The electrons moving inside the atoms behave as de Broglie waves and produce stationary waves because they are trapped within the atoms. As a result, the electrons are only allowed a discrete set of energy values – their energy is quantised.

An **energy level** is one of the discrete set of energies a bound electron can have.

Quantised energy levels

Figure 1 shows a typical energy level diagram for the electrons in gas atoms. The vertical scale has the energy of the electron. The horizontal lines indicate the permitted energy of the electrons.

- An electron cannot exist between energy levels.
- An electron can move from one energy level to another.
- All energy levels have negative values. This simply means that external energy is required to pull away the electrons from the attractive electrostatic forces of the positive nuclei.
- The **ground state** is the most negative energy level.

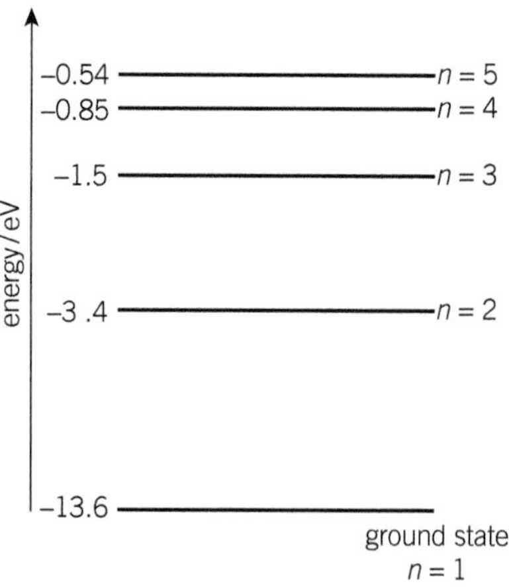

▲ **Figure 1** *Energy levels for the hydrogen atom. The levels are labelled with the principal quantum number n*

Emission lines

What happens when an electron makes a transition from a higher energy level to a lower energy level? Energy must be conserved. The transition of the electron produces a photon of a unique wavelength, see Figure 2. The energy of the photon is equal to the difference between the energy levels ΔE. Therefore

$$hf = \Delta E \qquad \text{or} \qquad \frac{hc}{\lambda} = \Delta E$$

where h is the Planck constant, f is the frequency of the electromagnetic radiation, and λ is the wavelength. An emission spectrum has sharp and bright lines, each corresponding to the unique wavelength originating from electron transitions between a pair of energy levels.

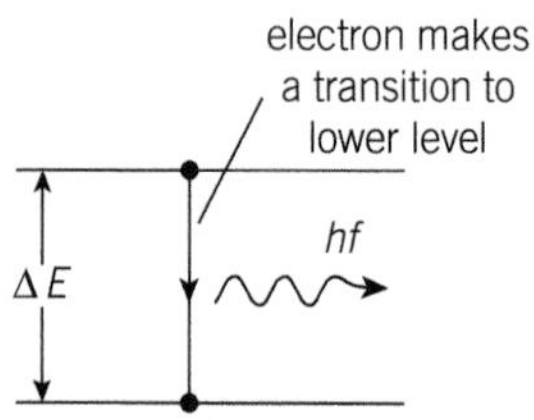

▲ **Figure 2** *Understanding emission spectrum*

19.5 Spectra

There are three types of spectra: **continuous**, **emission**, and **absorption**. The origin of **emission line spectrum** has already been outlined above.

Synoptic link

You studied photons in Topic 13.1, The photon model, and Topic 13.2, The photoelectric effect.

Continuous and absorption spectra

The atoms of a heated solid metal, such as a lamp filament, produce a continuous spectrum. This type of spectrum will have all the wavelengths of visible light from blue to red – effectively all the rainbow colours.

Absorption line spectrum is characterised by a series of dark lines against the background of a continuous spectrum. An absorption spectrum is produced when the electrons within cooler gas atoms absorb photons. Figure 3 shows a photon being absorbed by an electron. The electron will make a transition (a jump) to a higher energy level only when the photon has the right amount of energy ΔE. The photon disappears. After some time, the electron will make a transition to the lower energy level and re-emit the photon. The photons are emitted in random directions, so the intensity in the original direction is greatly reduced.

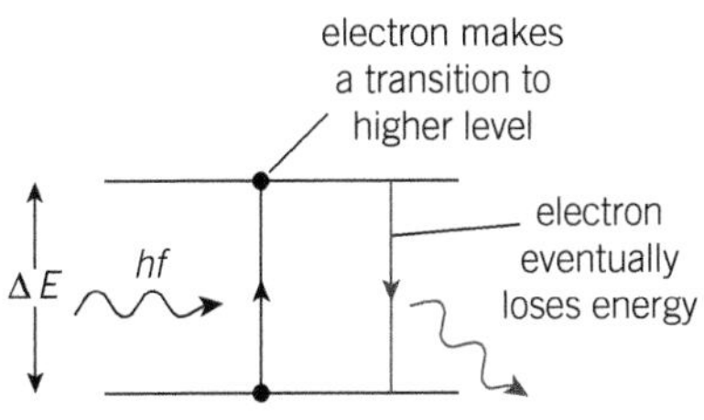

▲ **Figure 3** *Understanding absorption spectrum*

Worked example: Emission line

An electron makes a transition from −5.0 eV level to −6.1 eV level.

Calculate the wavelength of the photon.

Step 1: Calculate the energy loss of the electron in J.

energy loss = 6.1 − 5.0 = 1.1 eV

energy loss = $1.1 \times 1.6 \times 10^{-19} = 1.76 \times 10^{-19}$ J

Step 2: Calculate the wavelength.

$$\frac{hc}{\lambda} = \Delta E$$

$$\lambda = \frac{6.63\times10^{-34}\times3.00\times10^{8}}{1.76\times10^{-19}} = 1.13\times10^{-6}\text{ m}$$

wavelength = 1.1×10^{-6} m (2 s.f.)

Revision tip

1 eV = 1.60×10^{-19} J

Summary questions

1 Explain what is meant by an energy level. *(1 mark)*

2 An electron makes a jump from the −2.0 eV energy level to the −6.0 eV energy level.
Calculate the energy in eV of the emitted photon. *(1 mark)*

3 An electron makes a transition from the −10 eV level to the −2.0 eV level. Explain whether this involves the emission or the absorption of a photon. *(1 mark)*

4 Calculate the wavelength of the photon emitted in **Q2**. *(3 marks)*

5 Determine the maximum number of possible emission spectral lines from 4 energy levels. *(2 marks)*

6 An electron is in the −12.2 eV level. It absorbs a photon of wavelength 400 nm.
Calculate the value of the new energy level in eV. *(4 marks)*

19.6 Analysing starlight
19.7 Stellar luminosity

Specification reference: 5.5.2

19.6 Analysing starlight

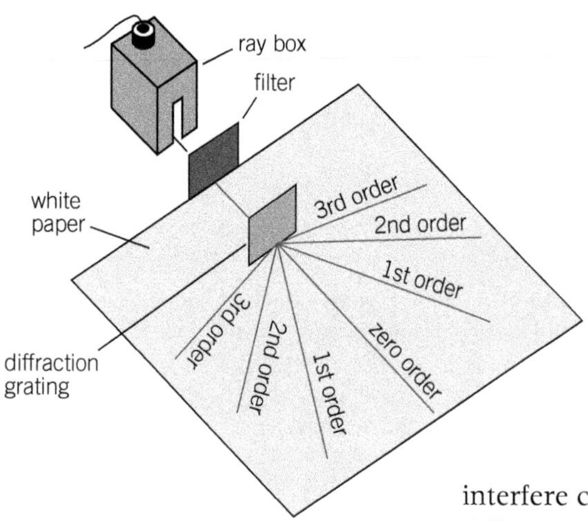

▲ **Figure 1** *A diffraction grating produces bright light in certain directions*

A **diffraction grating** consists of a large number of regularly spaced lines on a glass or plastic slide. Each line behaves like a narrow slit. A parallel beam of monochromatic light directed normally at the grating will be diffracted at each slit (line) and the diffracted light will then interfere beyond the slits. Bright light is only transmitted by the grating in certain directions.

Figure 1 shows the maxima (bright light) and the numbering (orders) of these maxima.

Transmission diffraction grating

Figure 2 shows two rays of diffracted light from adjacent slits of the grating. At the angle θ, the light from the slits P and Q interfere constructively. The phase difference must be zero and the path difference must be a whole number of wavelengths. Therefore

$$QY = n\lambda = d\sin\theta$$

This gives us the grating equation

$$d\sin\theta = n\lambda$$

where d is the separation between adjacent slits (known as **grating spacing**), λ is the wavelength of the monochromatic light, and n is the order.

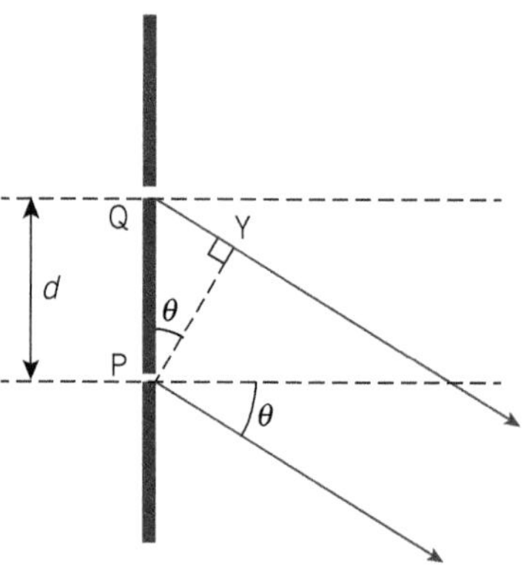

▲ **Figure 2** *Formation of the nth order maxima*

Worked example: Stellar spectral line

The light from a star is analysed using a diffraction grating with 500 lines per mm.

A bright spectral line is observed at an angle of 28° in the second-order spectrum.

Calculate the wavelength of this spectral line in nm.

Step 1: Calculate the grating spacing *d*.

$$d = \frac{1.0 \times 10^{-3}}{500} = 2.0 \times 10^{-6}\,\text{m}$$

Step 2: Write down the quantities and then use the grating equation to calculate λ.

$n = 2$ $\quad d = 2.0 \times 10^{-6}\,\text{m}$ $\quad \theta = 24°$ $\quad \lambda = ?$

$d\sin\theta = n\lambda$

$$\lambda = \frac{2.0 \times 10^{-6} \times \sin 28°}{2} = 4.7 \times 10^{-7}\,\text{m}$$

wavelength = 470 nm (2 s.f.) $\qquad$ ($1\,\text{nm} = 10^{-9}\,\text{m}$)

Synoptic link

You studied path and phase differences in Topic 12.1, Superposition of waves, 12.2, Interference, and 12.3, The Young double-slit experiment.

19.7 Stellar luminosity

Stars and many other hot objects can be modelled as a **black body**. Figure 3 shows a typical graph of intensity against wavelength of electromagnetic waves from a black body. The shape of the graph is unique to the surface temperature of the body. The wavelength λ_{max} at the peak of the graph is related to the thermodynamic temperature T of the surface of the body.

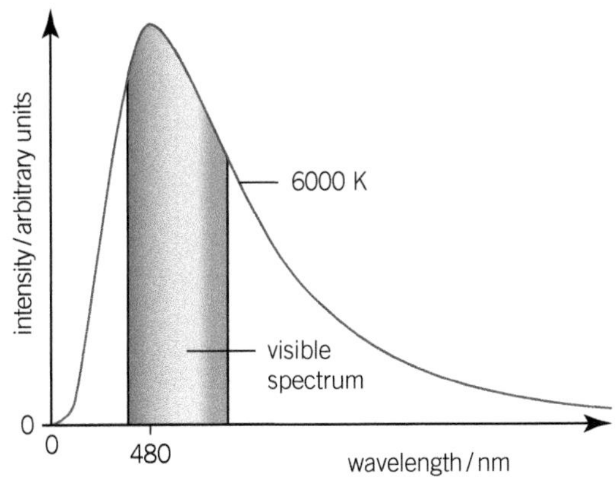

▲ **Figure 3** *Intensity–wavelength graph for a black body at 6000 K*

Wien's displacement law

Wien's displacement law: the black body radiation curve peaks at a wavelength that is inversely proportional to the thermodynamic temperature of the body.

Therefore

$$\lambda_{max} \propto \frac{1}{T} \qquad \text{or} \qquad \lambda_{max} T = \text{constant}$$

The value of the constant is 2.9×10^{-3} m K.

Stefan's law

The Stefan–Boltzmann law (also known as Stefan's law): the total power radiated per unit surface area of a black body is directly proportional to the fourth power of its thermodynamic temperature.

The luminosity L of a star is the total power it emits. For a star of radius r, the total surface area is $4\pi r^2$, and the luminosity is given by the relationship

$$L \propto 4\pi r^2 T^4$$

With the **Stefan constant** σ ($5.67 \times 10^{-8}\,\text{W m}^{-2}\text{K}^{-4}$), the relationship above can be written as an equation.

$$L = 4\pi r^2 \sigma T^4 \qquad \text{(Note: } L \propto r^2 \text{ and } L \propto T^4\text{)}$$

How big is a star?

The radiation from our Sun peaks at a wavelength of 5.0×10^{-7} m and the surface temperature of the Sun is about 5800 K. The temperature of a star can be determined using Wien's law. If its luminosity is known, then the equation $L = 4\pi r^2 \sigma T^4$ can be used to determine its radius r.

Summary questions

1 Calculate the grating spacing in metre (m) for a grating with 80 lines per mm. *(1 mark)*

2 State how the luminosity of a star depends on its surface temperature. *(1 mark)*

3 Monochromatic light of wavelength 6.4×10^{-7} m is incident normally at a grating with 800 lines per mm.
Calculate the angle θ for the first-order maxima. *(3 marks)*

4 The surface temperature of the Sun is 5800 K and its peak wavelength is 500 nm.
Calculate the peak wavelength for Rigel which has a temperature of 12 000 K. *(2 marks)*

5 Explain why red giants are very luminous compared with similar temperature main sequence stars. *(2 marks)*

6 Rigel has a radius 79 times that of the Sun.
Use information in **Q4** to calculate the luminosity of Rigel in term of the luminosity $L_{\odot}$ of the Sun. *(4 marks)*

Chapter 19 Practice questions

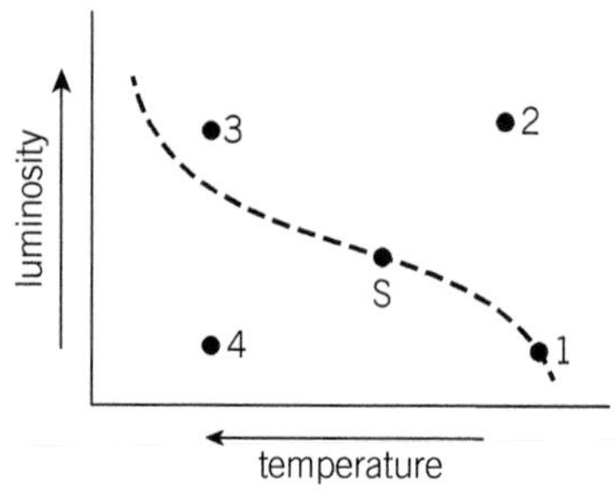

▲ Figure 1

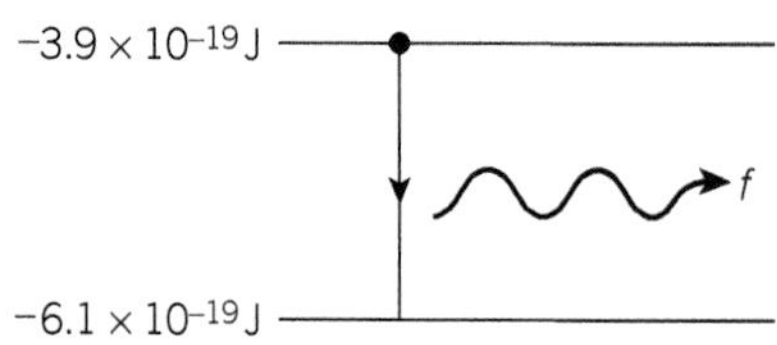

▲ Figure 2

1 Figure 1 shows an incomplete Hertzsprung–Russell (HR) diagram. The position of the Sun is marked by the letter S. The numbers 1, 2, 3, and 4 are positions on this HR diagram.

What is the likely evolution of the Sun?

A S → 2 → 1 **C** S → 4 → 1

B S → 3 → 4 **D** S → 2 → 4 *(1 mark)*

2 Figure 2 shows an electron making a transition between two energy levels.

What is the frequency f of the emitted photon?

A 3.3×10^{14} Hz **C** 9.2×10^{14} Hz

B 5.9×10^{14} Hz **D** 1.5×10^{15} Hz *(1 mark)*

3 Star X has radius R, surface temperature T, and luminosity L. Star Y has radius $2R$ and surface temperature $2T$.

What is the luminosity of star Y in terms of L?

A $4L$ **C** $32L$

B $8L$ **D** $64L$ *(1 mark)*

4 **a** Describe some of the characteristics of a white dwarf. *(2 marks)*

b **i** Sketch a graph to show the variation of intensity of a star with wavelength. *(1 mark)*

ii State the relationship between the wavelength at maximum intensity and the surface temperature of the star. *(1 mark)*

c The surface temperature of the Sun is 5800 K and its luminosity is 3.8×10^{26} W.

Sirius-B is a white dwarf with a surface temperature of about 25 000 K and a radius that is about 120 times smaller than the radius of the Sun. Estimate the luminosity of Sirius-B. *(3 marks)*

5 The visible spectrum of a distant star is analysed using a diffraction grating with 800 lines mm^{-1}.

a Explain how elements can be identified in the atmosphere of stars by analysing the spectrum of the light they emit. *(2 marks)*

b The diffraction grating is used to view the hydrogen emission line of wavelength 656 nm.

Calculate the maximum number of orders that can be observed using light of this wavelength. *(3 marks)*

c Calculate the angular separation between two emission lines observed in the first-order spectrum that have wavelengths 589 nm and 587 nm. *(3 marks)*

6 **a** Explain what is meant by the luminosity of a star. *(1 mark)*

b A star has radius 8.5×10^{5} km and a surface temperature of 5800 K. It is 3.9×10^{16} m from the Earth. Calculate:

i the luminosity of the star; *(2 marks)*

ii the intensity of the light from this star at the Earth. *(2 marks)*

c Suggest why the actual intensity of the light from the star in **b** at the Earth will be less than your answer to **b(ii)**. *(1 mark)*

Chapter 20 Cosmology (The Big Bang)

In this chapter you will learn about ...

- ☐ Astronomical unit (AU)
- ☐ Light year (ly)
- ☐ Parsec (pc)
- ☐ Doppler effect
- ☐ Doppler equation
- ☐ Hubble's law
- ☐ Cosmological principle
- ☐ The Big Bang theory
- ☐ The evolution of the Universe

20 COSMOLOGY (THE BIG BANG)
20.1 Astronomical distances
20.2 The Doppler effect

Specification reference: 5.5.3

20.1 Astronomical distances

In order to visualise the scale of space and separation between objects in the Universe, astronomers use astronomical unit, light-year and parsec to measure distances.

Astronomical unit (AU)

The **astronomical unit** is the average distance between the Earth and the Sun.

$$1\,\text{AU} = 1.50 \times 10^{11}\,\text{m}$$

The AU is used to measure the distances within our own Solar System. For example, the distance of Saturn is 9.58 AU from the Sun.

Light-year (ly)

The light-year is the distance travelled by light in a vacuum in a time of one year.

$$1\text{ ly} = \text{speed of light} \times \text{time}$$

$$1\text{ ly} = 2.9979 \times 10^{8} \times (365.25 \times 24 \times 3600)$$

$$1\text{ ly} = 9.46 \times 10^{15}\,\text{m} \approx 9.5 \times 10^{15}\,\text{m}$$

The light-year is used to measure distance between stars and galaxies. For example, the bright star Sirius is 8.60 ly from the Sun.

Parsec (pc)

The distance between stars and galaxies can also be measured in parsec (which comes from the abbreviation **par**allax of one **sec**ond of arc). In order to understand this unit, you need to first understand **stellar parallax** and angles measured in seconds of arc (or just arcseconds).

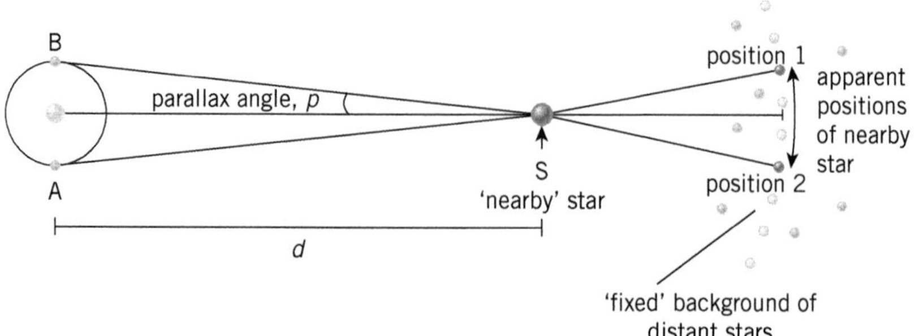

▲ **Figure 1** *Stellar parallax*

Stellar parallax is the apparent shift in the position of a nearby star when viewed against the background of very distant stars.

Figure 1 shows a nearby star observed from position A and then six months later from position B. The **parallax angle** p (also just known as parallax) is the angle subtended by a length of 1 AU at the position of the star. The parallax angle p is always significantly smaller than 1 degree. Very small angles are measured in seconds of arc – there are 3600 seconds of arc in 1°.

The parsec is the distance at which one astronomical unit (AU) subtends an angle of one arcsecond (see Figure 2).

Therefore

$$\tan\left(\frac{1}{3600}^{\circ}\right) = \frac{1\ \text{AU}}{1\ \text{pc}} = \frac{1.50\times10^{11}}{1\ \text{pc}} \quad \text{(See Figure 2)}$$

$$1\ \text{pc} = \frac{1.50\times10^{11}}{\tan\left(\frac{1}{3600}^{\circ}\right)} = 3.09\times10^{16}\text{m} \approx 3.1\times10^{16}\text{m}$$

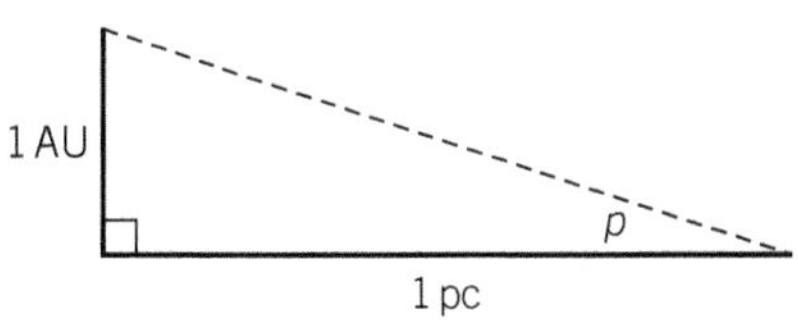

▲ Figure 2

The parallax angle p in arcsconds and the distance d in pc of a star are related by the equation $d = \frac{1}{p}$.

20.2 The Doppler effect

Doppler effect is the change in frequency or wavelength of a wave when there is relative velocity between a source and the observer.

You can observe Doppler effect with all waves, including sound and light. Doppler effect can be used to determine the relative speed of stars in our galaxy and of other galaxies too.

Light from moving sources

- If a source of light is moving towards the Earth, the entire spectrum observed from the source is shifted to shorter wavelengths. This is known as **blue-shift**.
- If a source of light is receding away the Earth, the entire spectrum observed from the source is shifted to longer wavelengths. This is known as **red-shift**.

The light from distant galaxies shows red-shift. All distant galaxies are receding away from each other. The light from Andromeda, our nearest galaxy, shows blue-shift. It is very slowly moving towards our galaxy

Doppler equation

The speed v of a star or a galaxy can be determined using the **Doppler equation** below:

$$\frac{\Delta\lambda}{\lambda} \approx \frac{v}{c}$$

where $\Delta\lambda$ is the change in wavelength, λ is the wavelength measured in the laboratory, and c is the speed of light in a vacuum. This equation can only be used when v is much smaller than c. You can also use the following equation:

$$\frac{\Delta f}{f} \approx \frac{v}{c}$$

where Δf is the change in frequency and f is the frequency.

Worked example: Speed of a distant galaxy

In the laboratory an absorption line of hydrogen is observed at a wavelength of 656.4 nm. The same spectral light from the distant galaxy is observed at 664.7 nm. Calculate the speed of this receding galaxy.

Step 1: Calculate the change in the wavelength.

$$\Delta\lambda = 664.7 - 656.4 = 8.3\ \text{nm}$$

Step 2: Substitute values into the Doppler equation and calculate the speed v.

$$\frac{\Delta\lambda}{\lambda} \approx \frac{v}{c}$$

$$v \approx \frac{c\times\Delta\lambda}{\lambda} \approx \frac{3.0\times10^{8}\times8.3}{656.4} = 3.8\times10^{6}\ \text{ms}^{-1}\ \text{(2 s.f.)}$$

Summary questions

1 Change the following distances into metres (m):
a 5.2 AU b 1.6 ly
c 2600 ly d 95 pc
(4 marks)

2 Convert 0.78 arcseconds into degrees. *(1 mark)*

3 The centre of our galaxy is about 8.0 kpc from the Earth. Calculate this distance in metres (m) and the time it takes light to travel this distance. *(3 marks)*

4 The star Sirius is travelling towards us at a speed of 7.6 km s^{-1}. Calculate the percentage change in the wavelength of a specific spectral line. *(2 marks)*

5 Sirius is 8.6 ly away from us. Calculate its parallax angle in arcseconds. *(3 marks)*

6 The wavelength of a specific spectral line in the laboratory is 119.5 nm. The same spectral line is observed in the spectrum of a star moving away from us at a speed of 5.3×10^{6} m s^{-1}. Calculate this observed wavelength. *(3 marks)*

20.3 Hubble's law
20.4 The Big Bang theory
20.5 Evolution of the Universe

Specification reference: 5.5.3

20.3 Hubble's law

All distant galaxies are receding from each other. This provides the evidence for the **Big Bang** theory of the **expanding Universe**. The galaxies are moving apart because the whole fabric of space has been, and is still, expanding.

Hubble constant

Hubble's law: The recession speed v of a galaxy is directly proportional to its distance d from us.

Therefore

$$v \propto d \qquad \text{or} \qquad v = H_0 d$$

where H_0 is the **Hubble constant**.

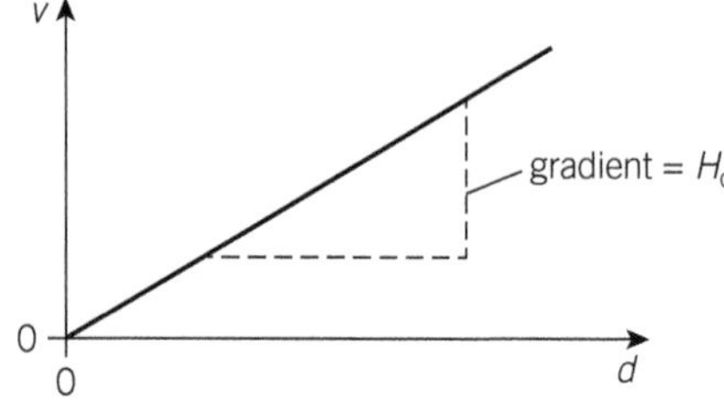

▲ **Figure 1** *A graph of v against d for galaxies*

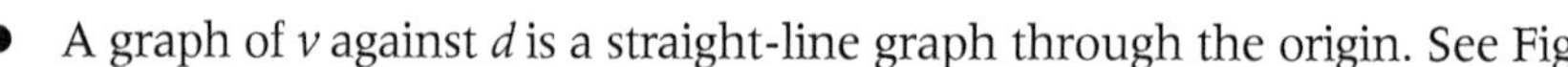

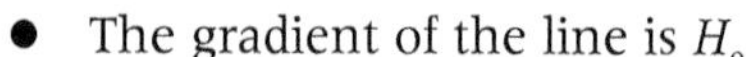

- A graph of v against d is a straight-line graph through the origin. See Figure 1.
- The gradient of the line is H_0.
- The SI unit for the Hubble constant is s^{-1}, but it is often quoted in $km\,s^{-1}\,Mpc^{-1}$ (kilometres per second per unit mega parsec). H_0 is about $2.2 \times 10^{-18}\,s^{-1}$ or $70\,km\,s^{-1}\,Mpc^{-1}$.
- The age of the Universe is given by the equation: age of Universe = H_0^{-1}.

Cosmological principle

The cosmological principle: When viewed on a large enough scale, the Universe is **homogeneous** and **isotropic**, and the laws of physics are universal.

The principle is simply stating that the Universe is the same for all observers in the Universe.

- The laws of physics are the same for all observers in the Universe.
- Homogeneous means that matter is distributed uniformly across the Universe.
- Isotropic means that the Universe looks the same in all directions to every observer.

20.4 The Big Bang theory

The Big Bang theory is the standard model that describes the origin of the Universe and its subsequent large-scale evolution.

- The Universe expanded from a singularity some 13.7 billion years ago.
- The Universe was extremely hot in the early stages.
- The expansion of the Universe over billions of years led to cooling to a temperature of 2.7 K.
- Treating the Universe as a black body at 2.7 K, the peak wavelength is about 1 mm, which is in the microwave region of the electromagnetic spectrum.
- There is alternative explanation for the microwave radiation – the early hot Universe mainly had energetic short-wavelength photons. The expansion of space stretched out the wavelength of these primordial photons, so they are now observed in the microwave region of the spectrum.

20.5 Evolution of the Universe

Figure 2 shows the composition of the Universe – matter is a very small contributor.

matter 4.9%
dark matter 26.8%
dark energy 68.3%

▲ **Figure 2** *Composition of the Universe*

Evolution

time ↓

Space-time expansion began 13.7 billion years ago from a singularity.

↓

Expansion of the hot Universe led to cooling.

↓

Universe has high-energy photons, quarks, and leptons.

↓

Hadrons (protons and neutrons) formed.

↓

Fusion produces primordial helium – about 25% of the matter.

↓

Electrons combine with nuclei to form atoms.

↓

Stars formed and eventually galaxies.

↓

The temperature of the Universe is now 2.7 K and we observe the same intensity of **microwave background radiation** in all directions (isotropic).

The receding galaxies, the existence of primordial helium in the very young galaxies, the temperature of 2.7 K, and the microwave background radiation all provide strong support for the Big Bang model of the Universe.

Worked example: How old?

The galaxy at a distance of 120 Mpc has a recession speed of 8000 km s^{-1}.

Determine an approximate age of the Universe based on this information.

Step 1: Calculate the Hubble constant in s^{-1}.

$$H_0 = \frac{v}{d} = \frac{8000 \times 10^3}{120 \times 10^6 \times 3.1 \times 10^{16}} = 2.15 \times 10^{-18}\,\text{s}^{-1}$$

(**Note:** 1 **km** s^{-1} = **10^3** m s^{-1} and 1 **M**pc = **10^6** × 3.1 × 10^{16} m)

Step 2: Determine the age of the Universe.

$$\text{age} \approx H_0^{-1} = (2.15 \times 10^{-18})^{-1} = 4.65 \times 10^{17}\,\text{s}$$

$$\text{age} \approx 4.7 \times 10^{17}\,\text{s (15 billion years) (2 s.f.)}$$

Summary questions

Use H_0 = 70 km s^{-1} Mpc^{-1} where required.

1 What is the current temperature of the Universe? *(1 mark)*

2 State the cosmological principle. *(1 mark)*

3 Calculate the speed in km s^{-1} of a galaxy at a distance of 200 Mpc. *(2 marks)*

4 Calculate the distance in Mpc of a galaxy receding at a speed of 5200 km s^{-1}. *(2 marks)*

5 Change 70 km s^{-1} Mpc^{-1} to s^{-1}. *(3 marks)*

6 Use Wien's displacement law and the information provided in Topic 19.7, Stellar luminosity, to show that the Universe is saturated with microwaves. *(3 marks)*

Chapter 20 Practice questions

1 When viewed on a large enough scale, the Universe is homogeneous and isotropic, and the laws of physics are universal.

This is a statement of which law or principle?

A Hubble's law

B Doppler's principle

C Cosmological principle

D Wien's displacement law *(1 mark)*

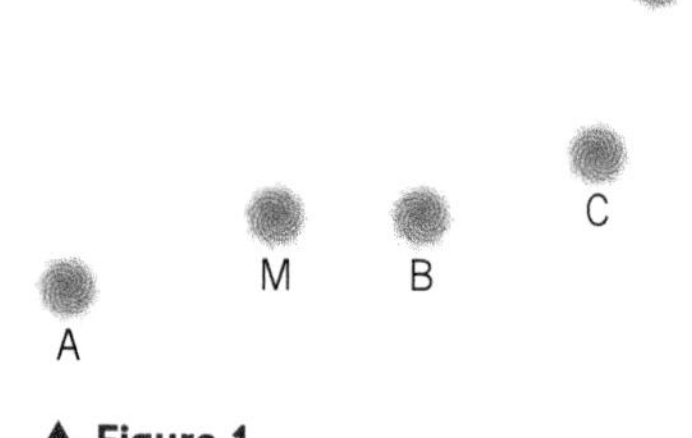

▲ **Figure 1**

2 Figure 1 shows five galaxies **A**, **B**, **C**, **D**, and **M**, where **M** is our galaxy (Milky Way).

Which galaxy will show the largest recession velocity when observed from **M**? *(1 mark)*

3 A star is at a distance of 2.1×10^{17} m from the Earth.

What is the distance of this star in light years (ly)?

A 4.2 ly **C** 14 ly

B 6.8 ly **D** 22 ly *(1 mark)*

4 The approximate wavelengths of the red-end and blue-end of the visible spectrum are 700 nm and 400 nm, respectively. The wavelength of an emission line in the blue part of the spectrum from a star shows a 0.058% increase.

What is the percentage increase in the wavelength of an emission line in the red part of the spectrum from the same star?

A 0% **C** 0.058%

B 0.033% **D** 0.102% *(1 mark)*

5 **a** With the help of a labelled diagram, show that a parallax of 1 arcsecond is equivalent to a distance of about 3.1×10^{16} m (1 pc).

1 AU = 1.5×10^{11} m *(3 marks)*

b The length of our galaxy is about 4.0×10^4 pc. Calculate the time it would take for light to travel the length of the galaxy. Write your answer both in seconds and years. *(3 marks)*

c The Hubble constant is about 70 km s^{-1} Mpc^{-1}.

i Calculate the Hubble constant in s^{-1}. *(2 marks)*

ii Estimate the farthest distance we can observe in the Universe. Explain your answer. *(3 marks)*

d List two observations that support the Big Bang model of the Universe. *(2 marks)*

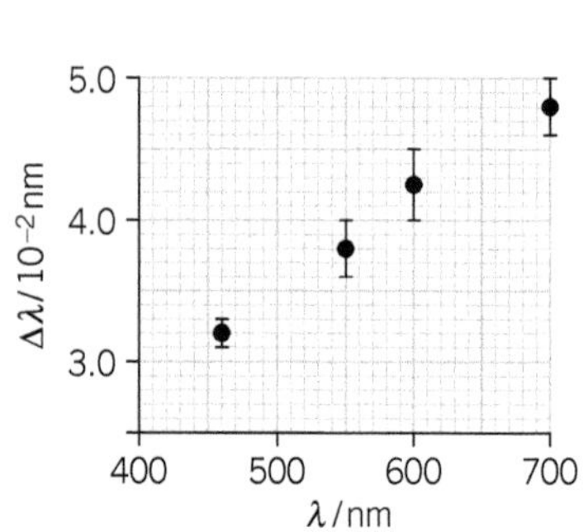

▲ **Figure 2**

6 Astronomers are observing the absorption lines in the visible spectrum from the star Alpha Centuari.

Figure 2 shows four data points plotted on a grid. The change in the wavelength of a spectral line is $\Delta\lambda$ and λ is wavelength of the same spectral line observed in the laboratory.

a Draw a line of best fit through the data points. *(1 mark)*

b Explain why a straight line graph is produced. *(1 mark)*

c Determine the gradient of the straight line and therefore estimate the speed of the star relative to the Earth. *(3 marks)*

Module 6 Particles and medical physics
Chapter 21 Capacitance

In this chapter you will learn about ...

- ☐ Capacitors
- ☐ Capacitance
- ☐ Capacitors in combination
- ☐ Energy stored by capacitors
- ☐ Discharge of a capacitor
- ☐ Time constant
- ☐ Charging of a capacitor
- ☐ Uses of capacitors

21 CAPACITANCE
21.1 Capacitors
21.2 Capacitors in circuits
21.3 Energy stored by capacitors

Specification reference: 6.1.1, 6.1.2

21.1 Capacitors

A capacitor is a component designed to store charge. It consists of two metal plates separated by an insulator (air, ceramic, paper, etc).

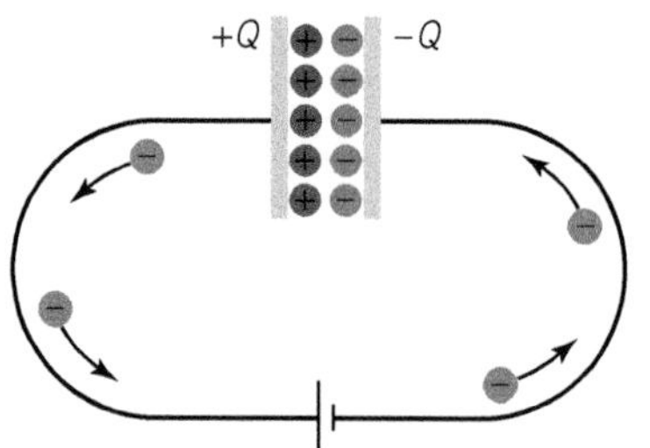

▲ **Figure 1** *The plates of a capacitor get oppositely charged*

Figure 1 shows a capacitor connected to a source of e.m.f. (a cell). Electrons are removed from the left-hand side plate and electrons are deposited onto the opposite plate. Each plate gains and loses the same number of electrons and the final charges on the plates are $+Q$ and $-Q$. The capacitor is charged fully when the potential difference (p.d.) V across it is equal to the e.m.f. of the cell.

Capacitance

Experiments show that the charge Q on one of the capacitor plates is directly proportional to the p.d. V across the capacitor. Therefore

$$Q \propto V \qquad \text{or} \qquad Q = VC$$

where C is the capacitance of the capacitor. Capacitance has the unit farad F, where $1\,\text{F} = 1\,\text{C}\,\text{V}^{-1}$.

The **capacitance** of a capacitor is defined as the charge stored per unit p.d. across it.

A capacitance of 1 **farad** is defined as 1 coulomb of charge stored per unit volt.

Revision tip

The farad F is a very large unit.
In practice, capacitors are marked in μF, nF, and pF.
($\mu = 10^{-6}$, $n = 10^{-9}$, and $p = 10^{-12}$)

Common misconception

The letter 'C' is used for both capacitance and coulomb. There is room for confusion – you just need to be vigilant.

21.2 Capacitors in circuits

In a circuit capacitors can be connected in many different combinations. You can simplify complex circuits by considering capacitors in series and parallel combinations.

Parallel

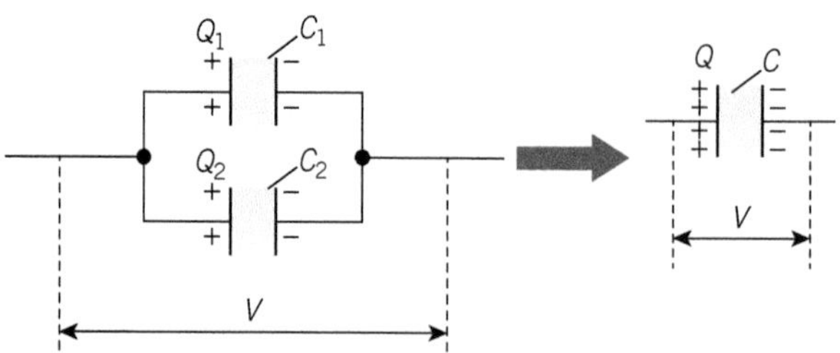

▲ **Figure 2** *Parallel combination*

Figure 2 shows capacitors of capacitance C_1 and C_2 connected in parallel. This combination is equivalent to a capacitor of capacitance C.

- The p.d. V across each capacitor is the same.
- The total charge Q stored is equal to the sum of the individual charges. $Q = Q_1 + Q_2 + \ldots$
- The total capacitance C is given by the equation $C = C_1 + C_2 + \ldots$

Series

Figure 3 shows capacitors of capacitance C_1 and C_2 connected in series. This combination is equivalent to a capacitor of capacitance C.

- The charge Q stored by each capacitor is the same.
- The total p.d. V across the combination is the sum of the individual p.d.s. $V = V_1 + V_2 + \dots$
- The total capacitance C is given by the equation

$$\frac{1}{C} = \frac{1}{C_1} + \frac{1}{C_2} + \dots \qquad \text{or} \qquad C = (C_1^{-1} + C_2^{-1} + \dots)^{-1}$$

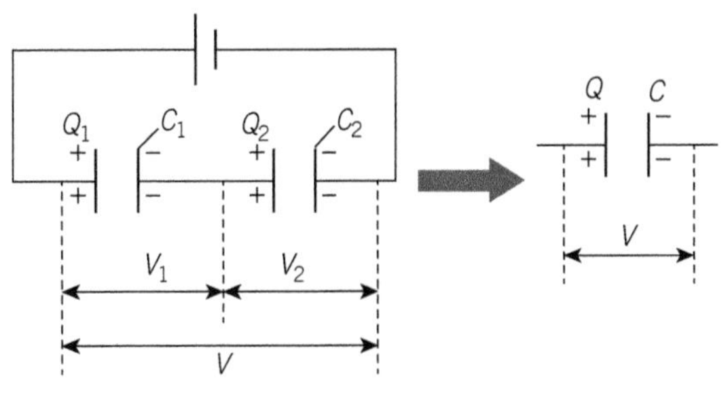

▲ **Figure 3** *Series combination*

21.3 Energy stored by capacitors

When a capacitor is charging up, work is done to push the electrons onto the negative plate and to pull the electrons away from the positive plate.

Revision tip

For capacitors in **series,** the total capacitance is always less than the *smallest* capacitance value.

Stored energy

The work done on the charges is equal to the area under a p.d. against charge graph. This area is also equal to the energy stored by the capacitor, see Figure 4.

work done W = energy stored = $\frac{1}{2}$ × charge × p.d.

$$W = \frac{1}{2}QV$$

Using $Q = VC$ gives two further equations.

$$W = \frac{1}{2}\frac{Q^2}{C} \text{ and } W = \frac{1}{2}V^2C.$$

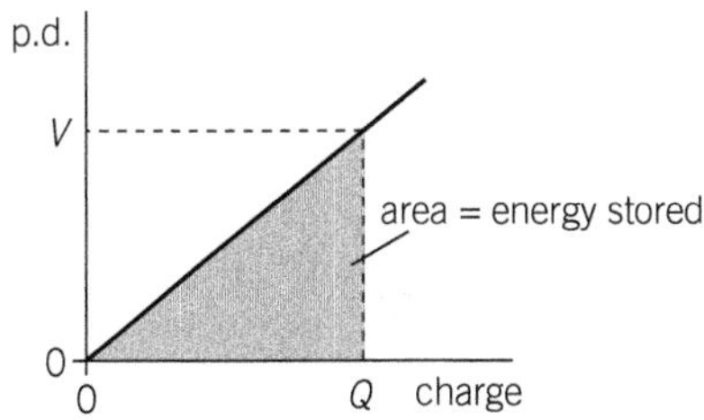

▲ **Figure 4** *The area under a p.d. against charge graph is the energy stored*

Worked example: Energy stored

A 100 μF capacitor and a 220 μF capacitor are connected in **series** to a 6.0 V supply.

Calculate the energy stored by the 100 μF capacitor.

Step 1: Calculate the total capacitance.

$C = (100^{-1} + 220^{-1})^{-1} = 68.75\ \mu F$

Step 2: Calculate the charge stored by the 100 μF capacitor.

Capacitors in series store the same charge.

$Q = VC = 6.0 \times 68.75 \times 10^{-6} = 4.125 \times 10^{-4}$ C

Step 3: Calculate the energy stored by the 100 μF capacitor.

$$\text{energy} = \frac{Q^2}{2C} = \frac{(4.125 \times 10^{-4})^2}{2 \times 100 \times 10^{-6}} = 8.5 \times 10^{-4}\text{ J (2 s.f.)}$$

Summary questions

1 A 500 μF capacitor is connected to a 9.0 V supply. Calculate the charge stored by the capacitor. *(2 marks)*

2 Calculate the energy stored by the capacitor in **Q1**. *(2 marks)*

3 A student is given three 100 μF capacitors. These capacitors can be connected in any combination.
Calculate the maximum and minimum capacitance. *(4 marks)*

4 Calculate the total capacitance of the circuit shown in Figure 5. *(3 marks)*

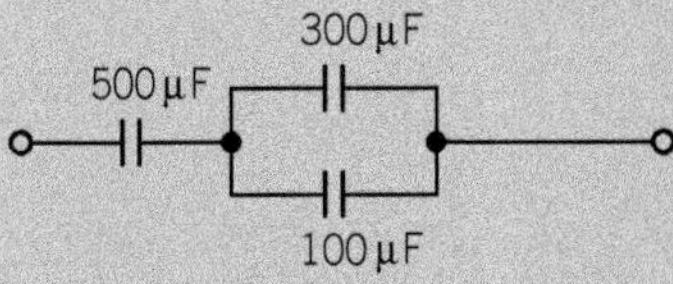

▲ **Figure 5**

5 The circuit of Figure 5 is now connected to a 6.0 V battery.
Calculate the p.d. measured by a digital voltmeter placed across the 500 μF capacitor. *(3 marks)*

6 A 1000 μF capacitor has a p.d. of 10 V. It is connected across an uncharged 500 μF capacitor.
Calculate the final p.d. across the 1000 μF capacitor. *(4 marks)*

21.4 Discharging capacitors
21.5 Charging capacitors
21.6 Uses of capacitors

Specification reference: 6.1.2, 6.1.3

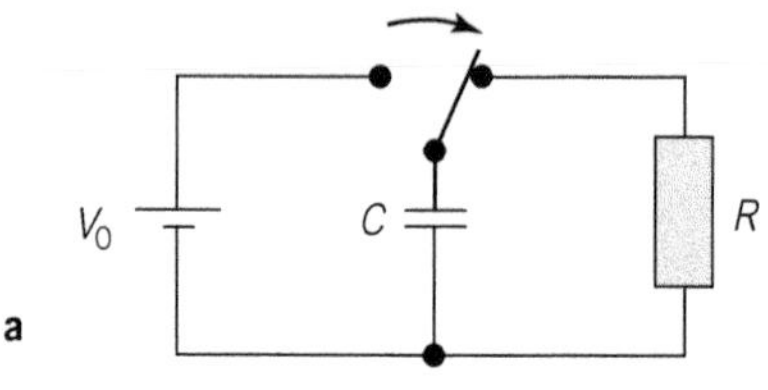

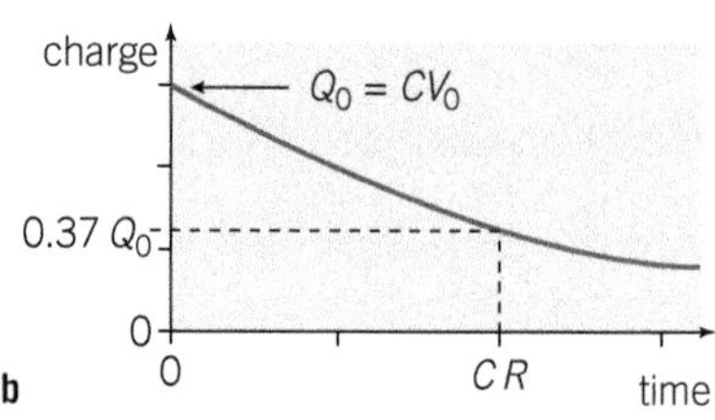

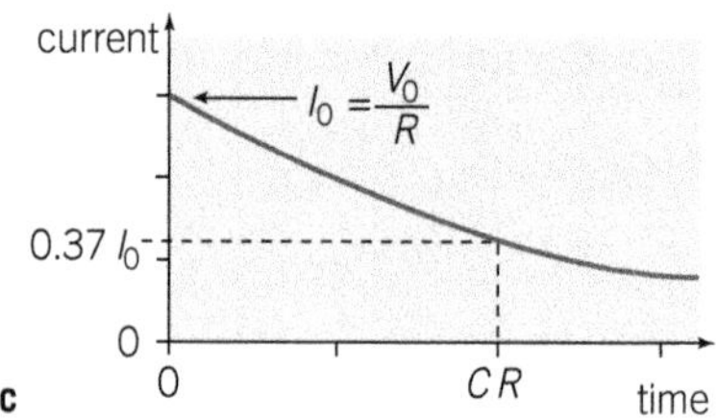

▲ **Figure 1** *(a) A capacitor discharging circuit, (b) charge-time graph, and (c) current-time graph*

21.4 Discharging capacitor

Knowledge of series and parallel circuits together with the three equations below can be used to analyse circuits where a capacitor discharges or charges through a resistor.

$$Q = VC \qquad V = IR \qquad I = \frac{\Delta Q}{\Delta t}$$

Exponential decay

Figure 1a shows a capacitor of capacitance C connected in **parallel** with a resistor of resistance R. The switch is connected to the left. The p.d. across the capacitor is V_0. At time $t = 0$, the switch is connected to the right. The capacitor starts to discharge through the resistor. The p.d. across the resistor and the capacitor is the same at all times.

- At $t = 0$, maximum current $I_0 = \frac{V_0}{R}$ and maximum charge $Q_0 = V_0C$.
- The p.d, current, and charge decrease *exponentially* with respect to time.
- At time t, the p.d. V across the capacitor (or resistor), the current I in the resistor, and the charge Q on the capacitor are given by the equations

$$V = V_0 e^{-\frac{t}{CR}} \qquad I = I_0 e^{-\frac{t}{CR}} \qquad Q = Q_0 e^{-\frac{t}{CR}}$$

These equations can be represented by the equation $x = x_0 e^{-\frac{t}{CR}}$, where e is the base of natural logarithms, which has a value of 2.718... Figures 1b and 1c show the variations of charge and current with time.

Time constant

The time constant τ for a discharging capacitor is equal to the time taken for the p.d. (or the current or the charge) to decrease to e^{-1} (about 37%) of its initial value.

- The time constant τ is also equal to the product CR.
- The unit of time constant is the second (s). Note: 1 s = 1 F Ω.
- A capacitor never discharges fully, but after a time of $5CR$ it is 'practically discharged' with less than 1% of the original charge left.

Modelling decay

For a discharging capacitor, the p.d. across the capacitor is the same as the p.d. across the resistor.

Therefore

$$IR = -\frac{Q}{C} \qquad \text{or} \qquad R\frac{\Delta Q}{\Delta t} = -\frac{Q}{C}$$

Therefore

$$\frac{\Delta Q}{\Delta t} = -\frac{Q}{CR}$$

The minus sign signifies the charge on the capacitor decreases with time.

This equation can be used to model the discharge of a capacitor over a period of time.

Example

$CR = 2.0\,\text{s}$ and $\Delta t = 0.10\,\text{s}$

The charge lost after every 0.10 s is $\Delta Q = 0.05\,Q$.

After every 0.10 s, 95% of the previous charge is left.

This constant-ratio property is characteristic of **exponential decay**.

21.5 Charging capacitors

Figure 2a shows a capacitor charging through a resistor. The charge Q stored by the capacitor and therefore the p.d. V_C across it *increases* with time. The p.d. V_R across the resistor (and therefore the current I) must *decrease* because $V_C + V_R = V_0$ (Kirchhoff's second law).

Important equations

- V_C and Q are given by the general equation $x = x_0(1 - e^{-\frac{t}{CR}})$
- V_R and I are given by the general equation $x = x_0 e^{-\frac{t}{CR}}$

Worked example: Charging capacitor

An uncharged capacitor is charged using the circuit shown in Figure 3.

The switch is closed at time $t = 0$. Calculate the p.d. across the 1.0 μF capacitor at time $t = 3.0$ s.

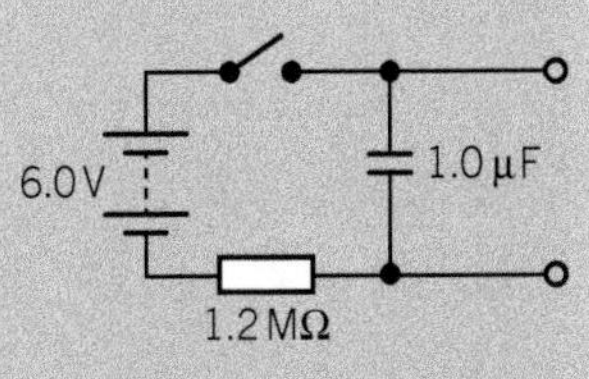

▲ **Figure 3**

Step 1: Calculate the time constant CR of the circuit.

$CR = 1.0 \times 10^{-6} \times 1.2 \times 10^{6} = 1.2\,\text{s}.$

Step 2: Calculate the p.d. V_C across the capacitor.

$V_C = V_0(1 - e^{-\frac{t}{CR}}) = 6.0 \times (1 - e^{-\frac{3.0}{1.2}}) = 5.51\,\text{V} \approx 5.5\,\text{V}$ (2 s.f.)

Maths: $\log_e$ or ln

ln is an abbreviation of $\log_e$

$V = V_0 e^{-\frac{t}{CR}}$. Taking $\log_e$ of both sides, we get $\ln V = \ln V_0 - \frac{t}{CR}$

A graph of $\ln V$ against t will be a straight line, with gradient $= -\frac{1}{CR}$.

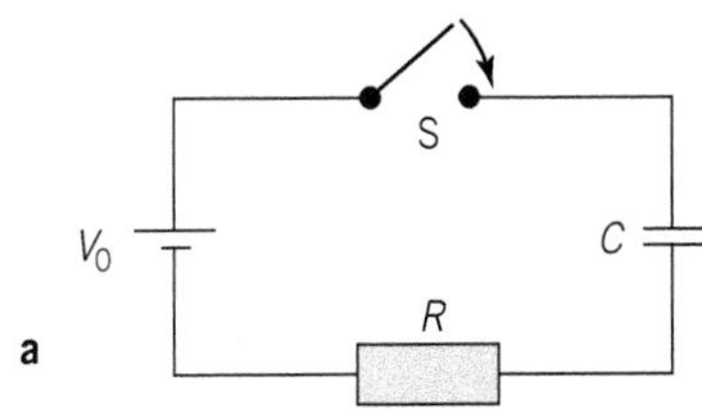

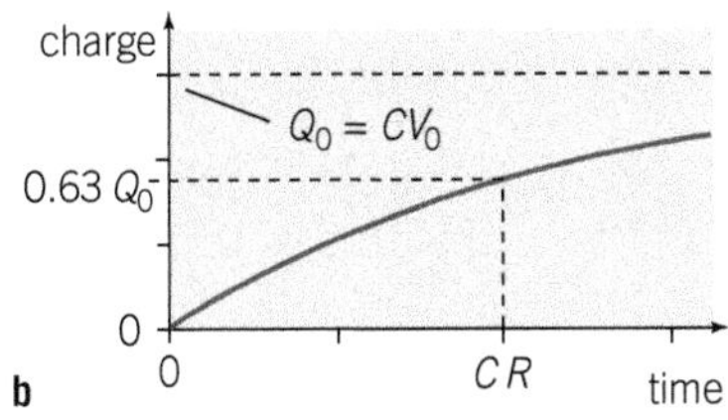

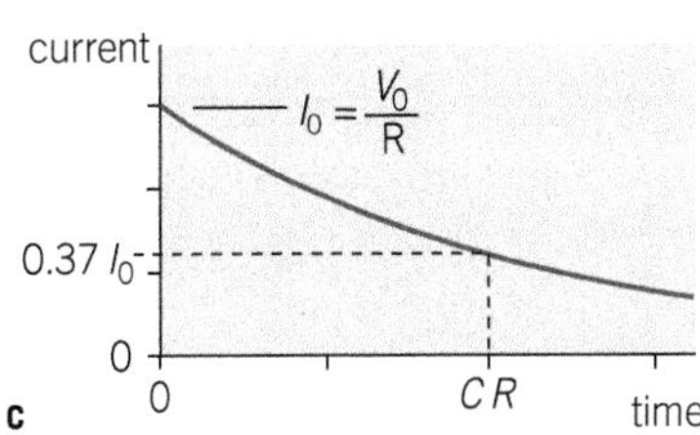

▲ **Figure 2** *(a) A capacitor charging circuit, (b) charge–time graph, and (c) current–time graph*

21.6 Uses of capacitors

Capacitors can store energy. This energy can be released in a very short period of time to produce large power. For example, releasing 1 J of stored energy is 1 μs can produce an output power of 1 MW. Capacitors are used in camera flashes, in particle accelerators, in electrical smoothing circuits, and so on.

Summary questions

1. Calculate the time constant of a circuit given $C = 100\,\mu\text{F}$ and $R = 150\,\text{k}\Omega$. *(1 mark)*
2. Explain why the p.d. across a charged capacitor decreases when it is connected across a resistor. *(2 marks)*
3. In the circuit of Figure 1a, $C = 100\,\mu\text{F}$, $R = 200\,\text{k}\Omega$, and $V_0 = 10\,\text{V}$. Calculate the p.d. across the capacitor after 38 s. *(3 marks)*
4. Show that the charge left on a discharging capacitor after five time constants is less than 1%. *(3 marks)*
5. In the circuit of Figure 2a, $C = 500\,\mu\text{F}$, $R = 100\,\text{k}\Omega$, and $V_0 = 10\,\text{V}$. Calculate the p.d. across the capacitor and resistor after 80 s. *(4 marks)*
6. For the circuit in **Q3**, how long would it take for the p.d. across the capacitor to halve? *(4 marks)*

Chapter 21 Practice questions

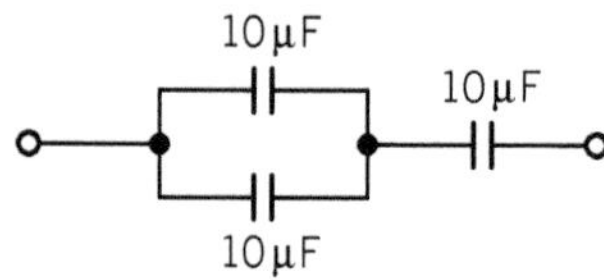

▲ **Figure 1**

1 Figure 1 shows a circuit with three identical capacitors.

What is the total capacitance of the circuit?

A 3.3 μF

B 6.7 μF

C 15 μF

D 30 μF *(1 mark)*

2 Which one of the following is **not** a unit for time constant?

A s

B FΩ

C $\mathrm{F\,V\,A^{-1}}$

D $\Omega\,\mathrm{V^{-1}}$ *(1 mark)*

3 A capacitor is discharging through a resistor. The time constant of the circuit is 2.0 s.

At time $t = 0$, the p.d. across the capacitor is 8.0 V.

What is the p.d. across the capacitor at time $t = 4.0$ s?

A 1.1 V

B 2.0 V

C 2.9 V

D 4.0 V *(1 mark)*

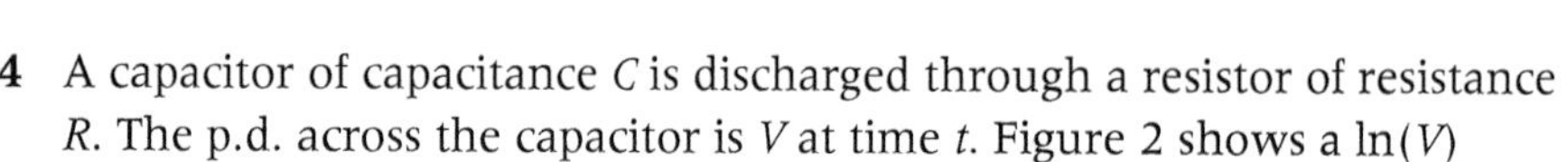

4 A capacitor of capacitance C is discharged through a resistor of resistance R. The p.d. across the capacitor is V at time t. Figure 2 shows a ln(V) against t graph.

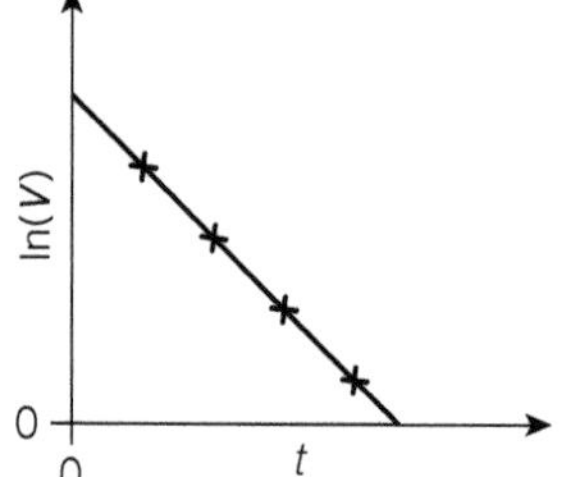

▲ **Figure 2**

What is the gradient of the graph equal to?

A −1

B −e

C $-CR$

D $-(CR)^{-1}$ *(1 mark)*

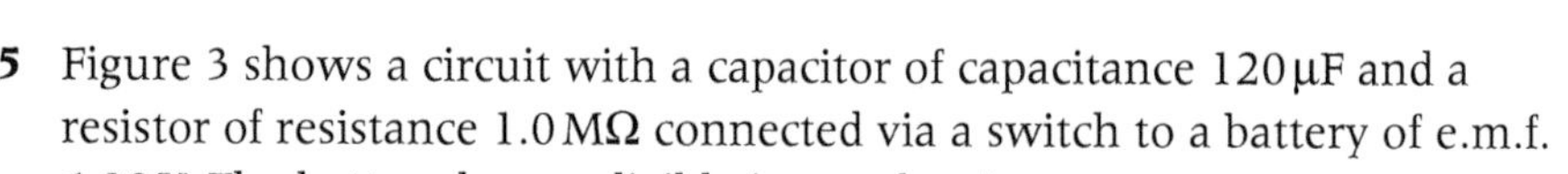

5 Figure 3 shows a circuit with a capacitor of capacitance 120 μF and a resistor of resistance 1.0 MΩ connected via a switch to a battery of e.m.f. 6.00 V. The battery has negligible internal resistance.

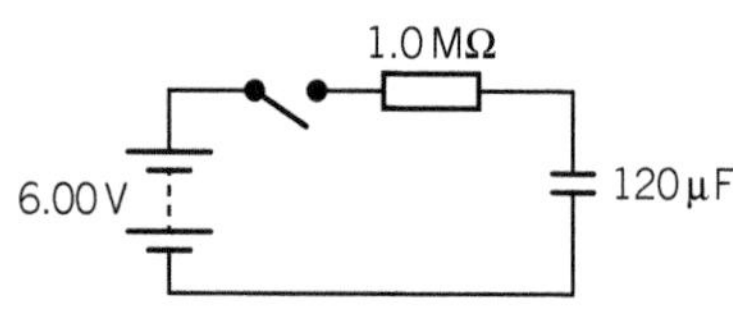

▲ **Figure 3**

At time $t = 0$, the switch is closed and the p.d. across the capacitor is zero.

a Describe and explain what happens to the p.d. across the resistor and the capacitor. *(4 marks)*

b Calculate the p.d. across the capacitor at time $t = 200$ s. *(3 marks)*

c Calculate the maximum current in the circuit in μA. *(1 mark)*

d Calculate the maximum energy stored by the capacitor. *(2 marks)*

6 Figure 4 shows a capacitor–resistor circuit.

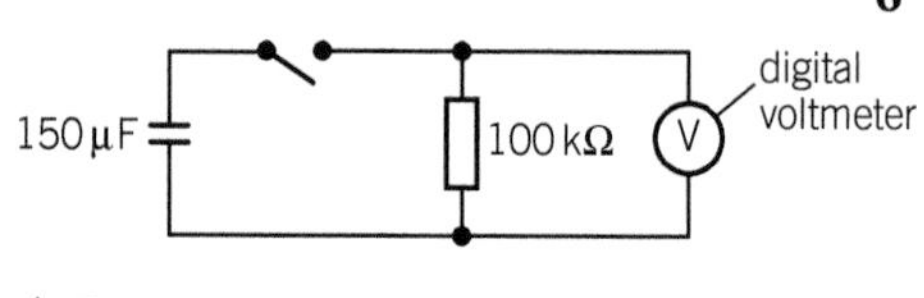

▲ **Figure 4**

The p.d. across the capacitor is 4.50 V. The switch is closed at time $t = 0$.

a Calculate the time constant of the circuit. *(1 mark)*

b Calculate the p.d. across the resistor at time $t = 35$ s. *(2 marks)*

c Calculate the total energy dissipated by the resistor between $t = 0$ and $t = 35$ s. *(3 marks)*

Chapter 22 Electric fields

In this chapter you will learn about ...

- ☐ Electric field lines
- ☐ Electric field strength
- ☐ Coulomb's law
- ☐ Radial electric field
- ☐ Uniform electric field
- ☐ Capacitance of a parallel plate capacitor
- ☐ Motion of charged particles
- ☐ Electric potential
- ☐ Electric potential energy
- ☐ Capacitance of an isolated sphere

22 ELECTRIC FIELDS
22.1 Electric fields
22.2 Coulomb's law

Specification reference: 6.2.1, 6.2.2

22.1 Electric fields

A charged particle creates an electric field in the space around it. Another charged particle in this electric field will experience either an attractive or a repulsive electrical force.

Field patterns

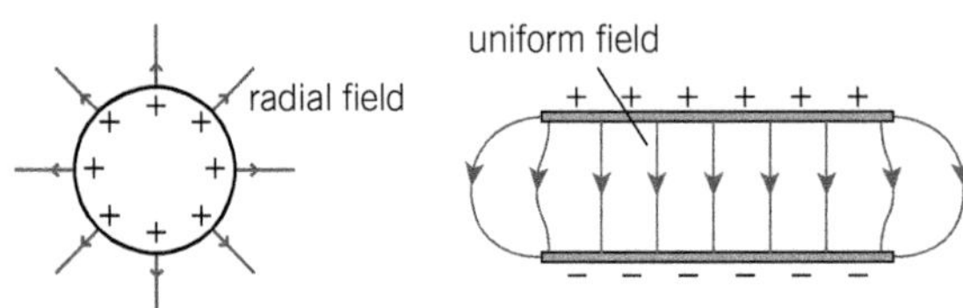

▲ **Figure 1** *Radial and uniform electric fields*

Electric field patterns can be mapped using electric field lines (or lines of force). The direction of the field at a point shows the direction of the force experience by a small *positive* charge placed at that point. The electric field strength (see later) is indicated by the separation between the field lines.

- A uniform electric field has equally spaced field lines. The electric field between two oppositely charged parallel plates is uniform.
- Electric field lines are always perpendicular to the surface of a conductor.
- A radial field has straight field lines converging to a point at the centre of the charged object. The field strength gets smaller with increased distance from the object.
- A uniformly charged sphere can be modelled as a point charge at its centre, see Figure 2.

Revision tip: Field strength

Closely spaced electric field lines indicate greater electric field strength *E*.

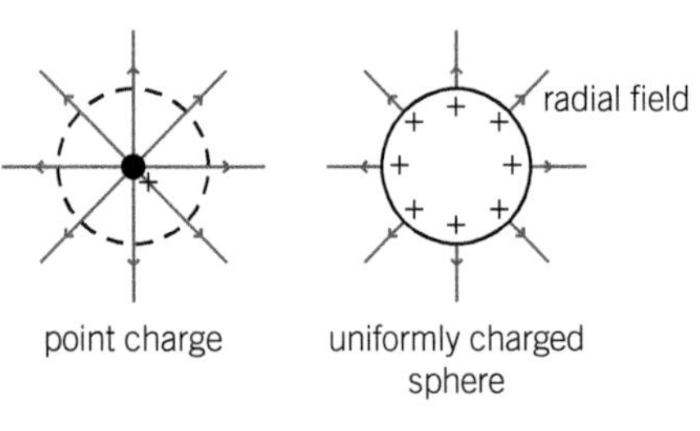

▲ **Figure 2** *A uniformly charged sphere with charge Q can be modelled as a point charge (particle) of charge Q*

Electric field strength *E*

The **electric field strength** E at a point is defined as the force experienced per unit *positive* charge at that point.

This can be written as

$$E = \frac{F}{Q}$$

where F is the force experienced by the positive charge Q. The SI unit for E is N C^{-1}. Electric field strength is a vector quantity.

22.2 Coulomb's law

Coulomb's law is a universal law that can be applied to all charged particles.

Coulomb's law: Two point charges exert an electrostatic (electrical) force on each other that is directly proportional to the product of their charges and inversely proportional to the square of the distance between them.

Equation for Coulomb's law

According to **Coulomb's law**

$$F \propto \frac{Qq}{r^2}$$

where F is the electrical force, Q and q are the charges, and r is the separation, see Figure 3. The law can be written as an equation using the permittivity of free space ε_0 (8.85×10^{-12} F m^{-1}) as follows:

$$F = \frac{Qq}{4\pi\varepsilon_0 r^2}$$

For two uniformly charged spheres, just remember that r is the centre-to-centre separation.

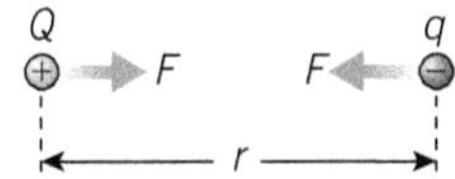

a *unlike charges attract*

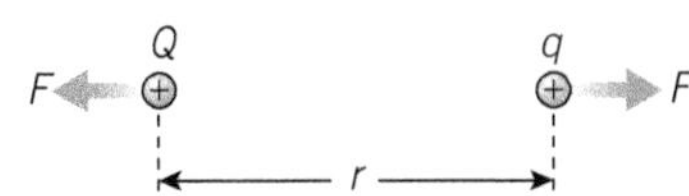

b *like charges repel*

▲ **Figure 3** *Electrical forces can be either attractive or repulsive*

Radial field

The **electric field strength** E at a distance r from the centre of a charge Q can be determined as follows:

$$E = F \div q = \frac{Qq}{4\pi\varepsilon_0 r^2} \div q$$

Therefore

$$E = \frac{Q}{4\pi\varepsilon_0 r^2}$$

The electric field strength E obeys an inverse square law with distance.

Gravitational and electric fields

Here are the main similarities and differences between the electric field of a point charge and the gravitational field of a point mass.

Similarities:

- The field strengths of both obey an inverse square law with distance.

$$g \propto \frac{1}{r^2} \text{ and } E \propto \frac{1}{r^2}$$

- Both produce a radial field pattern.

Differences:

- Gravitational field depends on the mass whereas electric field depends on charge.
- A gravitational field always produces an attractive force, whereas an electric field can produce either an attractive or a repulsive force.

Worked example: Electron acceleration

Calculate the acceleration of an electron at a distance of 7.0×10^{-11} m from a proton.

Step 1: Write down the quantities needed to calculate the force on the electron.

$r = 7.0 \times 10^{-11}$ m $Q = q = 1.60 \times 10^{-19}$ C $\varepsilon_0 = 8.85 \times 10^{-12}$ F m^{-1}

Step 2: Calculate the force experienced by the electron.

$$F = \frac{Qq}{4\pi\varepsilon_0 r^2} = \frac{1.60 \times 10^{-19} \times 1.60 \times 10^{-19}}{4\pi \times 8.85 \times 10^{-12} \times (7.0 \times 10^{-11})^2} = 4.70 \times 10^{-8}\text{ N}$$

Step 3: The mass of the electron is 9.11×10^{-31} kg. Use $F = ma$ to calculate the acceleration.

$$a = \frac{F}{m} = \frac{4.70 \times 10^{-8}}{9.11 \times 10^{-31}} = 5.2 \times 10^{22}\text{ m s}^{-2}\text{ (2 s.f.)}$$

Summary questions

1. The electrostatic force on an electron is 8.0×10^{-14} N. Calculate the electric field strength. *(2 marks)*
2. The electric field strength at a point is 6.0×10^4 N C^{-1}. Calculate the force experienced by an electron. *(2 marks)*
3. Two electrons are separated by 2.0×10^{-10} m. Calculate the electrostatic force on one of the electrons. *(3 marks)*
4. Calculate the force on the electron in **Q3** when the separation is halved. *(2 marks)*
5. A metal sphere has a positive charge of 3.8×10^{-9} C. Calculate its radius given the field strength on the surface is 5.0×10^4 N C^{-1}. *(3 marks)*
6. Charge density σ is defined as the charge per unit area. Derive an equation for σ and surface electric field strength E for a charged metal sphere. *(3 marks)*

22.3 Uniform electric fields and capacitance
22.4 Charged particles in uniform electric fields

Specification reference: 6.2.3

22.3 Uniform electric fields and capacitance

There is a uniform electric field between a pair of oppositely charged parallel plates. The plates store charge – they are equivalent to a capacitor.

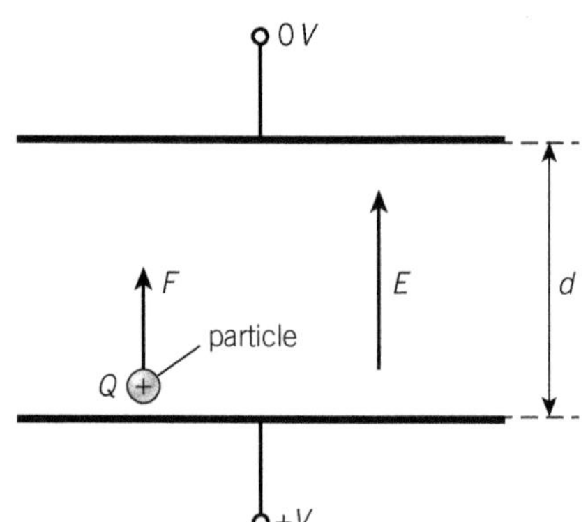

▲ **Figure 1** *A charged particle in a uniform electric field*

Parallel plates

A charged particle of charge $+Q$ will experience a force F in the direction of the field, see Figure 1. Consider this particle moving from the positive (where it is stationary) to the negative plate. The work done on the particle is VQ, where V is the p.d. across the plates. This work done is also Fd, where d is the separation between the plates. Therefore

$$Fd = VQ$$

or

$$\frac{F}{Q} = \frac{V}{d}$$

The electric field strength is defined as $\frac{F}{Q}$ therefore the uniform field strength of the electric field between oppositely charged plates is given by the equation

$$E = \frac{V}{d}$$

The unit for electric field strength can be either $\mathrm{N\,C^{-1}}$ or $\mathrm{V\,m^{-1}}$.

Revision tip: Field strength

The defining equation for electric field strength is $E = \frac{F}{Q}$ and *not* $E = \frac{V}{d}$.

Capacitor with parallel plates

The capacitance C of a capacitor made from two parallel metal plates in a vacuum with separation d and area of overlap A is given by the equation

$$C = \frac{\varepsilon_0 A}{d}$$

where ε_0 is the permittivity of free space ($8.85 \times 10^{-12}\,\mathrm{F\,m^{-1}}$). With an insulator (or dielectric) other than a vacuum between the plates, the capacitance increases and is given by the equation

$$C = \frac{\varepsilon A}{d}$$

where ε is the permittivity of the insulator. The permittivity of insulators is always greater than ε_0, so sometimes the term relative permittivity ε_r is also used, where $\varepsilon = \varepsilon_r \varepsilon_0$,

ε_r has no unit. For vacuum $\varepsilon_r = 1$ (by definition), for air $\varepsilon_r = 1.0006 \approx 1.0$, and for paper $\varepsilon_r \approx 4.0$.

Synoptic link

Capacitance was defined in Topic 21.1, Capacitors.

22.4 Charged particles in uniform electric fields

A charged particle experiences an electrical force in a uniform electric field, and therefore will have acceleration. The motion of a charged particle in a uniform electric field can be analysed using the equations of motion and the following equations:

$$F = ma \qquad F = EQ \qquad E = \frac{V}{d}$$

Motion parallel to field

A positive charge moving in the direction of the electric field will accelerate, whereas a negative charge will decelerate when moving in the direction of the field. See Figure 2.

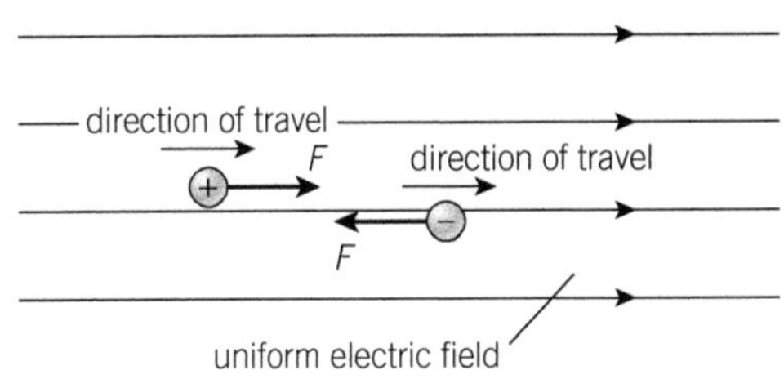

▲ **Figure 2** *Charged particles moving to the right. The positive particle will accelerate and the negative particle will decelerate*

Motion perpendicular to field

A charged particle will describe a parabolic path when its initial velocity is perpendicular to the direction of the field. You can use your knowledge of projectiles to analyse the motion of particles. Here are a couple of important reminders:

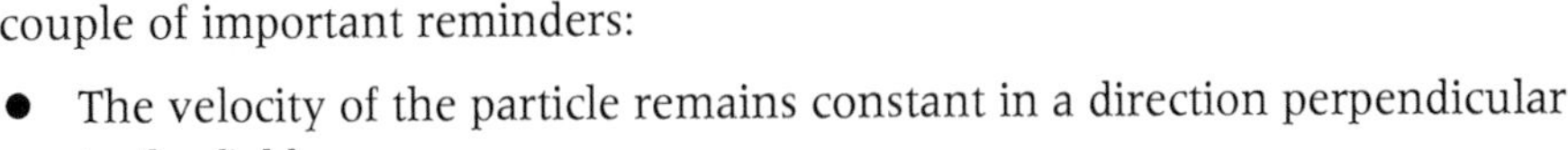

- The velocity of the particle remains constant in a direction perpendicular to the field.

- The particle will accelerate (or decelerate) in the direction of the field.

Worked example: Electron deflection

An electron, with a velocity of $2.0 \times 10^7\,\mathrm{m\,s^{-1}}$ in the horizontal direction, enters a region of uniform electric field of strength $4.0 \times 10^4\,\mathrm{V\,m^{-1}}$. The direction of the electric field is vertical. Calculate the vertical deflection of the electron after it has travelled 8.0 cm in the horizontal direction.

Step 1: Calculate the time taken to travel a horizontal distance of 8.0 cm.

$$\text{time } t = \frac{\text{distance}}{\text{speed}} = \frac{8.0 \times 10^{-2}}{2.0 \times 10^{7}} = 4.0 \times 10^{-9}\,\mathrm{s}$$

Step 2: Calculate the vertical acceleration of the electron.

$E = 4.0 \times 10^4\,\mathrm{V\,m^{-1}}$ $\quad Q = e = 1.60 \times 10^{-19}\,\mathrm{C}$ $\quad m = 9.11 \times 10^{-31}\,\mathrm{kg}$

$$a = \frac{F}{m} = \frac{EQ}{m} = \frac{4.0 \times 10^4 \times 1.60 \times 10^{-19}}{9.11 \times 10^{-31}} = 7.025 \times 10^{15}\,\mathrm{m\,s^{-2}}$$

Step 3: Use the equation of motion $s = ut + \frac{1}{2}at^2$ to calculate vertical deflection s.

$u = 0$ $\quad a = 7.025 \times 10^{15}\,\mathrm{m\,s^{-2}}$ $\quad t = 4.0 \times 10^{-9}\,\mathrm{s}$

$$s = \frac{1}{2} \times 7.025 \times 10^{15} \times (4.0 \times 10^{-9})^2 = 0.056\,\mathrm{m}\ (5.6\,\mathrm{cm})\ (2\ \text{s.f.})$$

Summary questions

1 Two parallel plates in air are separated by 1.0 cm. The plates are connected to a 2.0 kV supply.
Calculate the electric field strength between the plates. *(2 marks)*

2 The capacitance of the capacitor in **Q1** is 10 pF. Calculate the charge stored by the capacitor. *(1 mark)*

3 Calculate the area of overlap of the plates in **Q1**. *(2 marks)*

4 A capacitor is made by inserting a sheet of paper of thickness 0.070 mm between A4 size metal sheets.
Calculate the capacitance in nF of this capacitor. *(3 marks)*
ε_r for paper = 4.0 and area of A4 = $6.24 \times 10^{-2}\,\mathrm{m^2}$

5 Two metal plates are separated by 1.0 cm and are connected to a 500 V supply.
Calculate the time it takes for an electron to travel from the negative plate to the positive plate. *(4 marks)*
(Assume the electron is initially at rest.)

6 Two parallel plates are charged oppositely using a power supply. The power supply is then disconnected. The energy stored by the capacitor is E_0. The separation between the charged plates is doubled.
Calculate the final energy stored in terms of E_0. *(3 marks)*

22.5 Electric potential and energy

Specification reference: 6.2.4

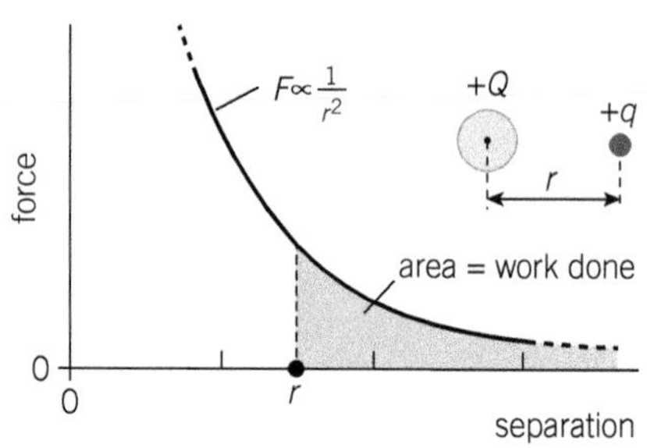

▲ **Figure 1** *The area under a force–separation graph is equal to the work done on the charges*

22.5 Electric potential and energy

You need the idea of **electric potential** to understand the gain and loss in the electric potential energy of charged particles.

Potential energy

Consider two charged particles of charges Q and q with a separation r. The force experienced by each particle is given by the equation for Coulomb's law

$$F = \frac{Qq}{4\pi\varepsilon_0 r^2}$$

where ε_0 is the permittivity of free space ($8.85 \times 10^{-12}\,\text{F m}^{-1}$).

External work must be done to bring like charged particles closer because they repel each other. The work done can be calculated from a force against separation graph, see Figure 1.

The area under a force–separation graph is the work done on the charged particles. The total work done on the charged particles to bring them from infinity to a separation r is known as electric potential energy.

The electric potential energy E for two charged particles separated by a distance r is defined as the work done to bring the charged particles from infinity to a separation r.

The equation for electric potential energy is $E = \frac{Qq}{4\pi\varepsilon_0 r^2}$.

- This equation can be used for uniformly charged spheres – just remember that r will be the separation between the centres of the spheres.
- The electric potential energy is negative for two particles with unlike charges. The negative potential energy simply means that external work has to be done to pull apart the particles.
- The *change* in electric potential energy ΔE is simply calculated as follows:

  $\Delta E = \text{final } E - \text{initial } E = \frac{Qq}{4\pi\varepsilon_0}\left(\frac{1}{r_f} - \frac{1}{r_i}\right)$ where r_f is the final separation and r_i is the initial separation.

Electric potential

The **electric potential** V at a point is defined as the work done per unit charge in bringing a positive charge from infinity to that point.

The electric potential V at a distance r away from a particle of charge Q can be determined by dividing the electric potential energy by the test charge q. This gives the following equation for electric potential:

$$V = \frac{Q}{4\pi\varepsilon_0 r}$$

- V has unit J C^{-1} or volt (V).
- V is defined to be zero at infinity.
- This equation can be used for a uniformly charged sphere – just remember that r will be the distance from the centre of the sphere.
- The *change* in electric potential ΔV is simply calculated as follows:

  $$\Delta V = \text{final } V - \text{initial } V = \frac{Q}{4\pi\varepsilon_0}\left(\frac{1}{r_f} - \frac{1}{r_i}\right)$$

  where r_f is the final distance and r_i is the initial distance.

Revision tip

Electric potential energy = electric potential × charge
Therefore, $E = Vq$.

Capacitance revisited

An isolated metal sphere can be charged using a power supply. The electric potential V on the surface of the sphere is related to the surface charge Q and the radius R of the sphere by the equation

$$V = \frac{Q}{4\pi\varepsilon_0 R}$$

The ratio $\frac{Q}{V}$ is the capacitance C of the sphere. Therefore

$$C = 4\pi\varepsilon_0 R$$

The value of $4\pi\varepsilon_0$ is $1.11 \times 10^{-10}\,\text{F m}^{-1}$. Therefore a sphere of radius 1.0 m will have capacitance of about 11 pF.

Worked example: Removing an electron

In the hydrogen atom, the electron is at a mean distance of about 5.0×10^{-11} m from the proton.

Estimate the energy required to completely remove the electron from the hydrogen atom in eV.

Step 1: Write down the information given.

$Q = +1.60 \times 10^{-19}\,\text{C}$ $q = -1.60 \times 10^{-19}\,\text{C}$ $r = 5.0 \times 10^{-11}\,\text{m}$ $\varepsilon_0 = 8.85 \times 10^{-12}\,\text{F m}^{-1}$

Step 2: Calculate the electric potential energy of the electron and proton.

$$E = \frac{Qq}{4\pi\varepsilon_0 r} = -\frac{(1.60 \times 10^{-19})^2}{4\pi \times 8.85 \times 10^{-12} \times 5.0 \times 10^{-11}} = -4.60 \times 10^{-18}\,\text{J}$$

Step 3: Calculate the energy required to completely remove the electron in eV.

$1\,\text{eV} = 1.60 \times 10^{-19}\,\text{J}$

$$\text{energy} = \frac{4.60 \times 10^{-18}}{1.60 \times 10^{-19}} = 29\,\text{eV (2 s.f.)}$$

Revision tip

The electron and the proton have a charge of the same magnitude, 1.60×10^{-19} C.

Summary questions

1. The electric potential on the surface of a sphere is −100 V.
   What is the energy required to remove a charge of 1 C from the surface of the sphere to infinity? *(1 mark)*
2. The electric potential on the surface of a sphere is + 2000 V.
   Calculate the potential at a distance equal to one radius from the *surface* of the sphere. *(2 marks)*
3. Calculate the electric potential at a distance of 1.2×10^{-10} m from a proton. *(2 marks)*
4. The electric potential on the surface of a metal sphere of radius 5.0 cm is 100 V.
   Calculate the charge on the sphere. *(3 marks)*
5. Calculate the capacitance of a metal sphere of radius 2.5 cm and the charge stored on its surface when it is connected to the positive terminal of a 5.0 kV supply. *(4 marks)*
6. Estimate the force experienced by each charged metal sphere shown in Figure 2. The electric potential on the surface of each sphere is + 3000 V. *(4 marks)*

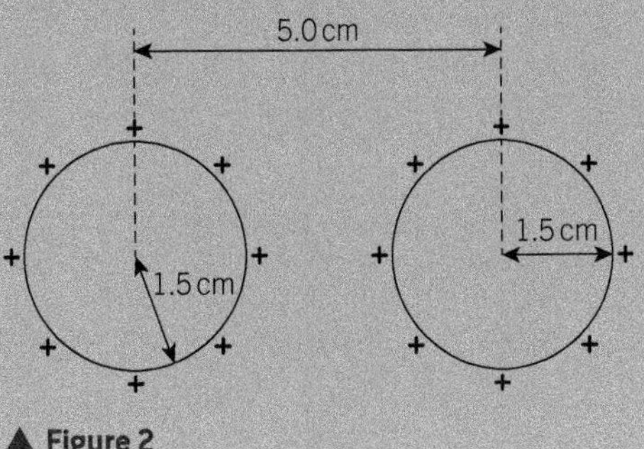

▲ **Figure 2**

Chapter 22 Practice questions

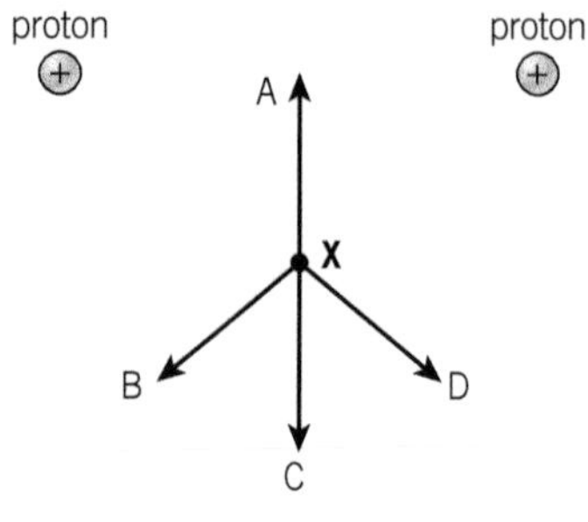

▲ **Figure 1**

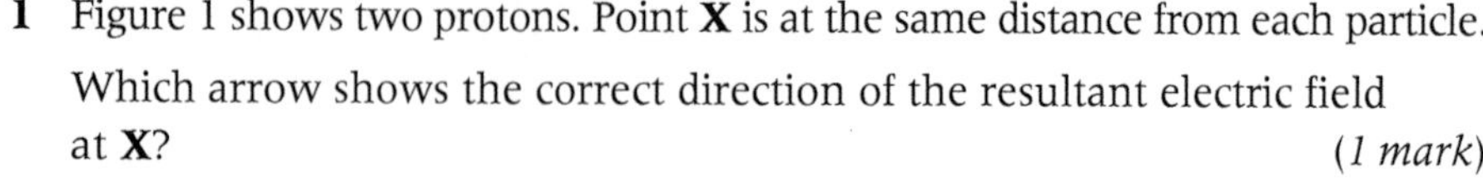

1 Figure 1 shows two protons. Point **X** is at the same distance from each particle.

Which arrow shows the correct direction of the resultant electric field at **X**? *(1 mark)*

2 What is electric potential measured in base units?

A $\mathrm{kg\,m^2\,A^{-1}s^{-3}}$

B $\mathrm{kg\,m^{-2}\,A^{-1}s^{-3}}$

C $\mathrm{kg\,m^2\,s^3\,A^{-1}}$

D $\mathrm{kg\,m^2\,A\,s^{-3}}$ *(1 mark)*

3 An electron experiences a force of magnitude F in a uniform electric field.

What is the force experienced by an alpha particle, in terms of F, in the same electric field?

A 0

B F

C $2F$

D $4F$ *(1 mark)*

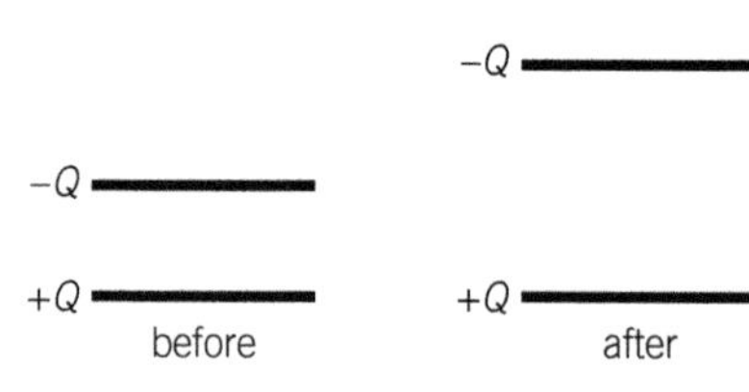

▲ **Figure 2**

4 Figure 2 shows opposite charged plates before and after the separation between the plates is doubled. The initial energy stored by this arrangement is E.

What is the final energy stored in terms of E?

A $\frac{E}{2}$

B E

C $2E$

D $4E$ *(1 mark)*

5 A metal sphere is momentarily connected to the positive terminal of a high-voltage power supply. The radius of the sphere is 5.0 cm. A scientist determines the electric field strength E at a distance r from the centre. The results are recorded in the table below.

r / cm	9.2	18.0	25.0
E / $10^3\,\mathrm{N\,C^{-1}}$	20.7	5.4	2.8

a Use the table to confirm the relationship between E and r. *(2 marks)*

b Calculate the charge Q on the surface of the sphere. *(3 marks)*

c Calculate the charge stored per unit surface area. *(2 marks)*

d Calculate the electric potential on the surface of the sphere. *(2 marks)*

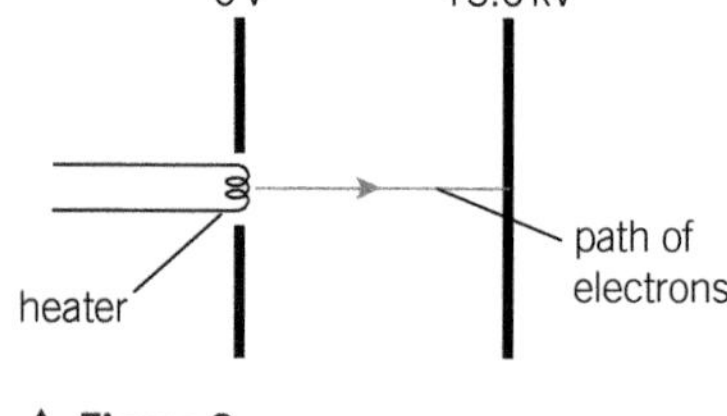

▲ **Figure 3**

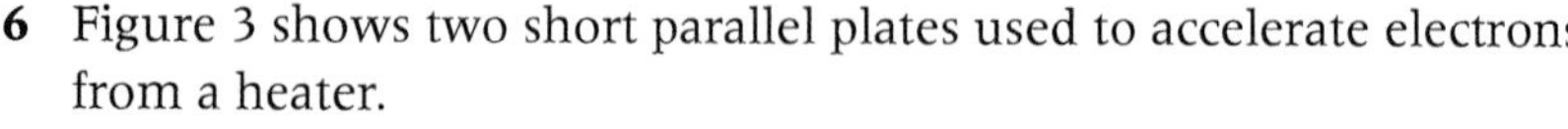

6 Figure 3 shows two short parallel plates used to accelerate electrons from a heater.

The potential difference between the plates is 5.0 kV and the separation between the plates is 7.2 cm.

a Describe the electric field and the field strength between the plates. *(2 marks)*

b Calculate the electric field strength between the plates. *(2 marks)*

c Calculate the maximum speed of an electron when at the positive plate. Assume the initial speed of the electron is zero. *(4 marks)*

d The p.d. across the plates is fixed at 5.0 kV.

Explain how the final kinetic energy of an electron arriving at the positive plate depends on the separation between the plates. *(2 marks)*

Chapter 23 Magnetic fields

In this chapter you will learn about ...

- [ ] Magnetic field lines
- [ ] Electromagnetism
- [ ] Magnetic flux density
- [ ] Fleming's left-hand rule
- [ ] Motion of charged particles
- [ ] Velocity selector
- [ ] Electromagnetic induction
- [ ] Faraday's law
- [ ] Lenz's law
- [ ] A.C. generator
- [ ] Transformers

23 MAGNETIC FIELDS
23.1 Magnetic fields
23.2 Understanding magnetic fields

Specification reference: 6.3.1

23.1 Magnetic fields

Magnetic fields are created in the space around a permanent magnet and a current-carrying conductor.

Field patterns

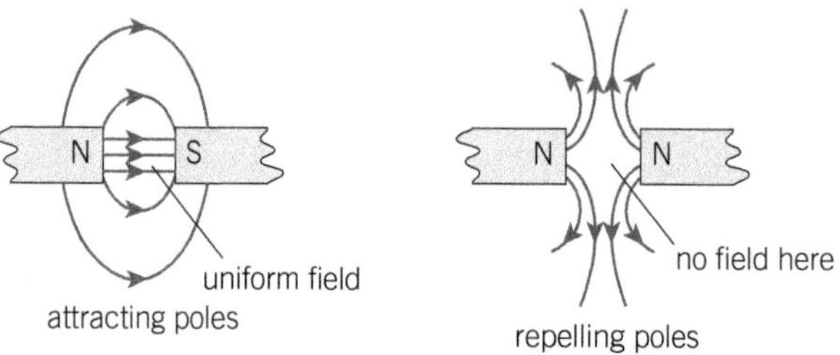

▲ **Figure 1** *Field patterns between unlike poles and like poles*

Revision tip: Field strength

A uniform magnetic field has equally spaced and parallel field lines. Closely spaced magnetic field lines indicate greater magnetic flux density *B*.

Magnetic field patterns can be mapped using **magnetic field lines**, see Figure 1. The direction of the field at a point shows the direction of the force experienced by a free north pole at that point.

Electromagnetism

A magnetic field is produced by moving charged particles, for example electrons moving within a wire.

Figure 2 shows the magnetic field patterns for a long straight current-carrying conductor, a flat coil, and a long solenoid.

- The magnetic fields lines are concentric circles for a long straight current-carrying conductor.
- The magnetic field pattern of a solenoid resembles the field pattern of a bar magnet. The magnetic field is uniform within the core of the solenoid.

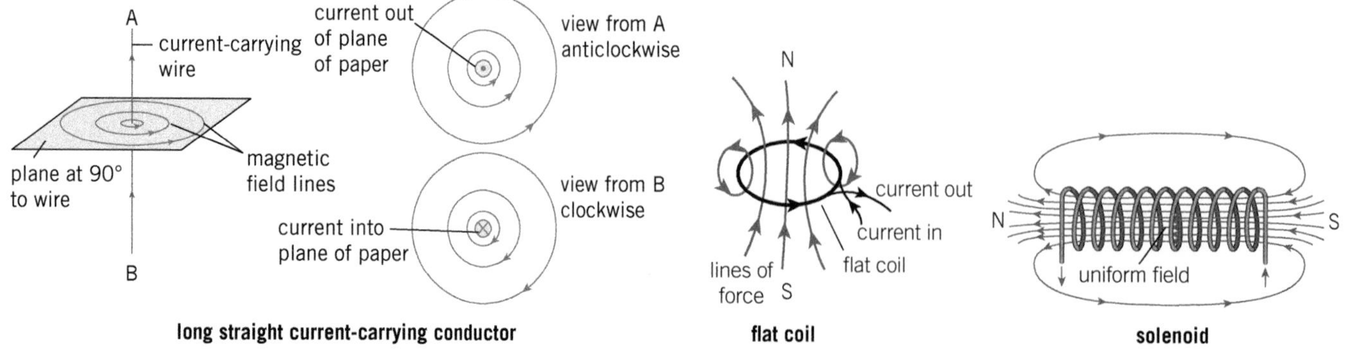

▲ **Figure 2** *Magnetic field patterns for a long straight current-carrying conductor, a flat coil, and a long solenoid*

23.2 Understanding magnetic fields

The strength of a magnetic field is called magnetic flux density. This important quantity is defined in terms of the force experienced by a current-carrying conductor.

Force on a current-carrying conductor

A current-carrying wire experiences a force when it is placed at right angles to the magnetic field of a magnet. According to Newton's third law the wire and the magnet experience equal but opposite forces.

The direction of the force experienced by the current-carrying wire in a uniform field can be determined using **Fleming's left-hand rule**, see Figure 3.

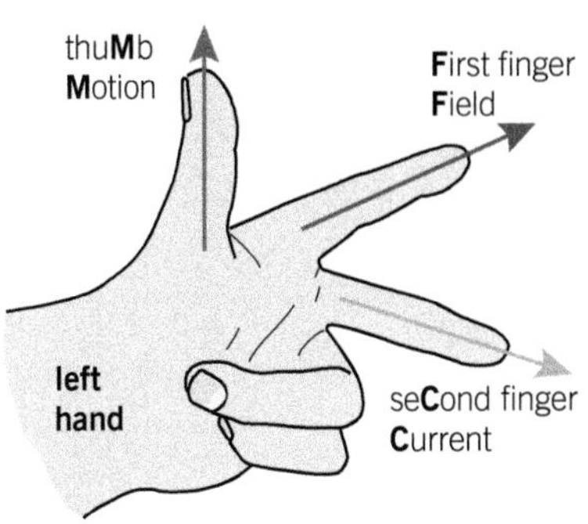

▲ **Figure 3** *Fleming's left-hand rule: The first finger points in the direction of the external field, the second finger points in the direction of the* ***conventional*** *current, and the thumb points in the direction of the motion (force)*

Magnetic flux density and $F = BIL$

The force F experienced by the wire placed in a uniform magnetic field is given by the relationship

$$F \propto IL\sin\theta$$

where L is the length of the wire in the field, I is the current in the wire, and θ is the angle between the field and the wire. The constant of proportionality is called the magnetic flux density B of the magnetic field. Therefore

$$F = BIL\sin\theta$$

The SI unit for magnetic flux density is the **tesla** (T) and $1\,\text{T} = 1\,\text{N}\,\text{m}^{-1}\,\text{A}^{-1}$. Magnetic flux density is a vector quantity.

The **magnetic flux density** is 1 tesla when a wire carrying a current of 1 ampere placed perpendicular to the magnetic field experiences a force of 1 newton per metre of its length.

When the wire is perpendicular to the magnetic field, $\theta = 90°$ and $\sin\theta = 1$. Therefore

$$F = BIL$$

You can determine the magnetic flux density between the poles of a magnet using the arrangement shown in Figure 4.

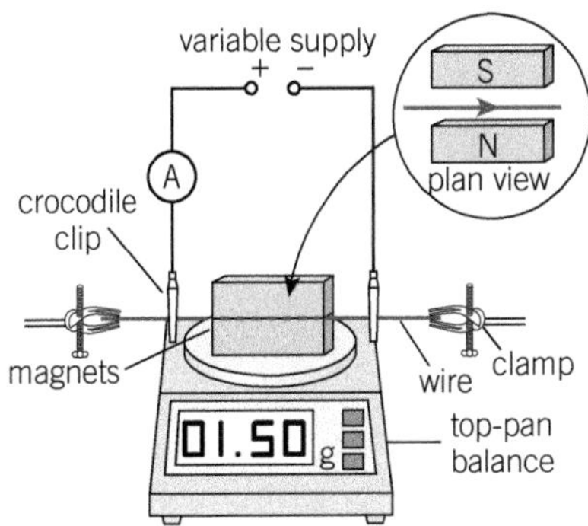

▲ **Figure 4** *An arrangement for determining magnetic flux density. The magnet experiences an equal but opposite force to the current-carrying wire placed between the poles*

Worked example: Magnetic flux density

A copper wire of length 6.0 cm and carrying a current of 8.0 A is placed in a uniform magnetic field. The maximum magnetic force experienced by the wire is 2.9×10^{-2} N. Calculate the magnetic flux density.

Step 1: Write the quantities given in the question.

$F = 2.9 \times 10^{-2}\,\text{N}$ $\quad B = ?$ $\quad I = 8.0\,\text{A}$ $\quad L = 0.060\,\text{m}$

Step 2: Use the equation $F = BIL$ to calculate B.

$$B = \frac{F}{IL} = \frac{2.9 \times 10^{-2}}{8.0 \times 0.060} = 6.0 \times 10^{-2}\,\text{T (2 s.f.)}$$

Summary questions

1. What can you infer about the magnetic flux density in a region that has equally spaced and parallel magnetic field lines? *(1 mark)*
2. Describe the magnetic field of a solenoid. *(2 marks)*
3. A wire of length 4.0 cm is placed perpendicular to a uniform magnetic field of flux density 0.020 T. The current in the wire is 5.0 A. Calculate the magnetic force experienced by the wire. *(2 marks)*
4. A current-carrying wire experiences a force F in a uniform magnetic field. The magnetic flux is trebled and the current in the wire is halved. Determine the new force on the wire in terms F. Explain your answer. *(2 marks)*
5. Two current-carrying copper wires are placed parallel to each other. Explain why each will experience a magnetic force. *(1 mark)*
6. The wire in **Q3** is now placed at an angle θ to the magnetic field. The magnetic force on the wire is 1.5 mN. Calculate θ. *(2 marks)*

23.3 Charged particles in magnetic fields

A current-carrying wire in a uniform magnetic field experiences a force because each electron moving in the wire experiences a tiny force.

Magnetic fields are used in many devices, from oscilloscopes to particle accelerators, to control the paths of charged particles.

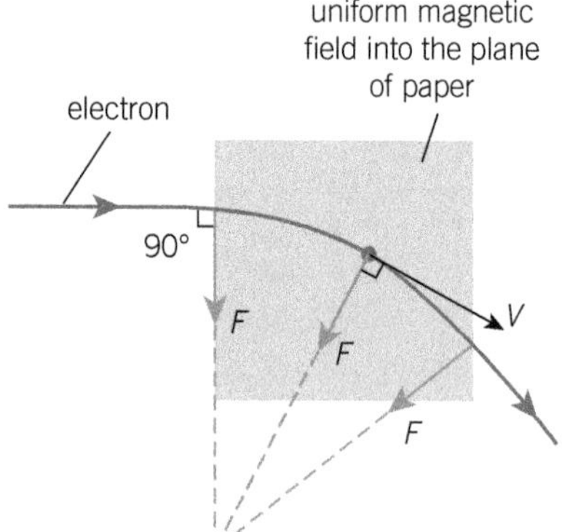

▲ **Figure 1 The** *force F on the electron is perpendicular to its velocity v*

Circular paths

Figure 1 shows the path of a single electron as it travels through a region of uniform magnetic field. The magnetic field is into the plane of the paper. The electron describes a circular path within the magnetic field because the magnetic force F it experiences is perpendicular to its velocity.

No work is done by the magnetic field on the electron because the force is perpendicular to the velocity – this force has no component in the direction of travel and therefore the speed of the electron remains constant.

The magnitude of the magnetic force F experienced by a particle of charge Q moving at right angles to the magnetic field is given by the equation

$$F = BQv$$

where B is the magnetic flux density and v is the speed of the particle.

For an electron, the charge Q is the elementary charge $e = 1.60 \times 10^{-19}$ C. Therefore, the equation for the force F for an electron becomes

$$F = Bev$$

Revision tip

When using Fleming's left-hand rule, do remember that the second finger shows the direction of *conventional* current and that this direction is opposite to the direction of electron flow.

Going round

Consider a charged particle of mass m and charge Q moving perpendicular to a uniform magnetic field of flux density B. The particle will describe a circular path of radius r. The magnetic force BQv provides the centripetal force, therefore

$$BQv = \frac{mv^2}{r} \qquad \text{or} \qquad r = \frac{mv}{BQ}$$

You can also use the equation $v = \frac{2\pi r}{T}$, where T is the period, to analyse the motion of the particle.

Synoptic link

Centripetal force and circular motion were covered in Topic 16.3, Exploring centripetal forces.

Worked example: Circular tracks

Electrons are travelling in a circular path at right angles to a uniform magnetic field.

Show that the period of revolution T of an electron is independent of the radius of the path or its speed, but depends on its mass m, the magnetic flux density B, and the elementary charge e.

Step 1: Write an expression for the centripetal force on the electron.

The centripetal force is provided by the BQv force. Therefore

$$BQv = \frac{mv^2}{r} \qquad \text{or} \qquad BQ = \frac{mv}{r}$$

Step 2: Substitute $v = \frac{2\pi r}{T}$ and $Q = e$ into the expression above.

$$Be = \frac{m}{r} \times \frac{2\pi r}{T} = \frac{2\pi m}{T}$$

Step 3: Simplify the expression and derive an equation for T.

$$T = \frac{2\pi m}{Be}$$

The period T just depends on m, B, and e.

Velocity selector

A velocity selector is a device used in instruments such as a mass spectrometer to select charged particles of a specific speed. Figure 2 shows a simple velocity selector.

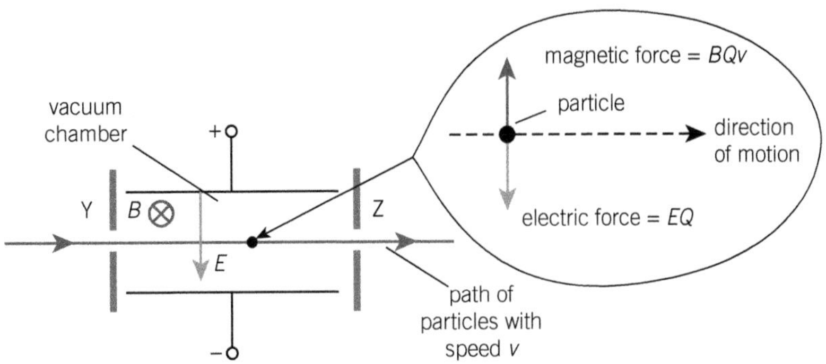

▲ **Figure 2** *Velocity selector has crossed electric and magnetic fields*

Two oppositely charged parallel plates provide a uniform electric field of field strength E. There is also a uniform magnetic field of flux density B at right angles to the electric field.

- The electric force on a charged particle of charge Q is equal to EQ.
- The magnetic force on the charged particle is equal to BQv, where v is the speed of the particle.
- The magnitude of either E or B is adjusted so that the magnetic force and the electric force are equal in magnitude and in opposite directions. Therefore

$$EQ = BQv \qquad \text{or} \qquad v = \frac{E}{B}$$

Any charged particle with the right speed v will travel in a straight line and emerge from the slit Z.

Summary questions

1. Explain why the speed of the electron in Figure 1 does not change. *(2 marks)*
2. Calculate the maximum magnetic force on an electron travelling at a speed of $4.0 \times 10^{6}\,\text{m s}^{-1}$ in a uniform magnetic field of flux density 0.080 T. *(2 marks)*
3. Calculate the acceleration of the electron in **Q2**. *(2 marks)*
4. Use your answer to **Q3** to calculate the radius of the circular path described by the electron. *(2 marks)*
5. An ion of mass 6.7×10^{-27} kg, charge $+2e$, and speed $6.0 \times 10^{5}\,\text{m s}^{-1}$ describes a circular path in a uniform magnetic field of flux density 720 mT. Calculate the radius of the path. *(3 marks)*
6. Calculate the period of revolution of a proton travelling at right angles to a magnetic field of magnetic flux density 900 mT. *(3 marks)*

23.4 Electromagnetic induction
23.5 Faraday's law and Lenz's law
23.6 Transformers

Specification reference: 6.3.3

23.4 Electromagnetic induction

An electromotive force (e.m.f.) is induced in a coil whenever there is relative motion between a magnet and the coil. This is an example of electromagnetic induction.

Important definitions

In order to understand electromagnetic induction, you need to know about magnetic flux density B (see Topic 23.1, Magnetic fields, and Topic 23.2, Understanding magnetic fields), magnetic flux ϕ, and magnetic flux linkage.

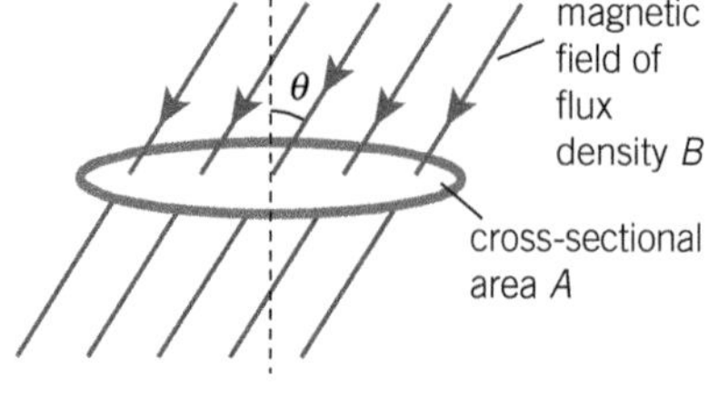

▲ **Figure 1** *Magnetic flux $\phi = BA\cos\theta$*

The **magnetic flux** ϕ is defined as the product of the component of the magnetic flux density B perpendicular to the area and the cross-sectional area A.

$$\phi = BA\cos\theta$$

where θ is the angle between the normal to the coil and the magnetic field, see Figure 1. When the magnetic field is normal to the area, $\cos 0° = 1$ and $\phi = BA$. The SI unit for magnetic flux is the weber (Wb) and $1\,\text{Wb} = 1\,\text{T}\,\text{m}^2$.

Magnetic flux linkage is defined as the product of the number of turns in the coil N and the magnetic flux.

magnetic flux linkage = $N\phi$

The SI unit of magnetic flux linkage is also the weber, but sometimes weber-turns (Wb-turns) is used to avoid confusion with magnetic flux.

23.5 Faraday's law and Lenz's law

An e.m.f. is induced in a circuit whenever there is a *change* in the magnetic flux linkage.

The laws

Faraday's law: The magnitude of the induced e.m.f. is directly proportional to the rate of change of magnetic flux linkage.

$$\varepsilon \propto \frac{\Delta(N\phi)}{\Delta t}$$

where ε is the induced e.m.f and $\Delta(N\phi)$ is the change in magnetic flux linkage in a time interval Δt.

This relationship can be written as an equation with the constant of proportionality equal to –1. Therefore

$$\varepsilon = -\frac{\Delta(N\phi)}{\Delta t}$$

The minus sign is a mathematical way of showing that energy is conserved as stated by Lenz's law.

Lenz's law: The direction of the induced e.m.f. or current is always such as to oppose the change producing it.

A.C. generator

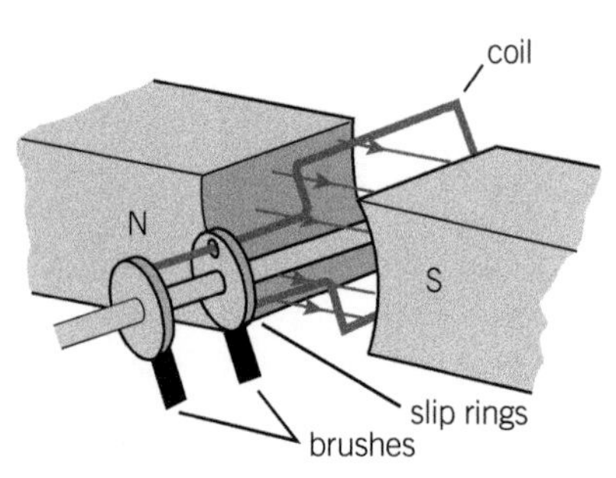

▲ **Figure 2** *A simple A.C. generator*

Figure 2 shows a simple alternating current generator. A coil is rotated at a steady speed in a magnetic field. The magnetic flux linking the coil varies sinusoidally with time. The induced e.m.f. ε in the coil can be determined using Faraday's law, see Figure 3.

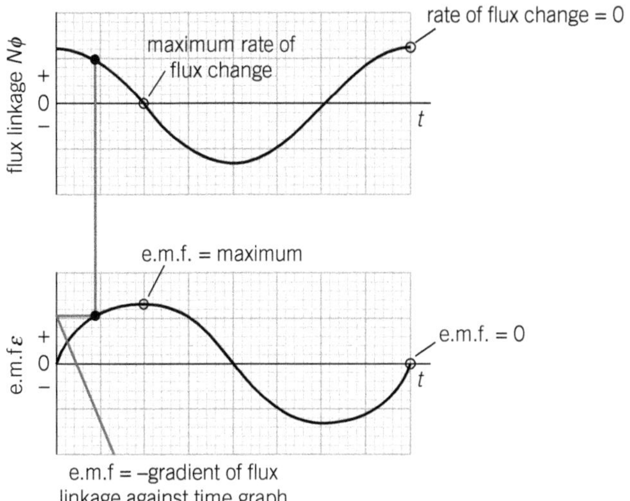

▲ **Figure 3** *The variation of flux linkage with time (above) and of the induced e.m.f. with time (below)*

Worked example: Induced e.m.f in a search coil

A search coil has 2000 turns and a cross-sectional area of $6.4 \times 10^{-3}\,m^2$. It is placed perpendicular to a magnetic field. The magnetic flux density changes from 60 mT to 20 mT in a time of 0.30 s. Calculate the magnitude of the induced e.m.f. in the coil.

Step 1: Write down the quantities given.

initial $B = 60\,mT$ final $B = 20\,mT$ $A = 6.4 \times 10^{-3}\,m^2$ $N = 2000$ turns $\Delta t = 0.30\,s$

Step 2: Use Faraday's law to determine the induced e.m.f. ε.

$$\varepsilon = -\frac{\Delta(N\phi)}{\Delta t} = \frac{[2000 \times 20 \times 10^{-3} \times 6.4 \times 10^{-3}] - [2000 \times 60 \times 10^{-3} \times 6.4 \times 10^{-3}]}{0.30 - 0}$$

$\varepsilon = 1.7\,V$ (2 s.f.)

23.6 Transformers

Figure 4 shows a simple transformer. An alternating current is supplied to the primary (input) coil. This produces a varying magnetic flux in the soft iron core. The iron core ensures that all the magnetic flux created by the primary coil links the secondary (output) coil and none is lost. A varying e.m.f. is produced across the ends of the secondary in accordance with Faraday's law.

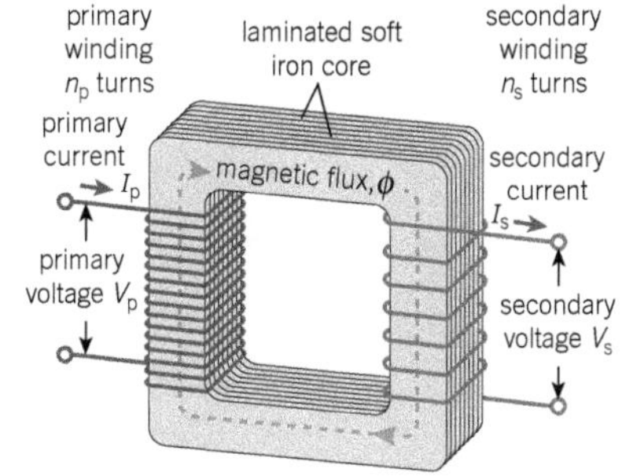

▲ **Figure 4** *A transformer has two coils mounted onto a laminated soft iron core*

Transformers are used to step-up and step-down voltages.

Important equations

The input voltage V_p and the output voltage V_s are related to the number of turns on the primary coil n_p and number of turns on the secondary coil n_s by the **turn-ratio equation**

$$\frac{n_s}{n_p} = \frac{V_s}{V_p}$$

For a 100% efficient transformer

output power = input power

$$V_s I_s = V_p I_p \qquad \text{or} \qquad \frac{V_s}{V_p} = \frac{I_p}{I_s}$$

where I_p is the current in the primary coil and I_s is the current in the secondary coil.

Summary questions

1. What is 1 Wb in terms of T and m? *(1 mark)*
2. The output voltage from a step-up transformer is greater than the input voltage. State which of the coils, primary or secondary, has the greater number of turns. *(1 mark)*
3. A bar magnet is placed inside the core of a coil and is at rest. Explain why there is no induced e.m.f. in the coil. *(2 marks)*
4. A transformer changes a voltage of 230 V to 5.0 V. Compare the number of turns used to make the primary and secondary coils. *(2 marks)*
5. Explain how increasing the frequency of rotation would affect the induced e.m.f. in a generator. *(3 marks)*
6. A coil with 500 turns and a cross-sectional area of $2.0 \times 10^{-4}\,m^2$ is placed perpendicular to a magnetic field. The magnetic flux density is reduced to zero in a time of 0.20 s. The average induced e.m.f. in the coil is 10 mV. Calculate the initial magnetic flux density through the coil. *(3 marks)*

Chapter 23 Practice questions

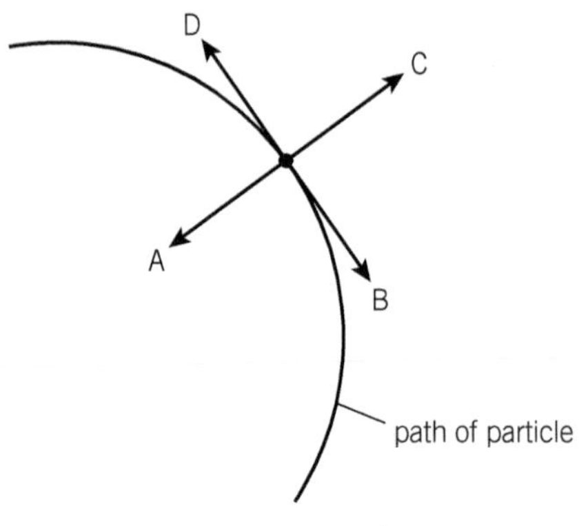

▲ **Figure 1**

1 A charged particle describes a circular path in a uniform magnetic field. Figure 1 shows part of this path.

Which arrow shows the correct direction of the magnetic force acting on the particle? *(1 mark)*

2 A lithium nucleus ${}^{7}_{3}$Li travelling at $2.0 \times 10^{5}\,\mathrm{m\,s^{-1}}$ enters a region of uniform magnetic field of flux density 0.10 T.

What is the maximum magnetic force experienced by this nucleus?

A 3.2×10^{-15} N

B 9.6×10^{-15} N

C 2.2×10^{-14} N

D 3.2×10^{-14} N *(1 mark)*

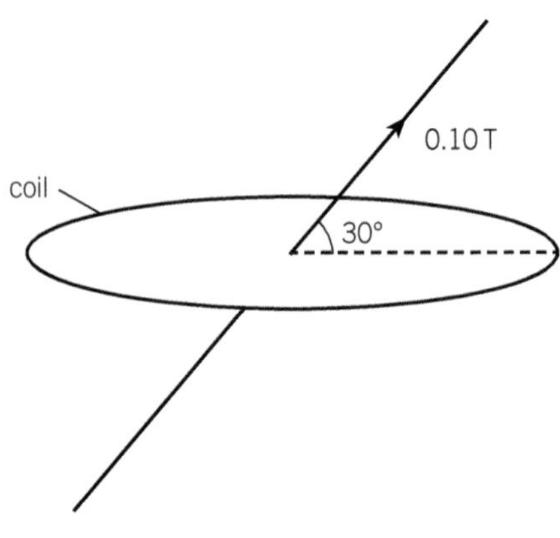

▲ **Figure 2**

3 Figure 2 shows a coil with 200 turns and cross-sectional area $2.0 \times 10^{-3}\,\mathrm{m^2}$ placed in a uniform magnetic field of flux density 0.10 T. The field makes an angle of 30° to the plane of the coil.

What is the magnetic flux linkage for this coil?

A 2.0×10^{-4} Wb

B 2.0×10^{-2} Wb

C 3.5×10^{-2} Wb

D 4.0×10^{-2} Wb *(1 mark)*

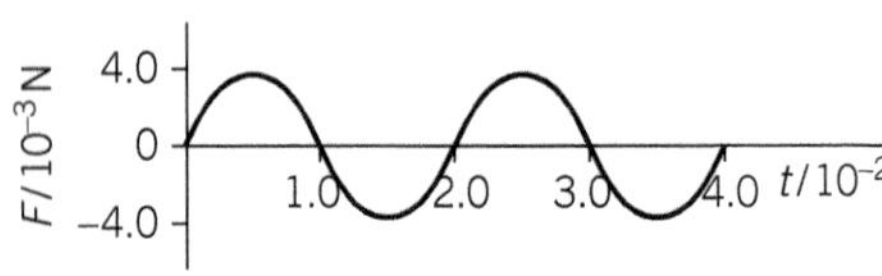

▲ **Figure 3**

4 A copper wire is placed in a uniform magnetic field. The magnetic field is at right angles to the wire. The magnetic flux density is 60 mT. The current in the wire is alternating at a frequency f. Figure 3 shows the variation of the force F experienced by the 1.0 cm length of the wire with time t.

a Determine the frequency f of the current in the wire. *(1 mark)*

b Calculate the maximum current in the wire. *(2 marks)*

c Explain how the graph in Figure 3 would change when the angle between the wire and the magnetic field is slowly reduced. *(2 marks)*

5 A charged particle enters a region of uniform magnetic field. The magnetic field is perpendicular to the velocity of the particle. The particle describes a circular arc in the magnetic field of radius r.

a Show that the momentum p of the charged particle is given by the equation

$$p = BQr$$

where B is the magnetic flux density and Q is the charge of the particle. *(2 marks)*

b An electron describes a circle of radius 3.2 cm in a magnetic field of flux density 0.012 T.

Calculate:

i the momentum of the electron; *(2 marks)*

ii the kinetic energy of the electron in eV. *(3 marks)*

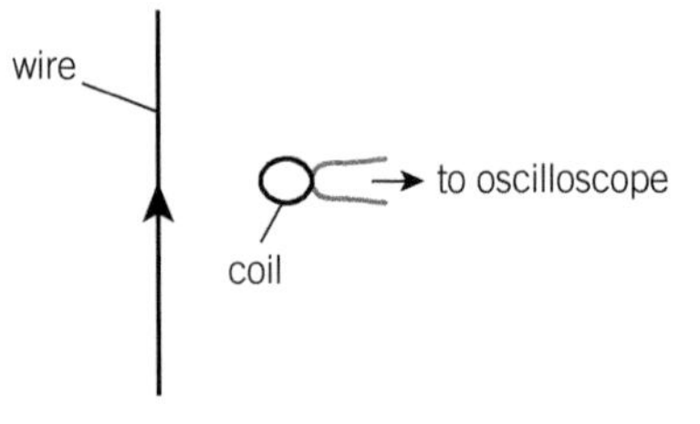

▲ **Figure 4**

6 Figure 4 shows a search coil placed close to a wire. The coil is connected to an oscilloscope.

a Explain why an e.m.f. lasting a short period of time is produced when the current in the wire is switched on. *(3 marks)*

b With the help of a diagram, sketch the magnetic field surrounding the current-carrying wire. *(2 marks)*

Chapter 24 Particle physics

In this chapter you will learn about ...

- ☐ Scattering of alpha particles
- ☐ The nucleus
- ☐ Model of the atom
- ☐ Radius of the nucleus
- ☐ Density of the atom and the nucleus
- ☐ Strong nuclear force
- ☐ Weak nuclear force
- ☐ Particles and antiparticles
- ☐ Hadrons and leptons
- ☐ Quarks
- ☐ Beta decay

24 PARTICLE PHYSICS
24.1 Alpha-particle scattering experiment
24.2 The nucleus

Specification reference: 6.4.1

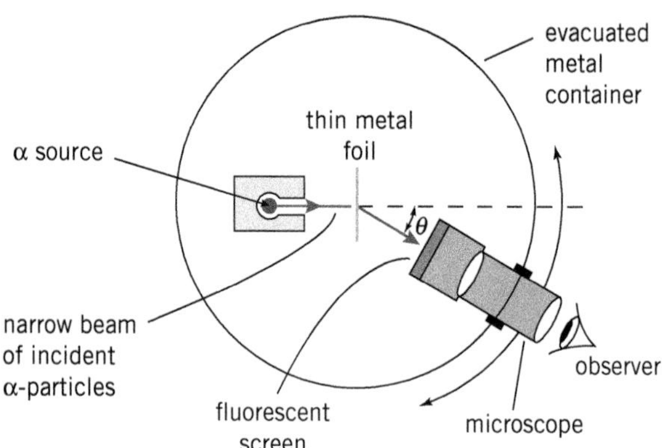

▲ **Figure 1** *Apparatus used to scatter alpha particles*

24.1 Alpha-particle scattering experiments

Before 1911 physicists knew about atoms and also knew their diameter to be about 10^{-10} m. Then, experiments carried out by Rutherford, Geiger, and Marsden on the scattering of alpha particles by thin metal foils provided evidence for the nuclear model of the atom.

Rutherford's scattering experiment

Figure 1 shows the arrangement used to perform the scattering experiments. A beam of alpha particles from a radioactive source were targeted towards a thin gold foil.

▼ **Table 1**

Observations	Conclusions
Most of the alpha particles went straight through the gold foil.	The gold atoms were mostly empty space (vacuum).
Some of the alpha particles were scattered through large angles (> 90°).	Each gold atom has a positive nucleus with a radius less than about 10^{-14} m*. (*A better modern approximation for the radius of the nucleus is about 10^{-15} m.)

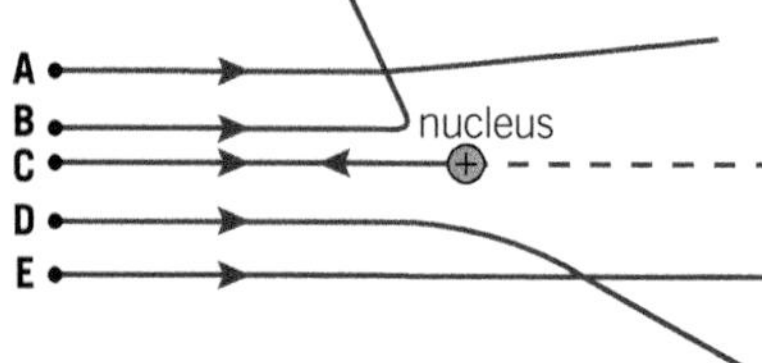

▲ **Figure 2** *Paths of alpha particles. The alpha particle at C is back-scattered and the alpha particle at E is too far away and therefore is not deflected*

An alpha particle is a helium nucleus with a charge of $+2e$, where e is the elementary charge. The charge on the gold nucleus is $+79e$. The scattering of an alpha particle can be explained in terms of electrostatic repulsion by the gold nucleus. The force on the particle can be calculated using the equation $F = \frac{Qq}{4\pi\varepsilon_0 r^2}$. Figure 2 shows the paths of some alpha particles.

24.2 The nucleus

The nucleus has neutrons and protons. The neutron has no charge and the proton has a positive charge $+e$. The term **nucleon** is used to refer to either a proton or a neutron. The nucleus is surrounded by a cloud of electrons. A neutral atom has an equal number of protons and electrons.

Model of the atom

The nucleus of an atom for a particular element is represented as ${}^{A}_{Z}X$, where X is the chemical symbol for the element (e.g., Pb for lead), A is the **nucleon number** (the total number of protons and neutrons), and Z is the **proton number** or the **atomic number**. The number of neutrons is equal to $(A - Z)$.

Isotopes are nuclei of the same element that have the same number of protons but different numbers of neutrons.

Some of the isotopes of oxygen: ${}^{14}_{8}O$ ${}^{15}_{8}O$ ${}^{16}_{8}O$ ${}^{17}_{8}O$ ${}^{18}_{8}O$ ${}^{19}_{8}O$

- The radius R of a nucleus is given by the equation $R = r_0 A^{\frac{1}{3}}$ where A is the nucleon number and r_0 is a constant with a value of about 1.2 fm or 1.2×10^{-15} m.

Revision tip

The diameter of the atom is about 10^{-10} m and the nucleus is 100 000 times smaller with a diameter of about 10^{-15} m.

- In nuclear physics, the mass of a particle is often quoted in **atomic mass units** u instead of kg.

$$1\,u = 1.661 \times 10^{-27}\,kg.$$

The masses of the electron, proton, and neutron are 0.00055 u, 1.00728 u, and 1.00867 u, respectively.

Density

The volume of the nucleus is about 10^{15} times smaller than the volume of the atom. The mass of the **atom** is approximately the same as the mass of the **nucleus** because of the miniscule mass of the electrons. The density of a material is $\frac{\text{mass}}{\text{volume}}$. Therefore, the mean density of the nucleus must be about 10^{15} times the mean density of atoms. The density of nuclear matter is about $10^{17}\,kg\,m^{-3}$.

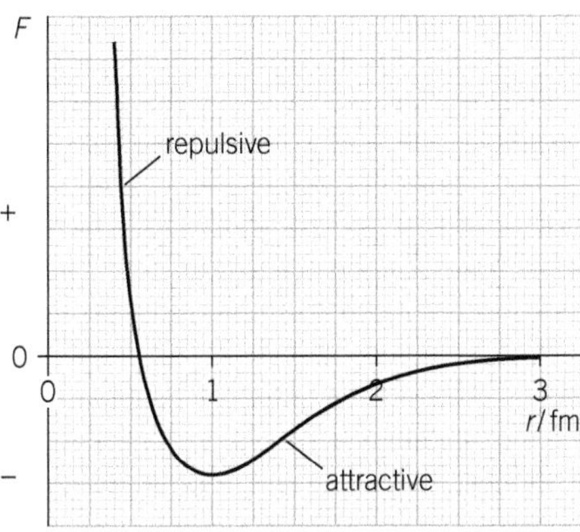

▲ **Figure 3** *Graph showing the variation of the nuclear force between two nucleons with separation r*

Strong nuclear force

The protons within a nucleus experience large electrostatic repulsive forces. The attractive gravitational force between the protons is too small to explain why they remain in the nucleus. All nucleons inside the nucleus experience the **strong nuclear force**. The strong force is attractive to about 3 fm and repulsive below about 0.5 fm, see Figure 3.

Worked example: Very dense

Estimate the density of a proton.

Step 1: Estimate the radius of the proton.

$R = r_0 A^{\frac{1}{3}}$; $A = 1$ for the proton therefore radius $= r_0 = 1.2 \times 10^{-15}\,m$

Step 2: Estimate the density.

mass of proton $\approx 1.7 \times 10^{-27}\,kg$ volume $= \frac{4}{3}\pi R^3$

$$\text{density} = \frac{\text{mass}}{\text{volume}} = \frac{1.7 \times 10^{-27}}{\frac{4}{3}\pi \times (1.2 \times 10^{-15})^3} \approx 2 \times 10^{17}\,kg\,m^{-3}\ (1\text{ s.f.})$$

Summary questions

1 State how many protons and neutrons there are in an oxygen-15 ($^{15}_{8}O$) nucleus. *(2 marks)*

2 Explain why very few alpha particles are scattered through large angles in the scattering experiment. *(2 marks)*

3 Calculate the radius in fm of the $^{4}_{2}He$ nucleus and the $^{235}_{92}U$ nucleus. *(2 marks)*

4 Estimate the number of atoms in a chain of atoms of length 1.0 cm. *(2 marks)*

5 An alpha particle with initial kinetic energy of 7.7 MeV makes a head-on collision with a nucleus of aluminium. The proton number for the aluminium nucleus is 13.
Calculate the minimum separation between these two particles. *(4 marks)*

6 In **Q5**, calculate the electrostatic force experienced by the alpha particle at the minimum separation. *(3 marks)*

24.3 Antiparticles, hadrons, and leptons
24.4 Quarks
24.5 Beta decay

Specification reference: 6.4.2

24.3 Antiparticles, hadrons, and leptons

Most of the matter in the Universe is made up of matter particles. Antimatter particles also exist, but in smaller numbers. An electron is a matter particle and the **positron** is its corresponding antimatter. The positron has the same mass as an electron but, unlike the electron, it has a positive charge of $+e$. When a particle and its **antiparticle** meet, they completely destroy each other and produce high-energy photons. Such an event is called **annihilation**.

The list below shows some of the particle-antiparticle pairs you will come across in this course:

electron–positron proton–antiproton neutron–antineutron neutrino–antineutrino

Synoptic link

There is more information on annihilation in Topic 26.1, Einstein's mass–energy equation.

Fundamental particles

A **fundamental particle** has no internal structure and therefore cannot be divided into smaller particles. The electron is a fundamental particle. Quarks *are* considered to be fundamental particles but protons and neutrons are *not* fundamental particles because they are composed of quarks.

Particles are classified into two families – **hadrons** and **leptons**.

- Hadrons are particles and antiparticles that are affected by the **strong nuclear force**. Examples include protons, neutrons, baryons, and mesons. Charged hadrons also experience the electromagnetic force. Hadrons decay by the **weak nuclear force**.
- Leptons are particles and antiparticles that are not affected by the strong nuclear force. Examples include electrons, neutrinos, and muons. Charged leptons will also experience the electromagnetic force.

Table 1 shows the characteristics of the four fundamental forces or interactions.

▼ **Table 1** *The four fundamental forces or interactions*

Fundamental force	Effect	Relative strength	Range
strong nuclear	experienced by nucleons	1	~ 10^{-15} m
electromagnetic	experienced by static and moving charged particles	10^{-3}	infinite
weak nuclear	responsible for beta-decay	10^{-6}	~ 10^{-18} m
gravitational	experienced by all particles with mass	10^{-40}	infinite

24.4 Quarks

Any particle made up of **quarks** is called a hadron. The standard model of elementary particles requires six quarks (up, down, charm, strange, top, and bottom or simply u, d, c, s, t, and b) and their six anti-quarks to explain the existence of all the hadrons. All quarks have a charge Q that is a fraction of the elementary charge e. For example, the down quark has a charge $-\frac{1}{3}e$ or for simplicity just $-\frac{1}{3}$.

▼ **Table 2** *A summary of the three quarks you need for this course*

Quark			Anti-quark		
Name	Symbol	Charge $\frac{Q}{e}$	Name	Symbol	Charge $\frac{Q}{e}$
up	u	$+\frac{2}{3}$	anti-up	$\bar{u}$	$-\frac{2}{3}$
down	d	$-\frac{1}{3}$	anti-down	$\bar{d}$	$+\frac{1}{3}$
strange	s	$-\frac{1}{3}$	anti-strange	$\bar{s}$	$+\frac{1}{3}$

Revision tip

A bar over the letter for the particle is used to denote most of the antiparticles.

Protons and neutrons

As mentioned earlier, protons and neutrons are hadrons. The hadron group is further divided into **baryons** and **mesons**, see Figure 1. Baryons have a combination of three quarks and mesons have a combination of a quark and an anti-quark.

- A proton has two up quarks and a down quark, or for simplicity, u u d. The total charge of the quarks adds up to $+e$.
- A neutron has one up quark and two down quarks, or u d d. The total charge of the quarks adds up to zero, so a neutron is uncharged.

24.5 Beta decay

Radioactivity is the decay of unstable nuclei. There is more detail on this in Topic 25.1, Radioactivity. Some nuclei decay by emitting beta radiation. There are two types of **beta decays**; beta-minus (β^-) and beta-plus (β^+). The weak nuclear force is responsible for this type of decay.

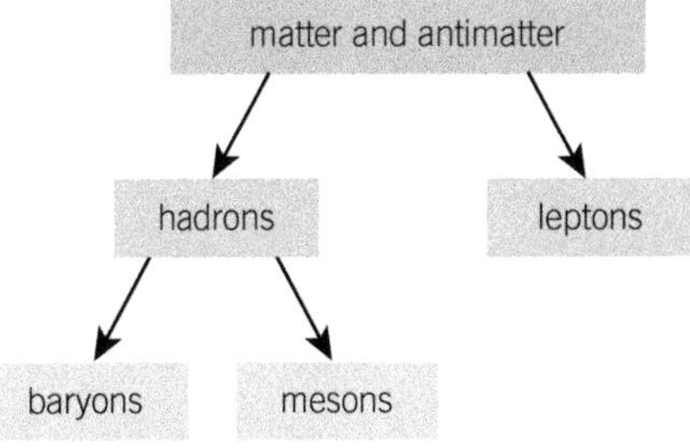

▲ **Figure 1** *Classification of particles*

Transformation of quarks

The two types of beta decays can be summarised by the following decay equations. The **neutrinos** have negligible mass and have no charge. Notice how the nucleon number and the proton number (and therefore charge) are conserved.

Beta-minus (β^-) decay: neutron → proton + electron + electron antineutrino

$$^{1}_{0}n \rightarrow {}^{1}_{1}p + {}^{0}_{-1}e + \overline{\nu_e}$$

At a quark level, one of the down quarks changes into an up quark:

$$u\,d\,d \rightarrow u\,u\,d + {}^{0}_{-1}e + \overline{\nu_e} \qquad \text{or} \qquad d \rightarrow u + {}^{0}_{-1}e + \overline{\nu_e}$$

Beta-plus (β^+) decay: proton → neutron + positron + electron neutrino

$$^{1}_{1}p \rightarrow {}^{1}_{0}n + {}^{0}_{+1}e + \nu_e$$

At a quark level, one of the up quarks changes into a down quark:

$$u\,u\,d \rightarrow u\,d\,d + {}^{0}_{+1}e + \nu_e \qquad \text{or} \qquad u \rightarrow d + {}^{0}_{+1}e + \nu_e$$

Summary questions

1 Explain what is meant by fundamental particles. Give two examples. *(2 marks)*
2 State the force experienced by quarks within a stable hadron. *(1 mark)*
3 Write the decay equation for a neutron decaying into a proton and state two quantities conserved. *(3 marks)*
4 Describe the nature of the gravitational force and the strong nuclear force acting between two protons. *(2 marks)*
5 Use Table 2 to show that the charge on a neutron is zero. *(2 marks)*
6 Complete the reaction: $uud \rightarrow ??d + {}^{0}_{+1}e + ?$ *(2 marks)*
7 Suggest how we know that there must be more matter than antimatter in the universe. *(2 marks)*
8 Calculate the following ratio for two protons inside a nucleus: $\frac{\text{gravitational force on a proton}}{\text{electrostatic force on a proton}}$. *(4 marks)*

Chapter 24 Practice questions

1 How many neutrons are there in the nucleus of $^{7}_{4}\text{Li}$?

A 3

B 4

C 7

D 10 *(1 mark)*

2 Which statement is correct about beta-minus decay?

A A proton transforms into a neutron.

B A neutron transforms into a proton.

C A neutron transforms into an up quark.

D A proton transforms into a down quark. *(1 mark)*

3 In the alpha-scattering experiment, most of the alpha particles fired into a foil went straight though without any deflection.

What can be deduced from this observation?

A Gold atoms are mainly vacuum.

B Alpha particles are small particles.

C Alpha particles have positive charge.

D Gold nucleus is surrounded by electrons. *(1 mark)*

4 The nucleus of $^{16}_{8}\text{O}$ has more nucleons than the nucleus of $^{4}_{2}\text{He}$.

Which statement is correct about the density of these two nuclei?

The density of the oxygen nucleus is:

A greater than the helium nucleus

B smaller than the helium nucleus

C the same as that of the helium nucleus

D four times that of the helium nucleus. *(1 mark)*

5 Figure 1 shows a graph of radius R of a nucleus and its nucleon number A.

a Use the graph to confirm the relationship between R and A. *(2 marks)*

b The mass of the proton and the neutron is about the same.

Explain how the mass, volume, and density of a nucleus depends on its nucleon number A. *(3 marks)*

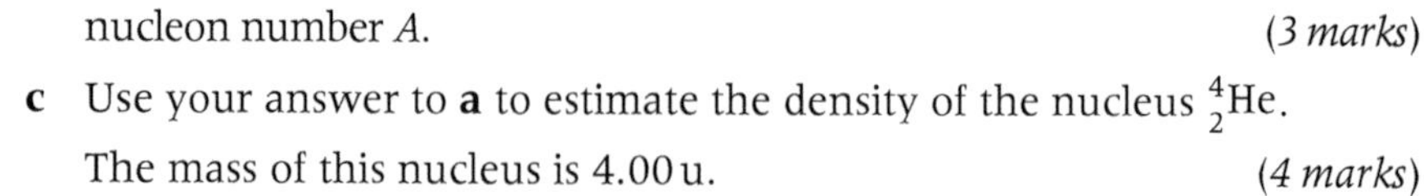

c Use your answer to **a** to estimate the density of the nucleus $^{4}_{2}\text{He}$.

The mass of this nucleus is 4.00 u. *(4 marks)*

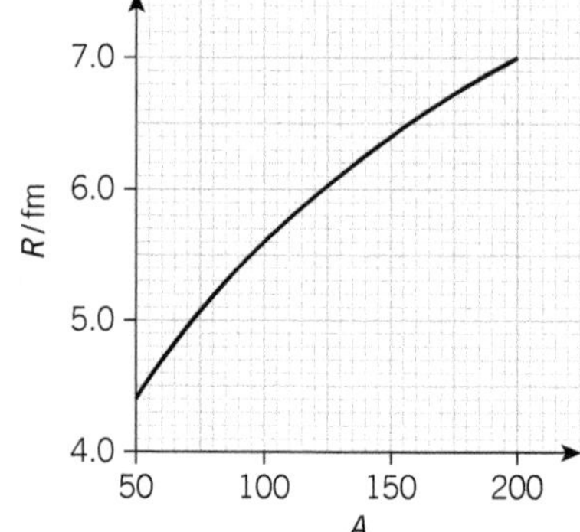

▲ **Figure 1**

6 Alpha particles, each with kinetic energy 8.2×10^{-13} J, are fired into a thin gold foil. The atomic number of gold is 79.

a Figure 2 shows the path of one alpha particle as it travels past a gold nucleus.

Explain why the path shows greater curvature when the alpha particle is closer to the gold nucleus. *(1 mark)*

b Show that the closest distance an alpha particle can get to the gold nucleus is about 4.4×10^{-14} m. *(3 marks)*

c Calculate the electrostatic force acting on the gold nucleus when the alpha particle is 4.4×10^{-14} m from the gold nucleus. *(2 marks)*

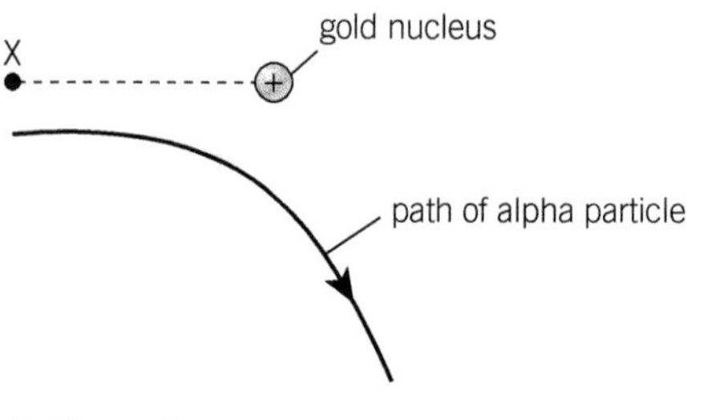

▲ **Figure 2**

Chapter 25 Radioactivity

In this chapter you will learn about ...

- [ ] Radioactivity
- [ ] Alpha, beta, and gamma radiations
- [ ] Nuclear transformation equations
- [ ] Activity
- [ ] Half-life
- [ ] Decay constant
- [ ] Exponential decay equation
- [ ] Modelling radioactive decay using spreadsheets
- [ ] Radioactive dating

25 RADIOACTIVITY
25.1 Radioactivity
25.2 Nuclear decay equations

Specification reference: 6.4.3

25.1 Radioactivity

Most nuclei of atoms are stable but some are unstable and consequently emit nuclear radiations in the form of particles or high-energy gamma photons. All nuclear radiations are described as **ionising radiations** because they can ionise atoms by removing some of their electrons, leaving behind positive ions.

There are four different types are radiation: alpha (α), beta-minus (β^-), beta-plus (β^+), and gamma (γ).

Revision tip
All types of radiation cause ionisation and therefore can damage living cells.

Properties and characteristics of the radiations

Alpha

- Each alpha particle is a helium nucleus (two proton and two neutrons).
- An alpha particle has mass 4.00151 u and a charge $+2e$ – this makes it very ionising.
- Alpha particles have a very short range in air and are easily stopped by a thin sheet of paper.

Beta-minus

- Each beta-minus particle is an electron of mass 0.00055 u and a charge $-e$.
- Beta-minus particles are less ionising than alpha particles.
- Beta-minus particles can be stopped by about 1–3 mm of aluminium.

Beta-plus

- Each beta-plus particle is a positron of mass 0.00055 u and a charge $+e$.
- Positrons get annihilated by electrons, and therefore cannot travel too far.

Gamma

- Gamma photons have no charge and travel at the speed of light.
- Gamma photons are poor ionisers.
- Most of the gamma photons can be stopped by lead of thickness of a few cm.

Revision tip
Alpha and beta particles are affected by electric and magnetic fields because they are charged particles.

25.2 Nuclear decay equations

The nucleon number and the proton number (and therefore charge) are conserved in all nuclear reactions and decays. The other quantities conserved are mass–energy (see Topic 26.1, Einstein's mass–energy equation) and momentum.

The original nucleus before the decay is known as the **parent nucleus** and the transformed nucleus after the decay is known as the **daughter nucleus**.

Alpha decay

- An unstable nucleus with more than 82 protons is likely to decay by emitting an alpha particle.
- A helium nucleus is emitted from the parent nucleus.
- The daughter nucleus is of a different element.
- Decay equation: $^{A}_{Z}\mathrm{X} \rightarrow {}^{A-4}_{Z-2}\mathrm{Y} + {}^{4}_{2}\mathrm{He}$

Beta-minus decay

- An unstable nucleus with 'too many' neutrons (neutron-rich) is likely to decay by emitting an electron.
- A neutron inside the parent nucleus changes into a proton, electron, and an electron antineutrino.
- Decay equation: $^{A}_{Z}X \rightarrow {}^{A}_{Z+1}Y + {}^{0}_{-1}e + \overline{\nu_e}$

Beta-plus decay

- An unstable nucleus with 'too many' protons (proton-rich) is likely to decay by emitting a positron.
- A proton inside the parent nucleus changes into a neutron, positron, and an electron neutrino.
- Decay equation: $^{A}_{Z}X \rightarrow {}^{A}_{Z-1}Y + {}^{0}_{+1}e + \nu_e$

Gamma decay

- Surplus energy within the nucleus is released as a gamma photon.
- Decay equation: $^{A}_{Z}X \rightarrow {}^{A}_{Z}X + \gamma$

Table 1 summaries the changes in the nucleon number A and proton number Z as a parent nucleus decays into a daughter nucleus.

▼ **Table 1** *Changes in nucleon and proton numbers*

Decay	Effect on A	Effect on Z
α	Decreases by 4	Decreases by 2
β^-	No change	Increases by 1
β^+	No change	Decreases by 1
γ	No change	No change

Worked example: A decay chain

A nucleus of uranium-238 ($^{238}_{92}U$) eventually transforms into a stable nucleus of lead (Pb) after emitting 8 alpha particles and 6 beta-minus particles. Predict this isotope of lead.

Step 1: Calculate the changes in A and Z due to the alpha-emissions.

decrease in $A = 4 \times 8 = 32$ decrease in $Z = 2 \times 8 = 16$

Step 2: Calculate the changes in A and Z due to the beta-emissions.

change in $A = 0$ increase in $Z = 1 \times 6 = 6$

Step 3: Determine the A and Z values of the lead isotope.

$A = 238 - 32 = 206$ $Z = 92 - 16 + 6 = 82$

The isotope of lead is $^{206}_{82}Pb$.

Summary questions

1 The nucleus of magnesium-28 decays as follows: $^{28}_{12}Mg \rightarrow {}^{28}_{13}Al + {}^{0}_{-1}e + \overline{\nu_e}$
Identify the particles $^{0}_{-1}e$ and $\overline{\nu_e}$. *(2 marks)*

2 For the decay shown in **Q1**:

a state two numbers conserved; *(1 mark)*

b calculate the number of neutrons in the nucleus of $^{28}_{12}Mg$ and in the nucleus of $^{28}_{13}Al$. *(2 marks)*

3 Complete the following decay equations:

a $^{204}_{82}Pb \rightarrow {}^{?}_{?}Hg + {}^{4}_{2}He$ *(1 mark)*

b $^{?}_{?}Cf \rightarrow {}^{245}_{96}Cm + {}^{4}_{2}He$ *(1 mark)*

4 Complete the following decay equations:

a $^{19}_{8}O \rightarrow {}^{?}_{?}F + {}^{0}_{-1}e + \overline{\nu_e}$ *(1 mark)*

b $^{21}_{?}Na \rightarrow {}^{?}_{10}Ne + {}^{0}_{+1}e + \nu_e$ *(2 marks)*

5 The count rate from a gamma source in air obeys an inverse square law with distance. A Geiger-Müller tube with an opening of area 2.0×10^{-4} m^2 placed at a distance of 30 cm detects 120 gamma ray photons (counts). Estimate the number of gamma photons emitted by the source per second. *(3 marks)*

6 A single alpha particle produces about 10^4 ions per mm in air. It takes about 10 eV to ionise an atom. Estimate the speed of an alpha particle that has a range of 3.0 cm in air.
mass of alpha particle = 6.6×10^{-27} kg *(4 marks)*

25.3 Half-life and activity
25.4 Radioactive decay calculations

Specification reference: 6.4.3

25.3 Half-life and activity

Natural **radioactivity** is the *random* and *spontaneous* decay of unstable nuclei. The decay is random because we cannot predict when a particular nucleus in a substance will decay or which one will decay next – each nucleus within a substance has the same chance of decaying per unit time. The decay is spontaneous because the decay of nuclei is not affected by the presence of other nuclei in the substance or external factors (e.g., pressure).

Half-life, activity, and decay constant

The **half-life** $t_{\frac{1}{2}}$ of an isotope is the average time it takes for half the number of active nuclei in the sample to decay.

Even isotopes of the same element can have a wide range of half-lives. This is illustrated in Table 1 for some of the isotopes of carbon.

▼ **Table 1** *Half-lives of some carbon isotopes*

Isotope	${}^{9}_{6}C$	${}^{10}_{6}C$	${}^{11}_{6}C$	${}^{12}_{6}C$	${}^{13}_{6}C$	${}^{14}_{6}C$	${}^{15}_{6}C$
Half-life	0.126 s	19.3 s	20.3 min	stable	stable	5730 y	2.45 s

The **activity** A of a substance is the rate at which the nuclei decay (disintegrate).

The SI unit of **activity** is the **becquerel** (Bq). 1 Bq = 1 decay per second.

What is meant by a beta-emitting sample having an activity of 2000 Bq? You could imagine 2000 nuclei in the sample decaying per second or 2000 beta-particles emitted per second.

Consider a source with a known isotope and where the parent nuclei decay into stable daughter nuclei. The activity A of a source is directly proportional to the number of undecayed nuclei N left in the source. The activity is given by the equation

$$A = \lambda N$$

where λ is the decay constant of the isotope. The unit for decay constant is normally s^{-1}, but min^{-1}, h^{-1}, and y^{-1} are frequently used.

The **decay constant** can be defined as the probability of decay of an individual nucleus per unit time.

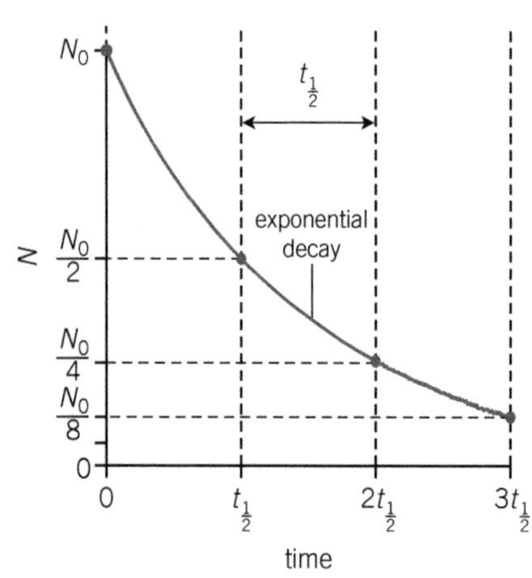

▲ **Figure 1** *A graph of number of undecayed nuclei N against time showing exponential decay*

25.4 Radioactive decay calculations

The decay of a radioactive substance follows an exponential pattern similar to the discharge of a capacitor. The rate of decay of a radioactive substance depends on the half-life and the decay constant of the isotope.

Exponential decay equation

The number of undecayed nuclei N left in the source at time t is given by the equation

$$N = N_0 e^{-\lambda t}$$

where N_0 is the number of undecayed nuclei at time $t = 0$ and e is the base of natural logarithms, 2.718. The number of undecayed nuclei decreases exponentially with time, see Figure 1.

The activity A of the source is directly proportional to N. Therefore

$$A = A_0 e^{-\lambda t}$$

where A_0 is the activity at time $t = 0$.

The decay constant λ and half-life $t_{\frac{1}{2}}$ are inversely proportional to each other and are given by the equation

$$\lambda t_{\frac{1}{2}} = \ln(2)$$

The half-life of the carbon-14 isotope is 5730 y. You can show that its decay constant is $1.21 \times 10^{-4}\,y^{-1}$ or $3.83 \times 10^{-12}\,s^{-1}$.

Revision tip

Instead of the equation $N = N_0 e^{-\lambda t}$ you can use $N = (0.5)^n \times N_0$ where n is the number of half-lives elapsed.

Worked example: Activity

A sample of a radioactive substance has 6.0×10^{10} nuclei of an isotope at time $t = 0$. The decay constant of the isotope is $2.0 \times 10^{-3}\,s^{-1}$. Calculate the activity at time $t = 6.0$ minutes.

Step 1: Write down the information given.

$N_0 = 6.0 \times 10^{10}$ $\quad \lambda = 2.0 \times 10^{-3}\,s^{-1}$ $\quad t = 6.0 \times 60 = 360\,s$ $\quad A = ?$

Step 2: Calculate the initial activity A_0.

$A_0 = \lambda N_0 = 2.0 \times 10^{-3} \times 6.0 \times 10^{10} = 1.2 \times 10^8\,Bq$

Step 3: Calculate the activity at time $t = 360\,s$.

$A = A_0 e^{-\lambda t} = 1.2 \times 10^8 \times e^{-2.0 \times 10^{-3} \times 360} = 5.8 \times 10^7\,Bq$ (2 s.f.)

Summary questions

1 What is the relationship between decay constant and half-life? *(1 mark)*

2 An alpha-emitting source has an activity of 120 Bq. Estimate the number of alpha particles emitted in a period of 2.0 s and state any assumption made. *(2 marks)*

3 A source has an isotope of half-life 2.0 minutes. The sample has 8.0×10^{15} undecayed nuclei of the isotope.
Calculate:
  a the number of nuclei left in the source after 4.0 minutes; *(2 marks)*
  b the total number of nuclei decayed in the source after 6.0 minutes. *(3 marks)*

4 The activity of an alpha-emitting source is 3.4×10^{10} Bq. The kinetic energy of each alpha particle is 1.5 MeV. Calculate the power of the source. *(3 marks)*

5 The isotope of thorium-234 has a half-life of 6.7 h. Calculate the activity of a sample of thorium of mass 1.5 mg. (The molar mass of thorium-234 is 0.234 kg mol^{-1}.) *(4 marks)*

6 Use the information given in Table 1 to calculate the time in years it takes for the activity of a small sample of carbon-14 to decrease to 72 % of its initial activity. *(4 marks)*

25.5 Modelling radioactive decay
25.6 Radioactive dating

Specification reference: 6.4.3

25.5 Modelling radioactive decay

The activity A of a source is given by the equation $A = \lambda N$, where λ is the decay constant of the isotope and N is the number of undecayed nuclei of the isotope in the source. This equation can be mathematically written as

$$\frac{\Delta N}{\Delta t} = -\lambda N$$

where the activity A, which is the rate of decay of nuclei, is written as $\frac{\Delta N}{\Delta t}$ and the minus sign is necessary because the number of active nuclei in the source decreases with time t. You have already met the 'solution' of the rate of decay equation, $N = N_0 e^{-\lambda t}$.

You can also use a spreadsheet method to estimate the number of nuclei left in the source.

Using spreadsheets

Imagine there are N nuclei in a sample. The number of nuclei decaying in a very small interval of time Δt is given by $\Delta N = (\lambda \Delta t)N$. The number of nuclei left in the sample after Δt will be ΔN subtracted from the initial number. This process can be repeated to determine the number of nuclei N at time t.

This is easily illustrated in the example below.

$N_0 = 6.00 \times 10^{10}$ nuclei $\quad t_{\frac{1}{2}} = 30$ s $\quad \lambda = \frac{\ln(2)}{30} = 0.0231\ \text{s}^{-1}$ $\quad \Delta t = 1.00$ s

$$\Delta N = (\lambda \Delta t) \times N = (0.0231 \times 1.00)\, N$$

or

$$\Delta N = 0.0231\, N$$

This means that 2.31 % of the nuclei have decayed leaving 97.69 % of the previous number of nuclei.

time /s	0.00	1.00	2.00	3.00	4.00	5.00	6.00	
$N/10^{10}$	6.00	5.86	5.72	5.59	5.46	5.34	5.21	....etc .

$N = 0.9769 \times 6.00 \times 10^{10}$

$N = 0.9769^2 \times 6.00 \times 10^{10}$

The number of nuclei N left in the sample can be determined accurately by making the interval Δt even smaller, and this is where a spreadsheet is helpful.

25.6 Radioactive dating

The isotope of carbon-14 has a half-life of 5730 years. The isotopes carbon-12 (stable) and carbon-14 are used in the technique of **carbon-dating** to determine the age of organic materials. The actual number of carbon-14 and carbon-12 nuclei in a small sample of organic material can be determined accurately using mass spectrometers

Carbon-dating

The ratio of carbon-14 nuclei to carbon-12 nuclei in atmospheric carbon is almost constant at about 1.3×10^{-12}.

The ratio is the same in all living (organic) things. Once the living thing (e.g., a tree or a person) dies, it stops taking in carbon. The carbon-14 nuclei within the dead material decreases exponentially with time. The age of the organic material can be determined by comparing ratios of carbon-14 nuclei to carbon-12 nuclei of the dead material and a similar living material. See the worked example below.

The ratio of carbon-14 nuclei to carbon-12 nuclei cannot be assumed to be constant over time. This ratio is affected by natural events such as volcanic eruptions and increased carbon-emission from cars and burning fossil fuels. The value of this ratio in the past is not known. This introduces a large uncertainty in the age of a relic.

Revision tip

The isotope of rubidium-87, with half-life of 49 billion years, is used to determine the age of ancient rocks on the Earth.

Worked example: Age of a relic

A mass spectrometer is used to determine the ratio r of carbon-14 nuclei to carbon-12 nuclei in a living tree and a wooden axe found in a cave. Use the information below to determine the age of the axe.

$r = 1.3 \times 10^{-12}$ for the living tree $\qquad r = 7.2 \times 10^{-13}$ for the axe

Step 1: Determine the percentage of carbon-14 nuclei left in the axe.

$$\text{\% of carbon-14 left in the axe} = \frac{7.2\times10^{-13}}{1.3\times10^{-12}}\times100 = 55.4\%$$

Step 2: Use $N = N_0 e^{-\lambda t}$ to determine the time t.

$$\lambda = \frac{\ln(2)}{5730} = 1.21\times10^{-4}\,\text{y}^{-1}$$

$$\frac{N}{N_0} = 0.554 = e^{-\lambda t}$$

$$t = -\frac{\ln(0.554)}{1.21\times10^{-4}} = 4880\,\text{y} = 4900\,\text{y (2 s.f.)}$$

Summary questions

1. A source contains an isotope of half-life 10 s. The source has 1000 nuclei of this isotope.
   Suggest why a time interval Δt of 5.0 s is not suitable when using the equation $\Delta N = (\lambda \Delta t)N$. *(1 mark)*
2. **a** Calculate the decay constant of the isotope in **Q1**. *(1 mark)*
   b Estimate the number of nuclei decaying in an interval of 0.10 s. *(2 marks)*
3. A sample of ancient wood has an activity of 0.10 Bq. A sample of living wood of the same mass has an activity of 0.40 Bq. Calculate the age of the ancient wood. *(2 marks)*
4. The activity of a living wood from the decay of carbon-14 nuclei is 1.7 Bq.
   Calculate the number of carbon-14 nuclei in this living wood. *(3 marks)*
5. A certain sample of dead wood is found to have an activity of 0.32 Bq. An identical mass of living wood has an activity of 1.6 Bq. Calculate the age of the dead wood. *(4 marks)*
6. The isotope of rubidium-87 has a half-life of 49 billion years. A sample of rock on the Earth is found to have 94% of its rubidium-87 nuclei left. Estimate the age of the Earth. *(4 marks)*

Chapter 25 Practice questions

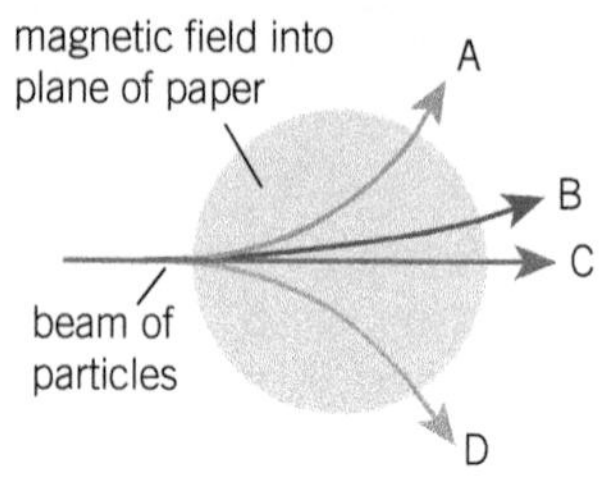

▲ **Figure 1**

1 A beam of radiation from a mixture of radioactive substances is passed through a uniform magnetic field, see Figure 1.

Which is the correct path described by the alpha particles? *(1 mark)*

2 The table below shows the number of nuclei N in the substance and the decay constant λ of the isotope of four substances A, B, C, and D.

Substance	A	B	C	D
$N / 10^{16}$	2.0	4.0	6.0	8.0
$\lambda / 10^{-6}\,s^{-1}$	7.0	5.0	3.0	1.0

Which substance has the greatest activity? *(1 mark)*

3 The nucleus of $^{233}_{91}Pa$ emits a beta-minus particle and then an alpha particle. It transforms into a nucleus of thorium Th.

Which is the correct transformed isotope of thorium?

A $^{228}_{89}Th$

B $^{229}_{89}Th$

C $^{229}_{90}Th$

D $^{233}_{90}Th$ *(1 mark)*

4 The activity of a radioactive substance is 700 Bq.

What is the activity of the substance after 3.5 half-lives?

A 57 Bq

B 62 Bq

C 88 Bq

D 200 Bq *(1 mark)*

5 This question is about the radioactive decay of carbon-10 nuclei in a source.

Figure 2 shows the activity A of the source from time $t = 30$ s to about $t = 45$ s.

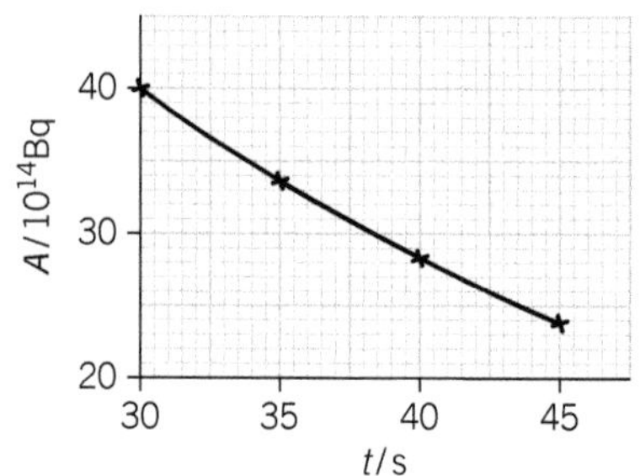

▲ **Figure 2**

a Use Figure 2 to show that the activity decreases exponentially with time. *(2 marks)*

b Use Figure 2 to show that the half-life of carbon-10 ($^{10}_{6}C$) is about 20 s. *(4 marks)*

c Carbon-10 is a beta-plus emitter.

Write the decay equation for the carbon-10 nucleus. *(2 marks)*

d The kinetic energy of each beta-plus particle is 3.4×10^{-13} J.

Calculate the power of the source at time $t = 30$ s. *(2 marks)*

6 **a** A student uses a Geiger-Müller tube and a counter to record the count rate from a radioactive source. Another student suggests the '*count rate is the activity of the source*'. *(2 marks)*

Discuss whether this suggestion is correct or not.

b A sample of a radioactive isotope contains 2.0×10^9 active nuclei. The half-life of the isotope is 6.0 hours. Calculate:

i the initial activity of the sample; *(3 marks)*

ii the number of active nuclei of the isotope remaining after 12 hours; *(2 marks)*

iii the number of active nuclei of the isotope remaining after 21 hours. *(2 marks)*

Chapter 26 Nuclear physics

In this chapter you will learn about ...

- ☐ Einstein's mass–energy equation
- ☐ Radioactivity
- ☐ Annihilation
- ☐ Creation of matter
- ☐ Binding energy
- ☐ Mass defect
- ☐ Binding energy per nucleon
- ☐ Nuclear fission
- ☐ Chain reaction
- ☐ Fission reactors
- ☐ Nuclear fusion

26 NUCLEAR PHYSICS
26.1 Einstein's mass–energy equation
26.2 Binding energy

Specification reference: 6.4.4

26.1 Einstein's mass–energy equation

Einstein's mass–energy equation is $\Delta E = \Delta mc^2$, where ΔE is the change in energy, Δm is the change in mass, and c is the speed of light in a vacuum. Mass and energy are equivalent. In nuclear reactions, it is 'mass–energy' that is conserved and not just 'energy'.

Two important outcomes for a system:

- Mass increases when energy increases.
- Mass decreases when energy decreases.

When you are travelling in a fast car, your mass is greater than your mass at rest (**rest mass**). The change in mass is incredibly small and therefore not noticeable. However, the mass of an electron in a particle accelerator can increase by a substantial amount.

Radioactivity

In natural radioactivity where does the energy of the emitted particles and photons come from?

The energy comes from matter being converted into energy. In all the nuclear decays below, the mass of the parent nucleus is *greater* than the total mass of the 'product' particles.

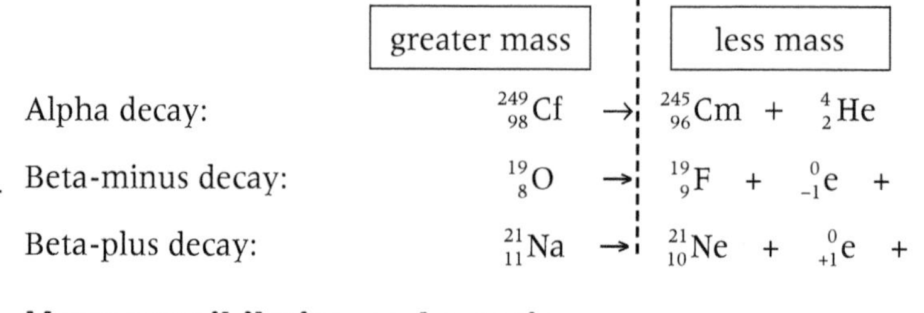

Matter annihilation and creation

Energy can be created from matter. When a particle and its corresponding antiparticle meet, they completely destroy each other and their entire mass is converted into two identical gamma photons. This process is called **annihilation**, see Figure 1a. The energy of each photon is mc^2, where m is the mass of the particle or the antiparticle. Electron–positron annihilation is used in PET scanners.

Matter can be created from energy. A single gamma photon of energy is equal to or greater than $2mc^2$ can produce a particle and an antiparticle, each of mass m. This process is known as **pair production.** See Figure 1b.

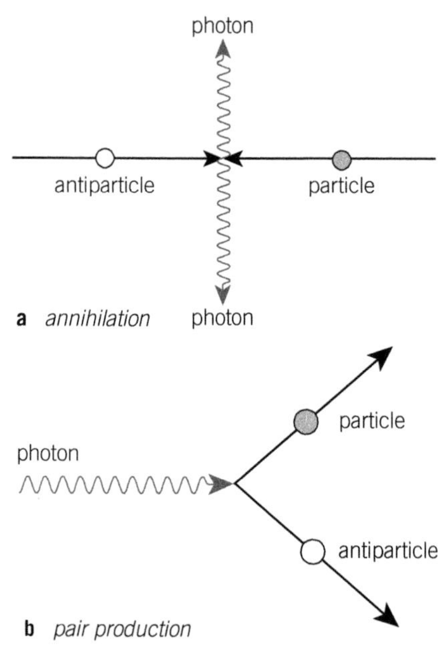

▲ **Figure 1** (*a*) *Annihilation.* (*b*) *Pair production*

Synoptic link

You will learn more about PET scanners in Topic 27.5, PET scans.

Worked example: Matter creation

Calculate the minimum energy in MeV of a photon capable of creating an electron–positron pair.

Step 1: Write down the known quantities.

mass of electron = mass of positron = 9.11×10^{-31} kg $c = 3.00 \times 10^8\ \text{m s}^{-1}$

Step 2: Use Einstein's mass–energy equation to calculate the energy of the photon.

$\Delta E = \Delta mc^2 = 2 \times 9.11 \times 10^{-31} \times (3.00 \times 10^8)^2 = 1.64 \times 10^{-13}$ J

Step 3: Convert the energy into MeV. 1 eV = 1.60×10^{-19} J.

$$\text{energy of photon} = \frac{1.64 \times 10^{-13}}{1.60 \times 10^{-19}} = 1.02 \times 10^{6}\ \text{eV}$$

energy of photon = 1.02 MeV (3 s.f.)

26.2 Binding energy

The mass of a nucleus is always *less* than the total mass of its separate nucleons (neutrons and protons). The nucleons are held together within the nucleus by the strong nuclear force. According to the mass–energy equation, external energy is required to pull apart these nucleons.

Mass defect and binding energy

Binding energy of a nucleus is related to its mass defect.

The **binding energy** of a nucleus is defined as the minimum energy required to completely separate a nucleus into its constituent protons and neutrons.

The **mass defect** of a nucleus is defined as the difference between the mass of the completely separated nucleons and the mass of the nucleus.

To calculate the binding energy of a nucleus, you can use the mass–energy equation.

$$\text{binding energy of nucleus} = \text{mass defect of nucleus} \times c^2$$

Binding energy per nucleon

The average **binding energy per nucleon** is useful when comparing different nuclei. The greater the binding energy per nucleon, the more tightly bound are the nucleons within the nucleus. Figure 2 shows the binding energy (BE) per nucleon against nucleon number A for nuclei.

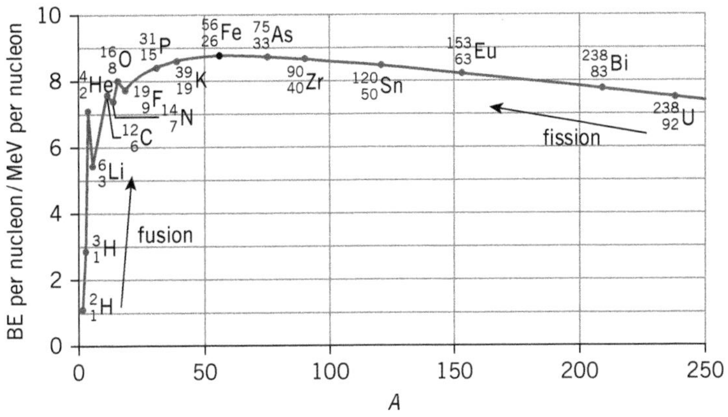

- The isotope of iron-56 has the greatest BE per nucleon – its nucleons are the most tightly bound together.
- In natural radioactive decay, the BE of the product particles is greater than the BE of the parent nucleus. The difference in the binding energies is the energy released.
- BE increases in fission and fusion processes and therefore energy is released. There is more detail in Topic 26.3, Nuclear fission, and Topic 26.4, Nuclear fusion.

▲ **Figure 2** *BE per nucleon against nucleon number graph for nuclei*

Summary questions

1. Convert a mass of 1.0 mg into energy in joules. *(2 marks)*
2. Convert the mass of an electron into energy in joules. *(2 marks)*
3. State and explain the change in the mass of the following systems:
   - **a** A lump of hot iron cooling in a room. *(2 marks)*
   - **b** An electron slowing down. *(2 marks)*
   - **c** A proton getting faster in an accelerator. *(2 marks)*
4. Use Figure 2 to estimate the binding energy of the following nuclei in MeV.
   a $^{2}_{1}$H **b** $^{4}_{2}$He **c** $^{238}_{92}$U *(6 marks)*
5. The mass of the $^{16}_{8}$O nucleus is 2.656×10^{-26} kg. Calculate its BE per nucleon in MeV. *(4 marks)*
6. There is a decrease in mass of 5.8×10^{-3} u when a polonium-210 emits an alpha particle and transforms to an isotope of lead. Calculate the total energy released by 1 mol (0.210 kg) of polonium-210. *(4 marks)*

26.3 Nuclear fission
26.4 Nuclear fusion

Specification reference: 6.4.4

26.3 Nuclear fission

Nuclear power stations use **fission** reactions to generate power. The energy released in fission reactions can be as much as a million times greater than burning a similar mass of fossil fuels.

Induced fission

Induced fission occurs when a nucleus absorbs a neutron and then splits into two approximately equal fragment nuclei and a few fast-moving neutrons.

A typical fission reaction of a uranium-235 nucleus is shown below:

$$^{1}_{0}\text{n} + ^{235}_{92}\text{U} \rightarrow ^{236}_{92}\text{U}^{*} \rightarrow ^{141}_{56}\text{Ba} + ^{92}_{36}\text{Kr} + 3^{1}_{0}\text{n}$$

A slow-moving neutron (**thermal neutron**) is absorbed by the uranium-235 nucleus. It temporarily forms the highly unstable nucleus of uranium-236 which very quickly splits into smaller nuclei of barium-141 and krypton-92 and also three fast-moving neutrons. In this single reaction about 180 MeV of energy is released as kinetic energy of the fragment nuclei and neutrons.

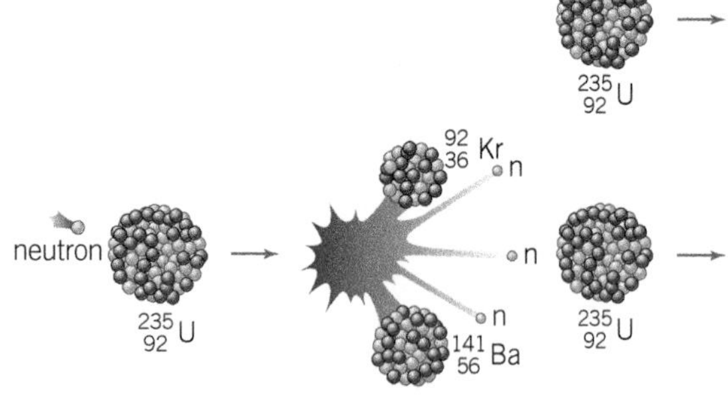

▲ **Figure 1** *Chain reaction*

- The chance of a fission reaction for a uranium-235 nucleus is greater with slow-moving neutrons than fast-moving neutrons.
- The total mass of the particles after the fission reaction is always less than the total mass of the particles before the reaction. The difference in the mass Δm is released as kinetic energy equal to Δmc^2.
- The total binding energy of the particles after fission is greater than the total binding energy before it. The difference in the binding energies is equal to the energy released.
- In a **chain reaction**, the neutrons released in a fission reaction can trigger further fission reactions, see Figure 1.

Fission reactor

The main components of a fission reactor are: fuel rods, control rods and moderator. See Figure 2.

- The fuel rods contain enriched uranium (uranium-238 with 2–3% uranium-235).
- The nuclei of the material of the **moderator** slow down the fast neutrons produced in the fission reactions without absorbing them. Water and graphite (carbon) are the two commonly used materials.
- The nuclei of the material of the **control rods** easily absorb neutrons. The control rods can be moved in and out of the reactor core to control the rate of the fission reactions. Boron and cadmium are the two commonly used materials.

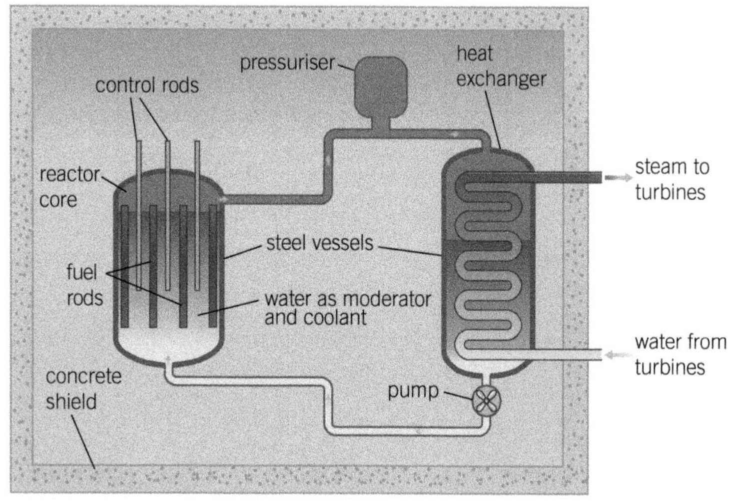

▲ **Figure 2** *The main components of a nuclear reactor. The water is a coolant; it removes thermal energy from the reactor.*

Storage and disposal of the by-products from fission reactors is difficult because of the toxicity and the long half-lives of many of the isotopes.

26.4 Nuclear fusion

Fusion is the process by which stars produce energy. The core of a star provides the high temperature (10^8 K) and high pressure necessary for the fusion of hydrogen nuclei into helium nuclei.

Fusion

Fusion occurs when two high-speed nuclei combine to form a bigger nucleus.

The positive nuclei repel each other. The nuclei therefore need to have large enough kinetic energy for them to get close enough (< 3 fm) for the strong nuclear force to bind them together.

A fusion reaction is shown below:

$$^{2}_{1}\text{H} + ^{1}_{1}\text{p} \rightarrow ^{3}_{2}\text{He}$$

- The mass of the $^{3}_{2}$He nucleus is less than the total mass the $^{2}_{1}$H nucleus and the proton $^{1}_{1}$p. The difference in the mass Δm is released as energy equal to Δmc^2.
- The binding energy of the $^{3}_{2}$He nucleus is greater than the binding energy of the $^{2}_{1}$H nucleus (the proton is a lone particle and has no BE). The difference in the binding energies is equal to the energy released.

Worked example: Energy from fusion

One of the many fusion reactions taking place in the core of a star is

$$^{2}_{1}\text{H} + ^{2}_{1}\text{H} \rightarrow ^{4}_{2}\text{He}$$

Use the binding energy (BE) per nucleon against nucleon number graph in Topic 26.2, Binding energy, to estimate the energy in joules released in this reaction.

Step 1: Write down the BE per nucleon of the nuclei.

BE per nucleon of $^{2}_{1}$H = 1.0 MeV   BE per nucleon of $^{4}_{2}$He = 7.1 MeV

Step 2: Calculate the BE of each nucleus.

BE of $^{2}_{1}$H = 2 × 1.0 = 2.0 MeV   BE of $^{4}_{2}$He = 4 × 7.1 = 28 MeV

Step 3: The energy released is the difference in the BE. Calculate this energy.

energy released = 28.4 − 2.0 = 26.4 MeV

energy released = $26.4 \times 1.60 \times 10^{-19}$ ($1\text{ eV} = 1.60 \times 10^{-19}$ J)

energy released = 4.2×10^{-12} J (2 s.f.)

Summary questions

1 State one similarity between fission and fusion reactions. *(1 mark)*

2 Show that the nucleon and the proton numbers are conserved in each reaction below.

a $^{2}_{1}\text{H} + ^{1}_{1}\text{p} \rightarrow ^{3}_{2}\text{He}$ *(2 marks)*

b $^{1}_{0}\text{n} + ^{235}_{92}\text{U} \rightarrow ^{141}_{56}\text{Ba} + ^{92}_{36}\text{Kr} + 3^{1}_{0}\text{n}$ *(2 marks)*

3 Explain why nuclei of hydrogen will only fuse together at high temperatures. *(2 marks)*

4 A single fission reaction of uranium-235 can produce energy of about 2.0×10^{-11} J. Estimate the energy produced from 1.0 kg of uranium-235.
molar mass of uranium-235 = 0.235 kg mol^{-1} *(3 marks)*

5 Use the binding energy (BE) per nucleon against nucleon number graph in Topic 26.2, Binding energy, to estimate the energy in joules released in the fusion reaction $^{1}_{1}\text{p} + ^{1}_{1}\text{p} \rightarrow ^{2}_{1}\text{H} + ^{0}_{+1}\text{e} + \nu$. *(3 marks)*

6 Use your answer to **Q5** to estimate the energy produced from 1.0 kg of hydrogen-1.
molar mass of hydrogen-1 = 1.0 g mol^{-1} *(3 marks)*

Chapter 26 Practice questions

1 Annihilation occurs when:

A electron and positron meet

B electron and proton interact

C electron and a photon collide

D electron and neutrino interact *(1 mark)*

2 In a fusion reaction, hydrogen nuclei $^{2}_{1}H$ and $^{1}_{1}H$ combine together to form a nucleus of helium $^{3}_{2}He$. These nuclei have binding energy BE.

The BE of $^{3}_{2}He$ nucleus is ____________ the total BE of the hydrogen nuclei.

What is the missing word or words?

A twice

B equal to

C less than

D greater than *(1 mark)*

3 What is the energy equivalent to the mass of an electron?

A 2.7×10^{-22} J

B 8.2×10^{-14} J

C 1.6×10^{-19} J

D 1.5×10^{-10} J *(1 mark)*

4 The binding energy per nucleon for the nucleus of $^{92}_{36}Kr$ is 8.7 MeV.

What is the binding energy of this nucleus?

A 310 MeV

B 490 MeV

C 800 MeV

D 1100 MeV *(1 mark)*

5 Use Figure 2 in Topic 26.2 to answer this question.

a Explain why the isotope of iron-56 cannot decay. *(1 mark)*

b **i** Use Figure 1 to explain why energy is released in the fusion reaction shown below.

$^{2}_{1}H + ^{2}_{1}H \rightarrow ^{4}_{2}He$ *(3 marks)*

ii Estimate the energy released when 1 mole of hydrogen-2 fuse together. *(4 marks)*

6 **a** Explain what is meant by the statement 'mass and energy are equivalent'. *(1 mark)*

b One of the many fusion reactions in the Sun is shown below:

$^{2}_{1}H + ^{1}_{1}H \rightarrow ^{3}_{2}He$

i Explain why high temperatures are necessary for this fusion reaction to occur. *(2 marks)*

ii In terms of the masses of the particles, explain why energy is released. *(2 marks)*

iii The energy released in the reaction above is about 5.5 MeV.

Calculate the change in mass responsible for this energy. *(3 marks)*

c Explain the difference between fission and fusion reactions. *(2 marks)*

Chapter 27 Medical imaging

In this chapter you will learn about ...

- [ ] X-rays
- [ ] Interaction of X-rays with matter
- [ ] Intensity of transmitted X-rays
- [ ] Contrast materials
- [ ] CAT scanner
- [ ] Gamma camera
- [ ] PET scanner
- [ ] Ultrasound
- [ ] A and B scans
- [ ] Acoustic impedance
- [ ] Doppler imaging
- [ ] Speed of blood

27 MEDICAL IMAGING
27.1 X-rays
27.2 Interaction of X-rays with matter

Specification reference: 6.5.1

27.1 X-rays

X-rays are electromagnetic waves with a wavelength in the range 10^{-8} m to 10^{-13} m. An X-ray photon has greater energy than a photon of visible light because of its shorter wavelength. X-ray photons are produced when fast-moving electrons are decelerated by the atoms of a metal. The kinetic energy of the electrons is transformed into X-ray photons.

Production of X-rays

Figure 1 shows the main components of an **X-ray tube**. Electrons are accelerated through a high potential difference, typically 100 kV. They travel from the cathode (heater) to the anode (**target metal**). Most of the kinetic energy of the electrons is transformed into heat in the anode. The anode is cooled by circulating water through it. About 1% of the kinetic energy of the electrons is transformed into X-ray photons. These X-ray photons have a range of wavelengths.

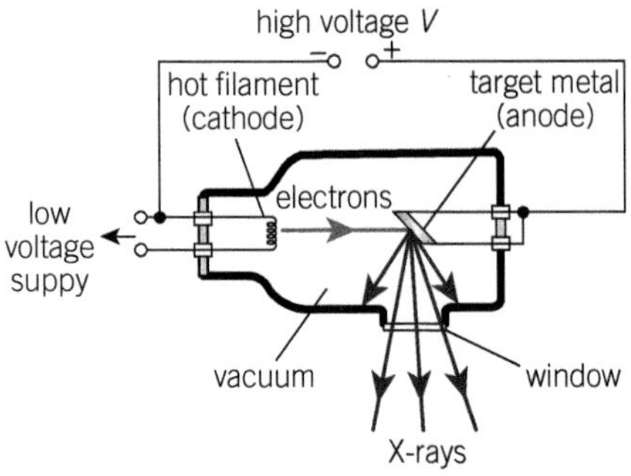

▲ **Figure 1** *The main components of an X-ray tube*

27.2 Interaction of X-rays with matter

The intensity of a beam of X-rays decreases as the X-ray photons are either stopped or scattered by the atoms of the material.

Attenuation mechanisms

Table 1 summarises the four **attenuation** mechanisms.

▼ **Table 1**

Attenuation mechanism	Diagram	Energy of photons	Description
Simple scatter	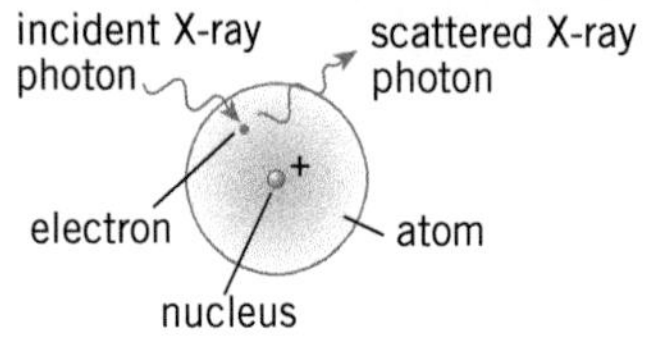 incident X-ray photon; scattered X-ray photon; electron; atom; nucleus	1 to 20 keV	The X-ray photon is scattered elastically by an electron.
Photoelectric effect	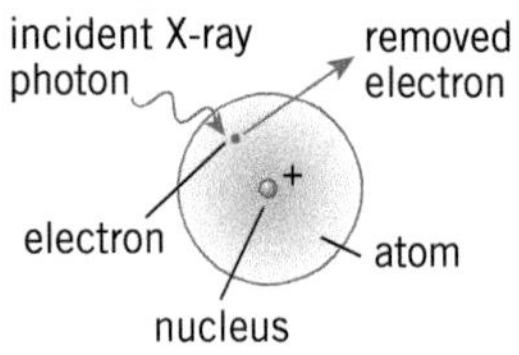 incident X-ray photon; removed electron; electron; atom; nucleus	< 0.1 MeV	The X-ray photon disappears and removes an electron from the atom.
Compton scattering	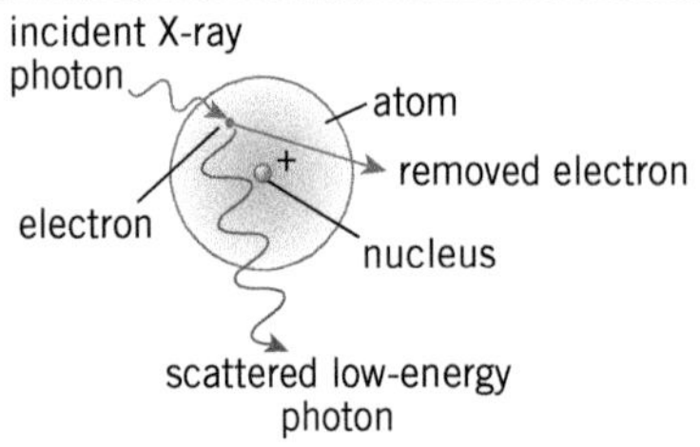 incident X-ray photon; atom; removed electron; electron; nucleus; scattered low-energy photon	0.5 to 5.0 MeV	The X-ray photon is scattered by an electron, its energy is reduced, and the electron is ejected from the atom.
Pair production	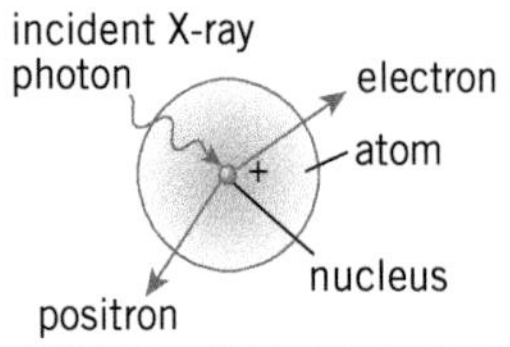 incident X-ray photon; electron; atom; nucleus; positron	> 1.02 MeV	The X-ray photon disappears to produce an electron–positron pair.

Intensity of transmitted X-rays

The **intensity** I of a parallel beam (collimated) of X-rays decreases exponentially with the thickness x of material and is given by the equation

$$I = I_0 e^{-\mu x}$$

where I_0 is the initial intensity and μ is the **attenuation coefficient** or the **absorption coefficient** of the material. The SI unit of μ is m^{-1}, but mm^{-1} and cm^{-1} are also used. Bone has a greater attenuation coefficient than soft tissues – this is why a simple X-ray scan (image) is used to identify broken bones in a patient. Figure 2 shows the variation of intensity I with thickness x of a material.

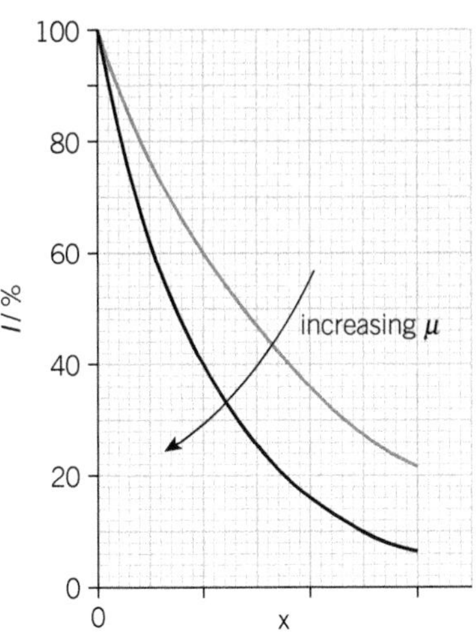

▲ **Figure 2** *The transmitted intensity I decreases exponentially with thickness x of the material*

Contrast materials

Soft tissues and muscles are poor absorbers of X-rays. A contrast medium is a material which is either injected into a patient or ingested by the patient in order to improve the visibility of soft tissues on an X-ray scan. A contrast medium must be harmless to patients.

The main interaction mechanism for the X-rays used in hospitals is the photoelectric effect. For this effect, $\mu \propto Z^3$, where Z is the atomic number of the element. Iodine and barium are the two most frequently used elements within contrast media. Both iodine and barium have larger Z values than soft tissues.

- Barium sulphate (barium meal) is used when imaging the digestive system of a patient.
- Iodine-based liquid is often injected into the blood vessels when diagnosing circulatory problems.

Synoptic link

Intensity was covered in Topic 11.5, Intensity.

Worked example: Thickness of bone

The attenuation coefficient of bone is 0.60 cm^{-1}. Calculate the thickness of bone that will halve the intensity of X-rays incident on the bone.

Step 1: Write down the quantities given.

$\mu = 0.60\ cm^{-1}$ $\quad I = 0.50 I_0$

Step 2: Use the equation $I = I_0 e^{-\mu x}$ to calculate the thickness x.

$0.50 I_0 = I_0 e^{-0.60x}$ $\quad$ or $\quad 0.50 = e^{-0.60x}$

$\ln(0.50) = -0.60x$

$$x = \frac{\ln(0.50)}{-0.60} = 1.16\ cm = 1.2\ cm\ (2\ s.f.)$$

Maths: $\log_e$ or ln

When $e^x = y$ then $x = \ln(y)$, where ln is an abbreviation for $\log_e$.

It is important that you use the ln button on your calculator and not the lg button (which is $\log_{10}$).

Summary questions

1. Name two attenuation mechanisms where the X-ray photon is scattered. *(1 mark)*
2. An X-ray tube operates using a 100 kV supply. Use Table 1 to state the most dominant attenuation mechanism for the X-ray photon. *(1 mark)*
3. The attenuation coefficient of muscle is about 0.21 cm^{-1}. Calculate the percentage of the original intensity of X-rays transmitted through 3.0 cm of muscle. *(3 marks)*
4. A single X-ray photon is produced from the kinetic energy of a single electron. Calculate the maximum energy of an X-ray photon in J from an X-ray tube connected to a 120 kV supply. *(3 marks)*
5. Calculate the shortest wavelength of an X-ray photon produced from an X-ray tube connected to a 200 kV supply. *(3 marks)*
6. The mean atomic number of elements within soft tissues is 7. The atomic number of barium is 56. Use this information to explain why barium is a useful contrast material. *(3 marks)*

27.3 CAT scans
27.4 The Gamma camera
27.5 PET scans

Specification reference: 6.5.1, 6.5.2

27.3 CAT scans

A computerised axial tomography (CAT) scanner digitally records a large number of cross-sectional scans (images) – very much like a loaf of bread that has been cut into many thin slices. Soft tissues can be distinguished but the patient is exposed to the ionising effects of X-rays for a longer period of time.

CAT scanner

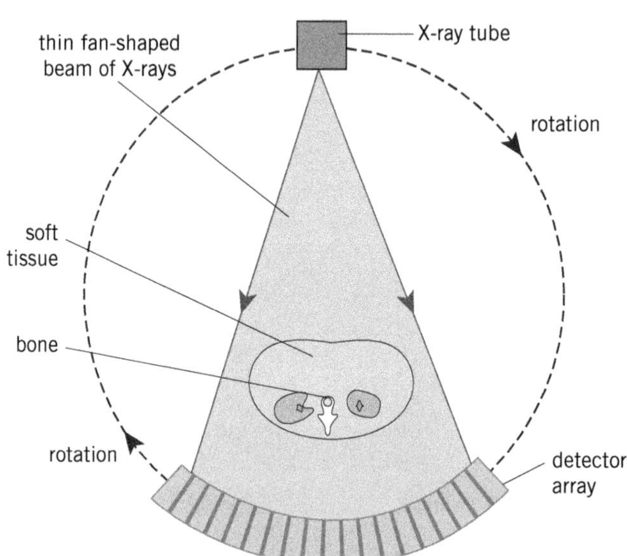

▲ **Figure 1** *The X-ray tube and the detectors can make three revolutions around the patient every second*

- The patient lies on a horizontal table that can move in and out of a large doughnut-shaped vertical ring (gantry).
- The ring contains a single X-ray tube and up to a thousand X-ray detectors on the opposite side. The detectors are all connected to a computer.
- The X-ray tube produces a thin fan-shaped beam of X-rays which irradiate a thin cross-section through the patient, see Figure 1.
- The X-ray tube and the detectors rotate around the patient. A two-dimensional image is digitally recorded by the computer.
- After every revolution, the X-ray tube and the detectors move a short distance (about 1 cm) along the length of the patient, so that in the next revolution, the X-ray beam irradiates the next slice through the patient.
- All the digital images can be manipulated by the computer software to produce cross-sectional images through the patient and also a three-dimensional image of the patient.

Revision tip

The term 'axial' in CAT refers to the images taken in the axial plane (cross-sections) through the patient. '*Tomos*' is Greek for '*slice*'.

27.4 The Gamma camera

The most common medical tracer contains the isotope of technetium-99m (Tc-99m). The Tc-99m isotope has a half-life of about 6.0 hours and emits 140 keV gamma photons. A compound containing Tc-99m is injected into a patient and this compound will target specific cells in the body. Tc-99m can be used to monitor the function of many major organs (brain, heart, liver, lungs, and kidney). The concentration of Tc-99m within the patient can be used to identify defects in the function of these organs.

Components of the camera

A gamma camera is placed above the patient. Figure 2 shows the main components of the gamma camera.

- **Collimator:** This consists of long and thin cylindrical tubes made from lead. Only gamma photons travelling along the axis of a tube will reach the scintillator.
- **Scintillator:** Each gamma photon hitting the scintillator (sodium iodide) can produce thousands of photons of visible light.
- **Photomultiplier tubes:** Each tube is an electrical device that produces an electrical pulse whenever a photon of visible light is incident on it.
- Computer and display: The electrical pulses from all the photomultiplier tubes are used by the computer and its software to accurately pinpoint the origin of the gamma photons, and therefore the technetium-99m within the patient. An image of the concentration levels of Tc-99m is displayed on a display screen.

▲ **Figure 2** *Components of a gamma camera*

27.5 PET scans

The most common **medical tracer** used in positron emission tomography (PET) scanning is fluorodeoxyglucose (FDG), which has radioactive atoms of fluorine-18. This isotope of fluorine-18 is a positron emitter with a half-life of about 110 minutes. FDG is injected into a patient and it gathers in tissues with a high rate of respiration. The *function* of organs such as the brain can be observed. PET scanning is an expensive technique because the medical tracers have to be produced on-site.

PET scanner

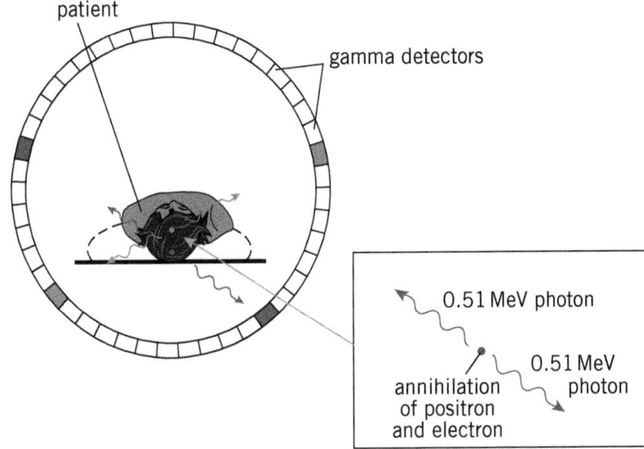

▲ **Figure 3** *The 0.51 MeV gamma photons are detected by the gamma detectors around the patient*

Figure 3 shows a patient surrounded by gamma detectors which are all connected to a computer.

- The positron emitted from a fluorine-18 nucleus does not travel too far before it gets annihilated by an electron.
- The annihilation produces two 0.51 MeV gamma photons emitted in opposite directions.
- The computer software can determine the exact point of annihilation within the patient from the arrival times of these gamma photons at diametrically opposite detectors and the speed of these photons ($3.00 \times 10^8\,m\,s^{-1}$).
- An image of the concentration levels of FDG within the patient can be displayed on a screen.

Summary questions

1. State one advantage of a CAT scan over traditional X-ray image. *(1 mark)*
2. State why a CAT scan can be harmful to a patient. *(1 mark)*
3. Explain why the collimators in a gamma camera need to be made of long and narrow lead tubes. *(2 marks)*
4. Discuss the technique that could be used to assess the effects of a drug on the function of the brain. *(2 marks)*
5. Distances as small as 0.5 cm need to be identified in a PET scan. Discuss why sophisticated computers are required for PET scanners. *(2 marks)*
6. Discuss the advantages of using Tc-99m as a medical tracer in a patient. *(3 marks)*

27.6 Ultrasound
27.7 Acoustic impedance
27.8 Doppler imaging

Specification reference: 6.5.3

27.6 Ultrasound

Ultrasound is any sound wave with frequency greater than 20 kHz. An **ultrasound transducer** is a device used both to generate and to receive ultrasound.

Ultrasound transducer

Some materials (e.g., quartz) produce an e.m.f. when they are either compressed or extended – this is the **piezoelectric effect**. The effect works in reverse too. When a p.d. is applied across the opposite ends of the material, it will either compress or expand.

The two common piezoelectric materials used in transducers are lead zirconate titanate and polyvinylidene fluoride.

- The piezoelectric material vibrates when a high-frequency alternating p.d. is applied between its opposite ends. The vibration of the material in the air produces ultrasound.
- The same piezoelectric material vibrates when ultrasound is incident on it. These vibrations induce an alternating e.m.f. in the material.

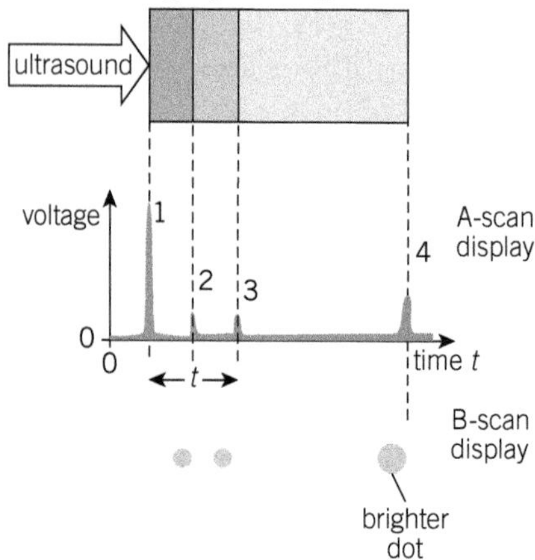

▲ **Figure 1** *Ultrasound A-scan. A, B, and C are different tissues*

A-and B-scans

In an A-scan, ultrasound pulses from a stationary transducer are sent into the patient. Figure 1 shows the voltage–time trace from the transducer. The voltage pulse 1 is responsible for sending in a pulse of ultrasound into the patient. The ultrasound is partially reflected at the boundaries between the soft tissues. The voltage pulses 2, 3, and 4 are produced from these reflections.

The thickness of tissues A and B can be determined from the average speed c of the ultrasound in A and B and the time t between the voltage pulses using the equation

$$\text{thickness} = \frac{ct}{2}$$

In a B-scan the ultrasound transducer is moved over the patient's skin. The output of the transducer is connected to a computer. For each position of the transducer, the computer produces a row of dots on the digital screen, where each dot corresponds to the boundary between two tissues. The brightness of the dot is proportional to the intensity of the reflected ultrasound pulse. These dots are used to form a two-dimensional image.

Revision tip

The A in the A-scan stands for 'amplitude' and the B in the B-scan stands for 'brightness'.

27.7 Acoustic impedance

The acoustic impedance Z of a material is defined as the product of its density ρ and the speed c of ultrasound in the material; $Z = \rho c$. The SI unit of acoustic impedance is $\text{kg m}^{-2}\,\text{s}^{-1}$.

Reflection at a boundary

For a parallel beam of ultrasound incident normally at a boundary between two materials of acoustic impedances Z_1 and Z_2, the ratio of the reflected intensity I_r to the incident intensity I_0 is given by the equation

$$\frac{I_r}{I_0} = \frac{(Z_2 - Z_1)^2}{(Z_2 + Z_1)^2}$$

This ratio is known as the **intensity reflection coefficient**. The acoustic impedance of air is very small compared with that of skin. Placing a transducer directly on the patient's skin would mean most of the ultrasound will be reflected off the skin. To reduce this reflection, a **coupling gel** is smeared on the transducer and the skin. The gel has acoustic impedance similar to that of skin and therefore most of the ultrasound is transmitted into the patient.

The terms impedance matching or acoustic matching are used when two materials have similar acoustic impedances.

27.8 Doppler imaging

The Doppler effect of ultrasound can be used to determine the speed of blood in blood vessels.

Speed of blood

The transducer is placed at an angle θ to the blood vessel, see Figure 2. It sends pulses of ultrasound into the patient. The frequency of the ultrasound reflected off the iron-rich blood cells is changed because of the moving blood cells. The change in frequency Δf is directly proportional to the speed v of the blood. The speed of the blood v can be determined using the equation

$$\Delta f = \frac{2fv\cos\theta}{c}$$

where f is the frequency of the ultrasound from the transducer and c is the speed of ultrasound in blood.

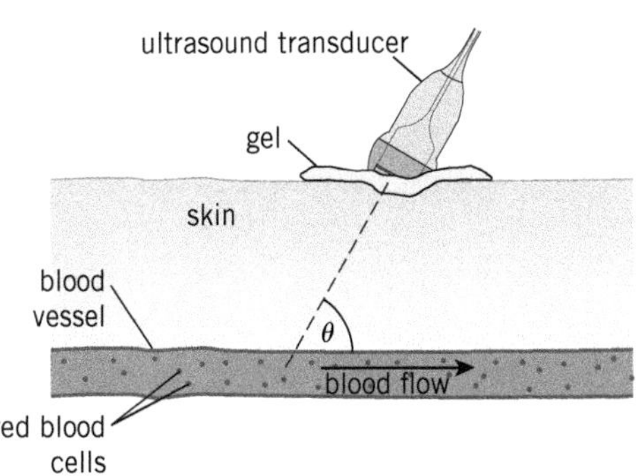

▲ **Figure 2** *An ultrasound transducer can be used to determine the speed of blood*

Worked example: Doppler shift

A transducer emitting ultrasound of frequency 15 MHz is held at an angle of 50° to a blood vessel. The speed of the blood is estimated to be 20 cm s^{-1}. Calculate the change in the frequency of the ultrasound reflected by the blood.

speed of the ultrasound in the blood = 1600 m s^{-1}

Step 1: Write down the quantities given.

$f = 15 \times 10^6$ Hz $\quad v = 0.20$ m s^{-1} $\quad c = 1600$ m s^{-1} $\quad \theta = 50°$

Step 2: Use the equation $\Delta f = \frac{2fv\cos\theta}{c}$ to calculate Δf.

$$\Delta f = \frac{2fv\cos\theta}{c} = \frac{2 \times 15 \times 10^6 \times 0.20 \times \cos 50°}{1600}$$

$\Delta f = 2.4 \times 10^3$ Hz (2 s.f.)

Summary questions

1. Describe how a transducer emits ultrasound. *(2 marks)*
2. Explain why the distance ct is divided by 2 in the equation thickness $= \frac{ct}{2}$ for an A-scan. *(1 mark)*
3. Suggest why the ultrasound transducer cannot be placed 90° to the blood vessel in Doppler imaging. *(1 mark)*
4. Ultrasound is incident at the boundary between two materials. The acoustic impedance of one of the materials is twice that of the other. Calculate $\frac{I_r}{I_0}$. *(3 marks)*
5. Explain why most of the ultrasound is reflected at the air–skin boundary. *(2 marks)*
6. A transducer emits ultrasound of frequency 18 MHz. It is placed at an angle of 60° to a blood vessel. The change in the frequency of the ultrasound reflected by the blood is 3.2 kHz. The speed of ultrasound in blood is 1600 m s^{-1}. Calculate the speed of the blood in cm s^{-1}. *(3 marks)*

Chapter 27 Practice questions

1 Which technique or device uses the annihilation of electron–positron pairs for imaging?

A CAT

B PET

C A-scan

D Doppler scan *(1 mark)*

2 The energy of a photon of visible light is about 2 eV.

What is a good estimate for the energy of an X-ray photon?

A 10^{-2} eV

B 10 eV

C 10^{4} eV

D 10^{8} eV *(1 mark)*

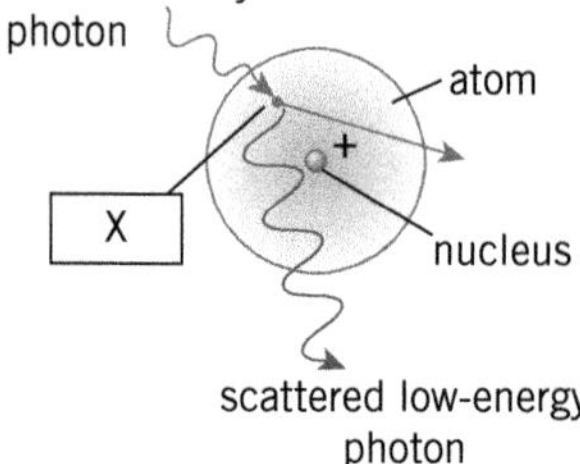

▲ **Figure 1**

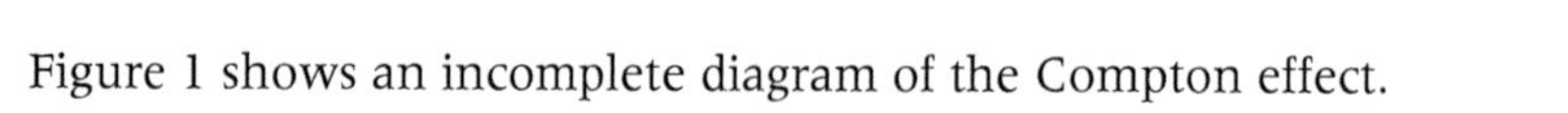

3 Figure 1 shows an incomplete diagram of the Compton effect.

What is **X** in this figure?

A electron

B positron

C proton

D vacuum *(1 mark)*

4 **a** Ultrasound can be used to investigate the internal structures of a patient.

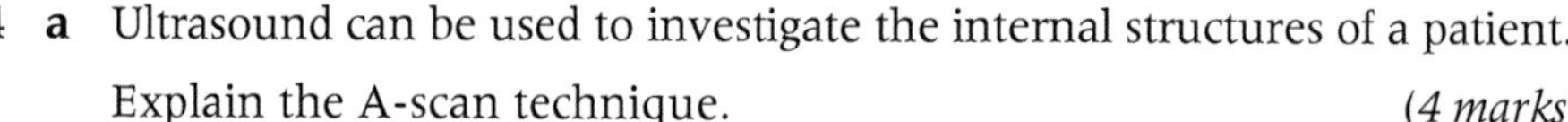

Explain the A-scan technique. *(4 marks)*

b Define the acoustic impedance of a material. *(1 mark)*

c The table below gives some information about two materials X and Y.

Material	Speed of ultrasound in material / m s^{-1}	Density of material / kg m^{-3}	Acoustic impedance /
X	4100	1500	
Y	1600	1100	

i Complete the table by inserting in the acoustic impedance values for X and Y. Include an appropriate unit. *(3 marks)*

ii A parallel beam of ultrasound is incident at the boundary of X and Y.

Calculate the percentage of ultrasound intensity *transmitted* at the boundary. *(3 marks)*

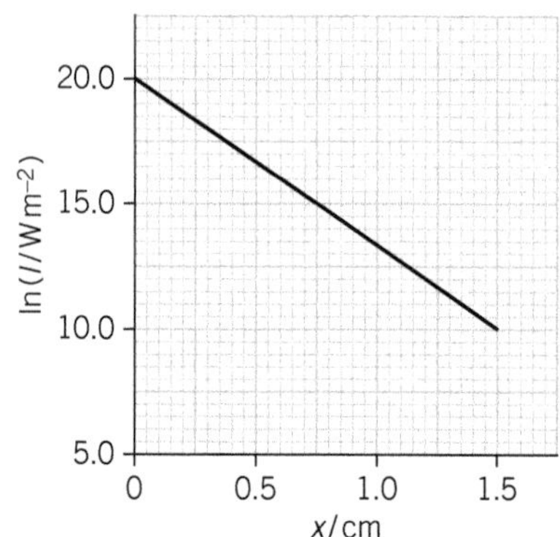

▲ **Figure 2**

5 **a** Explain why prolonged exposure to X-rays can be dangerous to a patient. *(2 marks)*

b A parallel beam of X-rays of intensity 2.0×10^{9} W m^{-2} is incident at the soft tissues of a patient for a total time of 1.0 minute.

Figure 2 shows the variation of $\ln(I/\mathrm{W\,m^{-2}})$ with thickness x of the tissue, where I is the intensity of the X-rays.

i Explain why the gradient of the straight-line graph in Figure 2 is $-\mu$, where μ is the attenuation (absorption) coefficient of the soft tissue. *(2 marks)*

ii Calculate μ in cm^{-1}. *(1 mark)*

iii Calculate the total energy incident on tissue of cross-sectional area 6.0×10^{-4} m^{2} at a depth of 1.8 cm. *(4 marks)*

A1 Physical quantities and units

Units

Most of the physical quantities you come across in your course can be expressed by combinations of six base units – kg, m, s, A, K, and mol.

All derived units can be worked out using an appropriate equation and then multiplying and/or dividing the base units.

Example:

density = mass/volume mass → kg volume → m^3

Therefore, density has units kg/m^3 or $kg\,m^{-3}$.

Homogeneous

An equation is homogeneous when the left-hand side has the same units as the right-hand side. A relationship between physical quantities can only be correct if the equation is homogeneous.

Example:

$s = \frac{1}{2}at^2$

right-hand side: $a \rightarrow m\,s^{-2}$, $t^2 \rightarrow s^2$, and $\frac{1}{2}$ has no unit.

Therefore $\frac{1}{2}at^2 \rightarrow m\,s^{-2} \times s^2$ or m

left-hand side: $s \rightarrow m$

The unit m is the same on both sides of the equal sign – the equation is homogeneous.

A2 Recording results and straight lines

Labelling and significant figures

In a table of results, each heading must have the quantity and its unit. The quantity can be represented by a symbol or in words. A slash (solidus /) is used to separate the quantity and its unit, for example, v / $m\,s^{-1}$ for speed. The labelling of graph axes follows the same rules as for table headings.

Be careful with significant figures in your table of results. The result of a calculation that involves measured quantities has the same number of significant figures as the measurement that has the *smallest* number of significant figures.

Example:

If the distance is 2.12 m (3 s.f.) and time is 3.2 s (2.s.f.), then the speed is written as $0.66\,m\,s^{-1}$ (2 s.f.).

Graphs

As a general rule, you plot the independent variable, the one you intentionally change in an experiment, on the x-axis, and the dependent variable, the variable which changes as a result, on the y-axis.

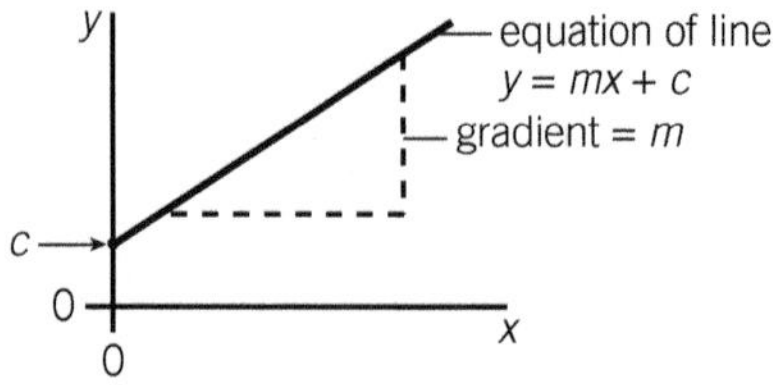

▲ **Figure 1** *Straight-line graph. The gradient of the line is* $\frac{\Delta y}{\Delta x}$

Figure 1 shows a straight-line graph. The equation for a straight line is $y = mx + c$, where m is the gradient and c is the y-intercept. Always use a large triangle to determine the gradient of the straight line.

Sometimes plotting the measured quantities does not produce a straight-line graph. You have to plot data carefully if you want to get a straight line and verify a relationship.

Example:

The equation is $v^2 = 2as$, and you have measured the velocity v and the displacement s. A graph of v against s will be a curve. Plotting v^2 against s should produce a straight line. The gradient of the line will be $2a$, therefore you can determine the acceleration a.

A3 Measurements and uncertainties

Definitions

Remember that no measurement can ever be perfect.

- **Error** (of measurement) is the difference between an individual measurement and the **true** value (or accepted reference value) of the quantity being measured.
- **Random errors** can happen when any measurement is being made. They are measurement errors in which measurements vary unpredictably.
- **Systematic errors** are measurement errors in which the measurements differ from the true values by a consistent amount each time a measurement is made.
- **Accuracy** is to do with how close a measurement result is to the true value.
- **Precision** is to do with how close repeated measurements are to each other.
- The **uncertainty** in the measurement is an interval within which the true value can be expected to lie. The **absolute uncertainty** is approximated as half the range.

Example:

$x = 52 \pm 3\,\text{mm}$

absolute uncertainty = 3 mm

$\%\ \text{uncertainty} = \frac{3}{52} \times 100 = 5.8\%$

Uncertainty rules

Rule 1: Adding or subtracting quantities

When you add or subtract quantities in an equation, you add the absolute uncertainties for each value.

Example:

$x = 4.2 \pm 0.3 \qquad y = 3.0 \pm 0.2 \qquad x - y = 1.2 \pm 0.5$

Rule 2: Multiplying or dividing quantities

When you multiply or divide quantities, you add the percentage uncertainties for each value.

Example:

$x = 2.0 \pm 0.3 \qquad y = 1.2 \pm 0.1$

% uncertainty in $xy = \left(\frac{0.3}{2.0} + \frac{0.1}{1.2}\right) \times 100 = 23\%$

Rule 3: Raising a quantity to a power

When a measurement in a calculation is raised to a power n, your percentage uncertainty is increased n times. The power n can be an integer or a fraction.

Example:

$x = 3.14 \pm 0.13$ % uncertainty in $x^4 = 4 \times \left(\frac{0.13}{3.14}\right) \times 100 \approx 17\%$

Graphs

You can show error bars for each plotted point, see Figure 2. The best-fit straight line must pass through all the error bars. You can determine the percentage uncertainty in the gradient by drawing a 'worst-fit' straight line and doing the following calculation:

$$\% \text{ uncertainty} = \frac{\text{gradient of worst-fit line} - \text{gradient of best-fit line}}{\text{gradient of best-fit line}} \times 100\%$$

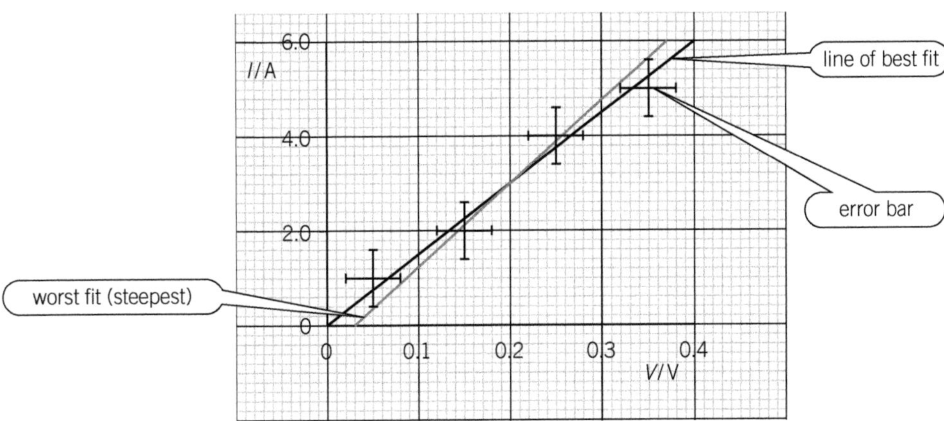

▲ **Figure 2** *Plotted points can have errors bars*

5d. Physics A data sheet

Data, Formulae, and Relationships

The data, formulae, and relationships in this datasheet will be printed for distribution with the examination papers.

Data

Values are given to three significant figures, except where more – or fewer – are useful.

Physical constants

acceleration of free fall	g	$9.81\ \mathrm{m\,s^{-2}}$
elementary charge	e	$1.60 \times 10^{-19}\ \mathrm{C}$
speed of light in a vacuum	c	$3.00 \times 10^{8}\ \mathrm{m\,s^{-1}}$
Planck constant	h	$6.63 \times 10^{-34}\ \mathrm{J\,s}$
Avogadro constant	N_A	$6.02 \times 10^{23}\ \mathrm{mol^{-1}}$
molar gas constant	R	$8.31\ \mathrm{J\,mol^{-1}\,K^{-1}}$
Boltzmann constant	k	$1.38 \times 10^{-23}\ \mathrm{J\,K^{-1}}$
gravitational constant	G	$6.67 \times 10^{-11}\ \mathrm{N\,m^2\,kg^{-2}}$
permittivity of free space	ε_0	$8.85 \times 10^{-12}\ \mathrm{C^2\,N^{-1}\,m^{-2}\ (F\,m^{-1})}$
electron rest mass	m_e	$9.11 \times 10^{-31}\ \mathrm{kg}$
proton rest mass	m_p	$1.673 \times 10^{-27}\ \mathrm{kg}$
neutron rest mass	m_n	$1.675 \times 10^{-27}\ \mathrm{kg}$
alpha particle rest mass	m_α	$6.646 \times 10^{-27}\ \mathrm{kg}$
Stefan constant	σ	$5.67 \times 10^{-8}\ \mathrm{W\,m^{-2}\,K^{-4}}$

Quarks

up quark	charge $= +\frac{2}{3}e$
down quark	charge $= -\frac{1}{3}e$
strange quark	charge $= -\frac{1}{3}e$

Conversion factors

unified atomic mass unit	$1\ \mathrm{u} = 1.661 \times 10^{-27}\ \mathrm{kg}$
electronvolt	$1\ \mathrm{eV} = 1.60 \times 10^{-19}\ \mathrm{J}$
day	$1\ \text{day} = 8.64 \times 10^{4}\ \mathrm{s}$
year	$1\ \text{year} \approx 3.16 \times 10^{7}\ \mathrm{s}$
light year	$1\ \text{light year} \approx 9.5 \times 10^{15}\ \mathrm{m}$
parsec	$1\ \text{parsec} \approx 3.1 \times 10^{16}\ \mathrm{m}$

Mathematical equations

arc length $= r\theta$

circumference of circle $= 2\pi r$

area of circle $= \pi r^2$

curved surface area of cylinder $= 2\pi rh$

surface area of sphere $= 4\pi r^2$

area of trapezium $= \frac{1}{2}(a + b)\,h$

volume of cylinder $= \pi r^2 h$

volume of sphere $= \frac{4}{3}\pi r^3$

Pythagoras' theorem: $a^2 = b^2 + c^2$

cosine rule: $a^2 = b^2 + c^2 - 2bc\cos A$

sine rule: $\frac{a}{\sin A} = \frac{b}{\sin B} = \frac{c}{\sin C}$

$\sin\theta \approx \tan\theta \approx \theta$ and $\cos\theta \approx 1$ for small angles

$\log(AB) = \log(A) + \log(B)$

(Note: $\lg = \log_{10}$ and $\ln = \log_e$)

$\log\left(\frac{A}{B}\right) = \log(A) - \log(B)$

$\log(x^n) = n\log(x)$

$\ln(e^{kx}) = kx$

Formulae and relationships

Module 2 Foundations of physics

vectors

$$F_x = F\cos\theta$$

$$F_y = F\sin\theta$$

Module 3 Forces and motion

uniformly accelerated motion

$$v = u + at$$

$$s = \frac{1}{2}(u + v)t$$

$$s = ut + \frac{1}{2}at^2$$

$$v^2 = u^2 + 2as$$

force

$$F = \frac{\Delta p}{\Delta t}$$

$$p = mv$$

turning effects

$$\text{moment} = Fx$$

$$\text{torque} = Fd$$

density

$$\rho = \frac{m}{V}$$

pressure

$$p = \frac{F}{A}$$

$$p = h\rho g$$

work, energy, and power

$$W = Fx\cos\theta$$

$$\text{efficiency} = \frac{\text{useful energy output}}{\text{total energy input}} \times 100\%$$

$$P = \frac{W}{t}$$

$$P = Fv$$

springs and materials

$$F = kx$$

$$E = \frac{1}{2}Fx;\ E = \frac{1}{2}kx^2$$

$$\sigma = \frac{F}{A}$$

$$\varepsilon = \frac{x}{L}$$

$$E = \frac{\sigma}{\varepsilon}$$

Module 4 Electrons, waves, and photons

charge

$$\Delta Q = I\Delta t$$

current

$$I = Anev$$

work done

$$W = VQ;\ W = \mathcal{E}Q;\ W = VIt$$

resistance and resistors

$$R = \frac{\rho L}{A}$$

$$R = R_1 + R_2 + \ldots$$

$$\frac{1}{R} = \frac{1}{R_1} + \frac{1}{R_2} + \ldots$$

power

$$P = VI;\ P = I^2R;\ P = \frac{V^2}{R}$$

internal resistance

$$\mathcal{E} = I(R + r);\ \mathcal{E} = V + Ir$$

potential divider

$$V_{out} = \frac{R_2}{R_1 + R_2} \times V_{in}$$

$$\frac{V_1}{V_2} = \frac{R_1}{R_2}$$

waves

$$v = f\lambda$$

$$f = \frac{1}{T}$$

$$I = \frac{P}{A}$$

$$\lambda = \frac{ax}{D}$$

refraction

$$n = \frac{c}{v}$$

$$n\sin\theta = \text{constant}$$

$$\sin C = \frac{1}{n}$$

quantum physics

$$E = hf$$

$$E = \frac{hc}{\lambda}$$

$$hf = \phi + KE_{max}$$

$$\lambda = \frac{h}{p}$$

Answers to practice questions

Chapter 2

1 C [1] **2** D [1] **3** B [1] **4** B [1]

5 a Velocity is a vector with both magnitude and direction. The direction of travel of the satellite changes therefore its velocity changes. [1]

b i distance $= \frac{1}{4} \times$ circumference

$= \frac{1}{4} \times 2\pi \times 2.0 \times 10^7$ [1]

distance $= 3.1 \times 10^7$ m [1]

ii displacement = direct distance from A to B [1]

displacement$^2 = 2 \times (2.0 \times 10^7)^2$ [1]

displacement $= 2.8 \times 10^7$ m [1]

6 a The force N is 90° to the ramp therefore its component along the ramp is zero. ($N\cos 90° = 0$) [1]

b component $= 8.0\cos 60° = 4.0$ N [1]

c Correct diagram with sides and one of the angles labelled, see below. [2]

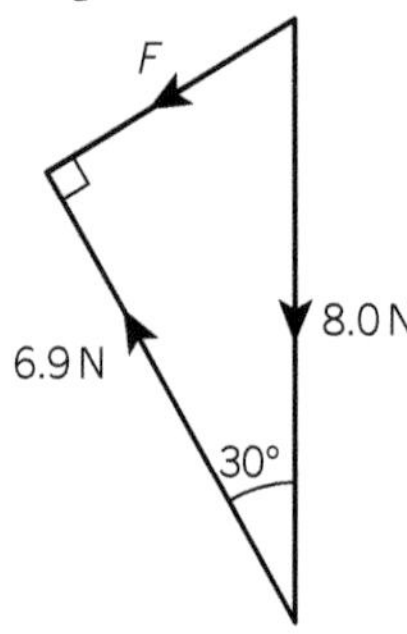

resultant force $F = (8.0^2 - 6.9^2)^{\frac{1}{2}}$ [1]

$F = 4.0$ N [1]

(Note: The resultant force is the same as the answer to **6b**.)

Chapter 3

1 D [1] **2** C [1] **3** D [1]

4 a Acceleration is the rate of change of velocity. [1]

b i The velocity of the ball is increasing with time. [1]

The gradient of the graph is equal to velocity and the gradient is increasing with time. [1]

ii velocity = gradient [1]

Tangent drawn at $t = 2.0$ s to determine the gradient. [1]

velocity $= 1.0\,\text{m s}^{-1}$ (allow $\pm 0.1\,\text{m s}^{-1}$) [1]

iii $a = \frac{1.0 - 0}{2.0}$ [1]

$a = 0.50\,\text{m s}^{-2}$ [1]

5 a $s = ut + \frac{1}{2}at^2$ and $u = 0$. Therefore, $h = \frac{1}{2}at^2$ [1]

With constant acceleration a, $h \propto t^2$ – therefore, a straight line passing through the origin. [1]

b gradient $= \frac{g}{2}$ [1]

gradient = 5.0 (allow ± 0.1) [1]

Therefore, $g = 2 \times 5.0 = 10\,\text{m s}^{-2}$ [1]

c No change to the gradient of the graph because g is the same for all falling objects on the Earth. [1]

Chapter 4

1 C [1] **2** A [1] **3** D [1]

4 a $a = \frac{200 - 180}{400}$ [1]

$a = 0.05\,\text{m s}^{-2}$ [1]

b Both are 90° to the horizontal and therefore each has no component in the horizontal plane. [1]

c pressure $= \frac{400 \times 9.81}{1.8}$ [1]

pressure $= 2.2 \times 10^3$ Pa [1]

5 a The sum of the clockwise moments about a point is equal to the sum of the anticlockwise moments about the same point. [1]

b i mass $= \frac{1800}{9.81}$ [1]

volume $= \frac{\text{mass}}{\text{density}} = \frac{1800}{9.81 \times 900}$ [1]

volume $= 0.20\,\text{m}^3$ [1]

ii sum of clockwise moments about X = sum of anticlockwise moments about X [2]

$1800 \times 1.0 = T\cos 45° \times 1.5$ [1]

$T = 1.7 \times 10^3$ N

iii The force at X must have a horizontal component that would cancel out the horizontal component of the tension. [1]

6 a The resultant force is zero because the ship is travelling at constant speed. [1]

b force $= 2 \times 8.6 \times 10^3 \cos 30°$ [2]

- force $= 1.5 \times 10^4$ N [1]

c The drag is equal to 1.5×10^4 N because the resultant force on the ship is zero. [1]

The direction of drag is opposite to the direction of travel. [1]

Chapter 5

1 B [1] **2** C [1] **3** D [1] **4** C [1]

5 a Energy can neither be created nor destroyed. It can just be transferred from one form into another. [1]

b i loss in GPE $= mgh = 0.040 \times 9.81 \times 0.080$
$= 3.14 \times 10^{-2}\,\text{J} \approx 3.1 \times 10^{-2}$ J [1]

ii KE $= \frac{1}{2}mv^2 = \frac{1}{2} \times 0.040 \times 0.90^2$
$= 1.62 \times 10^{-2}\,\text{J} \approx 1.6 \times 10^{-2}$ J [1]

iii work done = $(3.14 - 1.62) \times 10^{-2}$ J [1]

work done = 1.52×10^{-2} J $\approx 1.5 \times 10^{-2}$ J [1]

iv $F \times 0.08 = 1.52 \times 10^{-2}$ [1]

$F = 0.19$ N [1]

6 **a** $a = (0.21 \times 9.81)/(0.80 + 0.21)$ [1]

$a = 2.04$ m s^{-2} [1]

b $v^2 = 2as = 2 \times 2.04 \times 0.50$ [1]

$\text{KE} = \frac{1}{2}mv^2 = \frac{1}{2} \times 0.80 \times [2 \times 2.04 \times 0.50]$ [1]

KE = 0.816 J ≈ 0.82 J [1]

c loss in GPE = $mgh = 0.21 \times 9.81 \times 0.50$

= 1.03 J ≈ 1.0 J [1]

d The loss in GPE of the 0.21 kg mass is equal to the KE of *both* the trolley and the falling mass – this is why the answers are not the same. [1]

Chapter 6

1 D [1] **2** D [1] **3** C [1] **4** D [1]

5 **a** **i** $k = \text{gradient} = \frac{5.0}{0.60}$ [1]

$k = 8.3$ N m^{-1} [1]

ii Assume that Hooke's law is obeyed. [1]

$F \propto x$ therefore $\frac{F}{x}$ = constant [1]

$x = 0.60 \times \frac{9.0}{5.0} = 1.08\text{ m} \approx 1.1\text{ m}$ [1]

b Force experienced by each spring = 10 N [1]

extension x of each spring $= \frac{10}{40} = 0.25$ m [1]

energy stored $= 2 \times \left(\frac{1}{2} \times 10 \times 0.25\right)$ [1]

energy stored = 2.5 J

6 **a** Young modulus $= \frac{\text{tensile stress}}{\text{tensile strain}}$ [1]

b **i** stress $= 0.4 \times 10^9 = \frac{16}{A}$ [1]

$A = 4.0 \times 10^{-8}$ m^2 [1]

ii E = gradient [1]

$E = \frac{0.3 \times 10^9}{2.0 \times 10^{-3}}$ [1]

$E = 1.5 \times 10^{11}$ Pa [1]

iii The shape does not depend on the dimensions of the material but just on the material itself. [1]

Therefore, there is no change to the shape of the graph. [1]

7 **a** strain $= \frac{x}{L} = \frac{0.54 \times 10^{-3}}{1.20}$ [1]

strain $= 4.5 \times 10^{-4}$ [1]

b stress $= E \times \text{strain} = 8.0 \times 10^{10} \times 4.5 \times 10^{-4}$ [1]

stress $= 3.6 \times 10^7$ Pa [1]

$F = \text{stress} \times A = 3.6 \times 10^7 \times 2.1 \times 10^{-7}$

= 7.56 N [1]

mass $= \frac{F}{g} = \frac{7.56}{9.81} = 0.77$ kg [1]

Chapter 7

1 B [1] **2** D [1] **3** A [1] **4** D [1]

5 **a** linear momentum = mass × velocity [1]

b The resultant force acting on an object is directly proportional to its rate of change of momentum. [1]

c **i** change in momentum

$= \Delta p = -0.800 \times (4.0 + 6.0)$

$\Delta p = (-)8.0$ kg m s^{-1} [1]

ii force $= \frac{\Delta p}{\Delta t} = \frac{8.0}{0.025}$ [1]

force = 320 N [1]

iii According to Newton's third law, the force acting on the ground is equal and opposite to the force experienced by the object. [1]

Therefore, the direction of the force experienced by the ground is vertically downwards. [1]

6 **a** impulse = force × time [1]

b **i** $F = ma$ therefore the acceleration a is directly proportional to the force F. [1]

Therefore, the acceleration is maximum at $t = 0$ and it decreases linearly to zero at $t = 0.4$ s. [1]

ii change in momentum = impulse = area under graph [1]

increase in momentum

$= \frac{1}{2} \times 6.0 \times 0.4 = 1.2$ kg m s^{-1} [1]

final momentum

$= 1.2 + (0.200 \times 3.0) = 1.8$ kg m s^{-1} [1]

final speed $v = \frac{1.8}{0.200} = 9.0$ m s^{-1} [1]

7 **a** Inelastic collision is one in which momentum is conserved but kinetic energy is not. Some of the KE is transferred to other forms. [1]

b **i** $p = 8.0 \times 10^4 \times 0.50 = 4.0 \times 10^4$ kg m s^{-1} [1]

ii initial momentum = final momentum [1]

$4.0 \times 10^4 = 12 \times 10^4 \times v$ [1]

$v = 0.33$ m s^{-1} [1]

iii **1** $\text{KE} = \frac{1}{2} \times 8.0 \times 10^4 \times (0.50^2 - 0.33^2)$ [1]

KE $= 5.64 \times 10^3$ J $\approx 5.6 \times 10^3$ J [1]

2 $\text{KE} = \frac{1}{2} \times 4.0 \times 10^4 \times 0.33^2 = 2.2 \times 10^3$ J [1]

iv The collision is inelastic with some of the kinetic energy being transferred to heat and sound. [1]

Chapter 8

1 B [1] **2** A [1] **3** C [1] **4** B [1]

5 a The direction of conventional current is the direction in which positive charges travel therefore the direction of this current is to the left. [1]

b i $Q = 0.040 \times 30$ [1]

charge = 1.2 C [1]

ii number of electrons $= \dfrac{1.2}{1.6 \times 10^{-19}}$ [1]

number of electrons $= 7.5 \times 10^{18}$ [1]

6 a $n = 8.5 \times 10^{22} \times 10^{6} = 8.5 \times 10^{28}\,\text{m}^{-3}$ $(1\,\text{cm}^3 = 10^{-6}\,\text{m}^3)$ [1]

b $I = Anev$ [1]

$v = \dfrac{9.5}{\pi \times (0.6 \times 10^{-3})^2 \times 8.5 \times 10^{28} \times 1.6 \times 10^{-19}}$ [1]

$v = 6.2 \times 10^{-4}\,\text{m s}^{-1}$ [1]

c distance $= 6.2 \times 10^{-4} \times 2.0 \times 60$ [1]

distance = 0.074 m

7 a current = rate of flow of charge therefore the gradient is equal to current. [1]

b current $= \dfrac{6.0}{4.0} = 1.5\,\text{A}$ [1]

c The current is constant at 1.5 A between $t = 0$ and $t = 4.0$ s. [1]

After $t = 4.0$ s, the current is smaller. [1]

Value of current after 4.0 s is 0.50 A. [1]

Chapter 9

1 C [1] **2** B [1] **3** D [1] **4** D [1]

5 a resistance $= \dfrac{\text{potential difference}}{\text{current}}$ [1]

b i $R = \dfrac{3.0}{2.0} = 1.5\,\Omega$ [1]

ii $\rho = \dfrac{RA}{L}$ [1]

$\rho = \dfrac{1.5 \times \pi \times (0.41 \times 10^{-3})^2}{0.09}$ [1]

$\rho = 8.8 \times 10^{-6}\,\Omega\text{m}$ [1]

iii current $= \dfrac{2.5}{1.5} = 1.7\,\text{A}$ [1]

6 a 12 W is the power of the lamp. [1]

12 J of electrical energy is transferred to heat (and light) per second. [1]

b $I_A = \dfrac{P}{V} = \dfrac{12}{230} = 0.052\,\text{A}$ [1]

$I_B = \dfrac{P}{V} = \dfrac{36}{230} = 0.157\,\text{A}$ [1]

total current $= 0.052 + 0.157 \approx 0.21\,\text{A}$ [1]

c energy in kWh $= \dfrac{12 + 36}{1000 \times 24} = 1.152\,\text{kWh}$ [2]

cost $= 1.152 \times 10.5 \approx 12\text{p}$ [1]

7 a The resistance of the LDR increases as the intensity of light decreases. [1]

Therefore, the current in the circuit decreases.

$\left(I = \dfrac{V}{R},\ V = \text{constant therefore } I \propto \dfrac{1}{R}\right)$ [1]

b For a given supply the power dissipated is greater when the resistance is smaller $\left(P = \dfrac{V^2}{R}\right)$. [1]

Therefore, greater power is dissipated when LDR is in sunlight and its resistance is smaller. [1]

Chapter 10

1 C [1] **2** C [1] **3** A [1]

4 a parallel: $R = (100^{-1} + 330^{-1})^{-1} = 76.7\,\Omega$ [2]

total $= 76.7 + 100 = 176.7\,\Omega \approx 180\,\Omega$ [1]

b p.d. across 330 Ω resistor:

$V = 2.7 \times 10^{-3} \times 330 = 0.891\,\text{V}$ [1]

current in the 100 Ω resistor:

$I = \dfrac{0.891}{100} = 8.91\,\text{mA}$ [1]

$\mathcal{E} = (8.91 + 2.7) \times 10^{-3} \times 176.7$ [1]

$\mathcal{E} = 2.05\,\text{V} \approx 2.1\,\text{V}$ [1]

5 a The sum of the p.d.s in a loop is equal to the sum of the e.m.f.s. [1]

b i $P = \dfrac{V^2}{R};\ R = \dfrac{3.0^2}{0.60}$ [1]

$R = 15\,\Omega$ [1]

ii $R = (15^{-1} + 20^{-1})^{-1}$ [1]

$R = 8.57\,\Omega \approx 8.6\,\Omega$ [1]

iii $\dfrac{8.57}{3.0} = \dfrac{r}{1.5}$ [1]

$r \approx 4.3\,\Omega$ [1]

6 a p.d. across resistor $= 0.015 \times 120 = 1.8\,\text{V}$ [1]

p.d. across thermistor $= 6.0 - 1.8 = 4.2\,\text{V}$ [1]

resistance of thermistor $= \dfrac{4.2}{0.015} = 280\,\Omega$ [1]

b The current in the circuit decreases. [1]

The p.d. across the variable resistor increases. [1]

The sum of the p.d.s must equal 6.0 V therefore the p.d. across the thermistor (voltmeter reading) will decrease. [1]

Chapter 11

1 D [1] **2** C [1] **3** B [1] **4** B [1]

5 a Any two properties from transverse waves: travel in a vacuum, travel at the speed of light c in a vacuum, etc. [2]

b i $\lambda = \dfrac{c}{f} = \dfrac{3.0 \times 10^8}{8.0 \times 10^9}$ [1]

$\lambda = 0.038\,\text{m}$ [1]

ii Microwaves. [1]

iii time $= \dfrac{6.3 \times 10^{11}}{3.0 \times 10^8}$ [1]

time = 2100 s (35 minutes) [1]

6 a The speed of light in glass is 1.54 times slower in glass than in vacuum (or air). [1]

b speed $= \dfrac{3.00 \times 10^8}{1.54} = 1.95 \times 10^8\,\text{m s}^{-1}$ [1]

c i $1.33 \times \sin 40 = 1.54 \sin r$ [1]

$\sin r = \frac{1.33 \times \sin 40}{1.54}$ [1]

$r = 34°$ [1]

ii Diagram showing refracted light into glass with angle of refraction less than the angle of incidence in water. [1]

Also, reflected light back into water. [1]

Chapter 12

1 D [1] **2** B [1] **3** D [1] **4** C [1]

5 a A progressive wave transfers energy from one place to another through vibrations. [1]

b i The waves emitted from L and R are coherent – they have a constant phase difference. [1]

ii The sound waves at P combine constructively. [1]

The phase difference between the waves is 0° or the path difference is a whole number of wavelengths. [1]

iii 1 $\lambda = \frac{ax}{D} = \frac{0.30 \times 1.4}{5.0}$ [2]

$\lambda = 0.084\,\text{cm}$ [1]

2 $v = f\lambda = 4.0 \times 10^3 \times 0.084$ [1]

speed = $340\,\text{m s}^{-1}$ [1]

6 a $\frac{\lambda}{2} = 30\,\text{cm}; \lambda = 60\,\text{cm}$ [1]

$v = f\lambda = 110 \times 0.60$ [1]

speed = $66\,\text{m s}^{-1}$ [1]

b The stable pattern of loops disappears. [1]

c $\frac{\lambda}{2} = 20\,\text{cm}; \lambda = 40\,\text{cm}$ [1]

$66 = f \times 0.40$ [1]

frequency = $165\,\text{Hz} \approx 170\,\text{Hz}$ [1]

Chapter 13

1 D [1] **2** B [1] **3** A [1] **4** B [1]

5 a This effect is the removal of electrons from the surface of a metal using electromagnetic radiation. [1]

b A single photon interacts with a single electron. [1]

Energy is conserved. [1]

energy of photon > work function (or frequency > threshold frequency) [1]

c i $hf = \phi + KE_{max}$ [1]

Equation for a straight-line graph is $y = mx + c$. Therefore, a graph of KE_{max} against f will be linear (with gradient = h and y-intercept = $-\phi$). [1]

ii $f = 9.0 \times 10^{13}$ Hz and $KE_{max} = 0.16\,\text{eV}$ [1]

$hf = \phi + KE_{max} = 6.63 \times 10^{-34} \times 9.0 \times 10^{13}$
$= \phi + (0.16 \times 1.60 \times 10^{-19})$ [1]

$\phi = 3.407 \times 10^{-20}\,\text{J}$ [1]

$\phi = \frac{3.407 \times 10^{-20}}{1.60 \times 10^{-19}} \approx 0.21\,\text{eV}$ [1]

iii threshold frequency
$= \frac{\phi}{h} = \frac{3.407 \times 10^{-20}}{6.63 \times 10^{-34}}$ [1]

threshold frequency = $5.1 \times 10^{13}\,\text{Hz}$ [1]

6 a This is the minimum energy required by an electron to escape from the surface of the metal. [1]

b $\frac{hc}{\lambda} = \phi + KE_{max}$ [1]

$6.63 \times 10^{-34} \times \frac{3.0 \times 10^8}{300 \times 10^{-9}}$
$= (3.7 \times 1.60 \times 10^{-19}) + KE_{max}$ [1]

$KE_{max} = 7.1 \times 10^{-20}\,\text{J}$ [1]

$KE_{max} = \frac{7.1 \times 10^{-20}}{1.60 \times 10^{-19}} \approx 0.44\,\text{eV}$ [1]

c $\frac{hc}{\lambda} = 3.7 \times 1.6 \times 10^{-19}$

$\lambda = 6.63 \times 10^{-34} \times \frac{3.0 \times 10^8}{3.7 \times 1.6 \times 10^{-19}}$ [1]

$\lambda = 3.4 \times 10^{-7}\,\text{m}$ [1]

7 a A moving electron has wave-like behaviour. [1]

The wavelength λ is given by the de Broglie equation: $\lambda = \frac{h}{p}$ [1]

(h = Planck and p = momentum of electron)

b $\lambda = \frac{h}{p}; p = \frac{h}{\lambda} = \frac{6.63 \times 10^{-34}}{2.0 \times 10^{-15}}$ [1]

momentum = $3.3 \times 10^{-19}\,\text{kg m s}^{-1}$ [1]

c Fast-moving electrons are made to travel through a (polycrystalline) metal or graphite. [1]

The electrons are diffracted by the spacing between the atoms. [1]

This is because their de Broglie wavelength is similar to atomic spacing. [1]

Chapter 14

1 C [1] **2** B [1] **3** A [1]

4 a $E = mc\Delta\theta = 0.120 \times 4200 \times 15$ [1]

rate of energy loss = $\frac{0.120 \times 4200 \times 15}{6 \times 60}$ [1]

rate of energy loss = 21 W [1]

b $E = 3.3 \times 10^5 \times 0.120$ [1]

time = $\frac{3.3 \times 10^5 \times 0.120}{21}$ [1]

time = 1900 s (31 mins) [1]

5 a A graph of m against t plotted with correct labelling. [1]

A line of best fit is drawn through the data points. [1]

b gradient = $9.3 \times 10^{-5}\,\text{kg s}^{-1}$ (Allow $\pm 0.1 \times 10^{-5}\,\text{kg s}^{-1}$) [1]

c $Pt = mL_f$ therefore gradient of m–t graph is $\frac{P}{L_f}$. [1]

$\frac{25}{L_f} = 9.3 \times 10^{-5}$ [1]

$L_f = 2.7 \times 10^5\,\text{J kg}^{-1}$ [1]

d The melting ice is also absorbing energy from the surroundings. [1]

6 a The temperature of the block is greater than that of the water, so there is a net transfer of thermal energy from the block to the water. [1]

b $E = mc\Delta\theta = 0.500 \times 4200 \times 17$ [1]

$E = 3.57 \times 10^4\,\text{J} \approx 3.6 \times 10^4\,\text{J}$ [1]

c $3.57 \times 10^4 = 0.210 \times 450 \times (\theta - 37)$ [1]

$\theta - 37 = 377.8$ [1]

$\theta = 414.8 \approx 410\,°\text{C}$ [1]

Chapter 15

1 B [1] **2** D [1] **3** C [1] **4** A [1]

5 a $pV = nRT$, with $n = 1$ therefore $pV = RT$ [1]

b $p \propto T$ therefore a graph of p against T will be a straight line (through the origin). [1]

c gradient $= \frac{R}{V} = \frac{N_A k}{V}$ [1]

gradient = $375\,\text{Pa K}^{-1}$ (Allow $\pm 15\,\text{Pa K}^{-1}$) [1]

$k = \frac{375 \times 2.2 \times 10^{-2}}{6.02 \times 10^{23}} = 1.37 \times 10^{-23}\,\text{J K}^{-1} \approx 1.4 \times 10^{-23}\,\text{J K}^{-1}$ [1]

6 a density $= \frac{M}{V} = \frac{0.65 \times 10^{-3}}{5.2 \times 10^{-4}} = 1.25\,\text{kg m}^{-3}$ [1]

b $n = \frac{0.65 \times 10^{-3}}{30 \times 10^{-3}} = 2.167 \times 10^{-2}\,\text{mol} \approx 2.2 \times 10^{-2}\,\text{mol}$ [1]

c $PV = nRT$ [1]

$T = \frac{1.2 \times 10^5 \times 5.2 \times 10^{-4}}{8.31 \times 2.167 \times 10^{-2}} = 346.5\,\text{K}$ [1]

temperature = $346.5 - 273 \approx 74\,°\text{C}$ [1]

d mean KE $= \frac{3}{2}kT = \frac{3}{2} \times 1.38 \times 10^{-23} \times 346.5$ [1]

mean KE = $7.2 \times 10^{-21}\,\text{J}$ [1]

Chapter 16

1 D [1] **2** D [1] **3** B [1] **4** D [1]

5 a Its velocity is changing. [1]

Acceleration is the rate of change of velocity – the object must therefore have acceleration. [1]

b i The arrow on the object is towards the centre of the disc. [1]

ii weight = $mg = 0.120 \times 9.81$ [1]

friction = $F = 0.50 \times 0.120 \times 9.81$ [1]

$F = \frac{mv^2}{r}$; $v = \sqrt{\frac{Fr}{m}} = \sqrt{\frac{0.50 \times 0.120 \times 9.81 \times 0.10}{0.120}}$ [1]

$v = 0.70\,\text{m s}^{-1}$ [1]

6 a $a = \frac{v^2}{r} = \frac{(4.2 \times 10^7)^2}{0.012}$ [1]

$a = 1.47 \times 10^{17}\,\text{m s}^{-2} \approx 1.5 \times 10^{17}\,\text{m s}^{-2}$ [1]

b $F = ma = 9.11 \times 10^{-31} \times 1.47 \times 10^{17}$ [1]

$F = 1.34 \times 10^{-13}\,\text{N} \approx 1.3 \times 10^{-13}\,\text{N}$ [1]

c $T = \frac{2\pi r}{v} = \frac{2\pi \times 0.012}{4.2 \times 10^7}$ [1]

$T = 1.8 \times 10^{-9}\,\text{s}$ [1]

7 a $T\cos 45° = 1.2$; $T = 1.697\,\text{N} \approx 1.7\,\text{N}$ [1]

b centripetal force = $T\sin 45° = 1.697 \times \sin 45° = 1.2\,\text{N}$ [1]

c $F = \frac{mv^2}{r}$; mass $m = \frac{1.2}{9.81} = 0.1223\,\text{kg}$ [1]

$v = \sqrt{\frac{Fr}{m}} = \sqrt{\frac{1.2 \times 0.15}{0.1223}}$ [1]

$v = 1.2\,\text{m s}^{-1}$ [1]

Chapter 17

1 D [1] **2** D [1] **3** B [1] **4** B [1]

5 a $a \propto -x$ [1]

b $A = 1.6\,\text{mm}$ [1]

c i $v_{max} = \omega A = 2\pi f A = 2\pi \times 840 \times 1.6 \times 10^{-3}$ [1]

$v_{max} = 8.446\,\text{m s}^{-1} \approx 8.4\,\text{m s}^{-1}$ [1]

ii $a = \omega^2 A = (2\pi \times 840)^2 \times 1.6 \times 10^{-3}$ [1]

$a = 4.5 \times 10^4\,\text{m s}^{-2}$ [1]

d The potential energy is zero at the equilibrium position. [1]

It increases as the displacement increases, reaching a maximum at the amplitude position. [1]

6 a $v_{max} = \omega A$ [1]

b $v_{max} = \omega A$, where ω is constant and therefore v_{max} is proportional to A.

This gives a straight line (through the origin). [1]

c i gradient = $8.0\,\text{s}^{-1}$ (Allow $\pm 0.1\,\text{s}^{-1}$) [1]

ii gradient = $\omega = 2\pi f$ [1]

$f = \frac{8.0}{2\pi}$ [1]

$f = 1.27\,\text{Hz} \approx 1.3\,\text{Hz}$ [1]

Chapter 18

1 C [1] **2** A [1] **3** C [1]

4 a gravitational field strength $g = \frac{\text{force}}{\text{mass}} = \frac{F}{m}$

and acceleration $a = \frac{\text{force}}{\text{mass}} = \frac{F}{m}$ [1]

Therefore $a = g$. [1]

b i g is inversely proportional to r^2 therefore gr^2 = constant. [1]

A minimum of two from:

$9.8 \times (6400 \times 10^3)^2 = 4.0 \times 10^{14}$

$6.3 \times (8000 \times 10^3)^2 = 4.0 \times 10^{14}$

$4.4 \times (9600 \times 10^3)^2 = 4.0 \times 10^{14}$ [1]

ii $gr^2 = GM = 4.0\times10^{14}$ [1]

$M = \frac{4.0\times10^{14}}{6.67\times10^{-11}}$ [1]

$M = 6.0\times10^{24}\,\text{kg}$ [1]

iii density $= \frac{M}{V} = \frac{6.0\times10^{24}}{\frac{4}{3}\pi\times(6400\times10^3)^3}$ [1]

density $= 5.5\times10^3\,\text{kg m}^{-3}$ [1]

5 a $\frac{GMm}{r^2} = \frac{mv^2}{r}$ [2]

$v = \sqrt{\frac{GM}{r}}$ [1]

b ratio $= \sqrt{\frac{r_{\text{Neptune}}}{r_{\text{Earth}}}}$ [1]

ratio $= \sqrt{30} = 5.5$ [1]

6 a 2.8 MJ [1]

b $V_g = -\frac{GM}{r} \propto \frac{1}{r}$ [1]

V_g at $3R = -\frac{2.8}{3} = -0.933\,\text{MJ kg}^{-1} \approx$ $-9.3\times10^5\,\text{J kg}^{-1}$ [1]

V_g at $4R = -\frac{2.8}{4} = -0.70\,\text{MJ kg}^{-1} \approx$ $-7.0\times10^5\,\text{J kg}^{-1}$ [1]

c energy $= \Delta V_g \times m = (0.933 - 0.70)\times10^6\times3500$ [1]

energy $= 8.2\times10^8\,\text{J}$ [1]

Chapter 19

1 D [1] **2** A [1] **3** D [1]

4 a Any <u>two</u> from:

Hot surface temperature, remnant of a low-mass star, mass less than 1.44 solar masses, no fusion takes place, etc. [2]

b i Correct shape of intensity against wavelength curve, see below. [1]

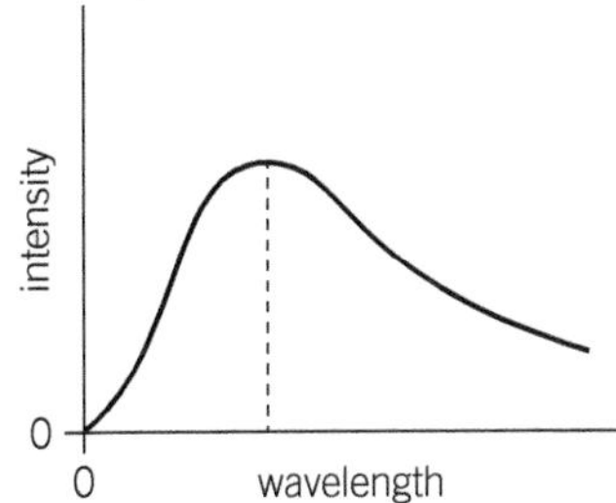

ii wavelength at maximum intensity × temperature in K = constant (Wien's displacement law) [1]

c $L = 4\pi r^2\sigma T^4$ [1]

$L = 3.8\times10^{26}\times\frac{1}{120^2}\times\left(\frac{25000}{5800}\right)^4$ [1]

$L = 9.1\times10^{24}\,\text{W}$ [1]

5 a The emission (or absorption) spectrum of each element is unique. [1]

By measuring the wavelengths using a grating, these elements can be identified in the stars. [1]

b $d\sin\theta = n\lambda$; $\theta = 90°$ and $d = \frac{10^{-3}}{800} = 1.25\times10^{-6}\,\text{m}$ [1]

$n = \frac{1.25\times10^{-6}}{656\times10^{-9}} = 1.905$ [1]

Therefore, maximum order n is 1. [1]

c $\sin\theta = \frac{1\times589\times10^{-9}}{1.25\times10^{-6}}$; $\theta = 28.11°$ [1]

$\sin\theta = \frac{1\times587\times10^{-9}}{1.25\times10^{-6}}$; $\theta = 28.01°$ [1]

angular separation = 0.10° [1]

6 a Luminosity is the total radiant power emitted from the surface of a star. [1]

b i $L = 4\pi r^2\sigma T^4 = 4\pi\times(8.5\times10^8)^2\times5.67\times10^{-8}\times(5800)^4$ [1]

$L = 5.83\times10^{26}\,\text{W}$ [1]

ii intensity $= \frac{5.83\times10^{26}}{4\pi\times(3.9\times10^{16})^2}$ [1]

intensity $= 3.05\times10^{-8}\,\text{W m}^{-2}$ [1]

c A sensible suggestion, e.g., absorption of radiation by dust or Earth's atmosphere. [1]

Chapter 20

1 C [1] **2** D [1] **3** D [1] **4** C [1]

5 a Correct diagram, see below. [1]

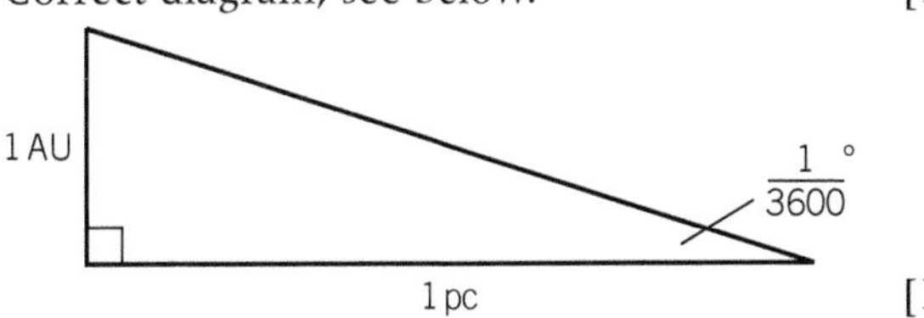

[1]

$\tan\left(\frac{1}{3600}°\right) = \frac{1.5\times10^{11}}{1\,\text{pc}}$ [1]

$1\,\text{pc} = \frac{1.5\times10^{11}}{\tan(2.78\times10^{-4})°} = 3.09\times10^{16}\,\text{m} \approx$ $3.1\times10^{16}\,\text{m}$ [1]

b distance $= 4.0\times10^4\times3.1\times10^{16} = 1.24\times10^{21}\,\text{m}$ [1]

time $= \frac{1.24\times10^{21}}{3.0\times10^8} = 4.13\times10^{12}\,\text{s} \approx 4.1\times10^{12}\,\text{s}$ [1]

time $= \frac{4.13\times10^{12}}{3.16\times10^7} = 1.3\times10^5\,\text{y}$ [1]

c i Hubble constant $= \frac{70\times10^3}{3.1\times10^{16}\times10^6}$ [1]

Hubble constant $H_0 = 2.26\times10^{-18}\,\text{s}^{-1} \approx 2.3\times10^{-18}\,\text{s}^{-1}$ [1]

ii age $= H_0^{-1} = (2.26\times10^{-18})^{-1} = 4.43\times10^{17}\,\text{s}$ [1]

The maximum distance is the product of the speed of light and the maximum time. [1]

distance $= 4.43\times10^{17}\times3.0\times10^8 = 1.33\times10^{26}\,\text{m} \approx 1.3\times10^{26}\,\text{m}$ [1]

d Any <u>two</u> from:

Microwave background radiation, 3K temperature, and abundance of helium. [2]

6 a Line of best fit drawn (must pass through all the error bars). [1]

b $\frac{\Delta\lambda}{\lambda} \approx \frac{v}{c}$, where c and v are constants therefore

$\Delta\lambda \propto \lambda$ and therefore a straight line (through the origin). [1]

c gradient = 6.8×10^{-5} (Allow $\pm 0.2 \times 10^{-5}$) [1]

$\frac{v}{c} = \frac{v}{3.00 \times 10^{8}} = 6.8 \times 10^{-5}$ [1]

$v = 2.0 \times 10^{4}\,\mathrm{m\,s^{-1}}$ [1]

Chapter 21

1 B [1] **2** D [1] **3** A [1] **4** D [1]

5 a Any four from:

The current in the circuit charges the capacitor.

The charge stored by the capacitor increases and therefore the p.d. across it increases.

The sum of the p.d.s across the capacitor and resistor must add to 6.00 V.

Therefore, as the p.d. across the capacitor increases, the p.d. across the resistor decreases.

The final p.d. across the capacitor is 6.00 V and zero across the resistor. [4]

b $V_c = V_0(1 - e^{-\frac{t}{CR}})$ and

$CR = 120 \times 10^{-6} \times 1.0 \times 10^{6} = 120\,\mathrm{s}$ [1]

$V_c = 6.00(1 - e^{-\frac{200}{120}})$ [1]

$V_c = 4.87$ V [1]

c maximum current = $\frac{6.00}{1.0 \times 10^{6}} = 6.00\,\mu\mathrm{A}$ [1]

d $E = \frac{1}{2}V^2C = \frac{1}{2} \times 6.00^2 \times 120 \times 10^{-6}$ [1]

$E = 2.2 \times 10^{-3}\,\mathrm{J}$ [1]

6 a $CR = 150 \times 10^{-6} \times 100 \times 10^{3} = 15\,\mathrm{s}$ [1]

b $V = V_0 e^{-\frac{t}{CR}} = 4.50 \times e^{-\frac{35}{15}}$ [1]

$V = 0.44$ V [1]

c energy dissipated = difference in the energy stored by the capacitor [1]

energy dissipated = $\frac{1}{2}C\left(V_1^2 - V_2^2\right) =$

$\frac{1}{2} \times 150 \times 10^{-6}\left(4.50^2 - 0.44^2\right)$ [1]

energy dissipated = $1.5 \times 10^{-3}\,\mathrm{J}$ [1]

Chapter 22

1 C [1] **2** A [1] **3** C [1] **4** C [1]

5 a $E \propto \frac{1}{r^2}$; therefore Er^2 = constant [1]

A minimum of two from:

$20.7 \times 10^{3} \times 0.092^2 = 175$

$5.4 \times 10^{3} \times 0.18^2 = 175$

$2.8 \times 10^{3} \times 0.25^2 = 175$ [1]

b $E = \frac{Q}{4\pi\varepsilon_0 r^2}$ therefore $Er^2 = \frac{Q}{4\pi\varepsilon_0}$ [1]

$Q = 4\pi \times 8.85 \times 10^{-12} \times 175$ [1]

$Q = 1.946 \times 10^{-8}\,\mathrm{C} \approx 1.9 \times 10^{-8}\,\mathrm{C}$ [1]

c charge per unit area = $\frac{Q}{4\pi r^2}$ [1]

charge per init area = $\frac{1.946 \times 10^{-8}}{4\pi \times 0.050^2} =$

$6.2 \times 10^{-7}\,\mathrm{C\,m^{-2}}$ [1]

d $V = \frac{Q}{4\pi\varepsilon_0 r} = \frac{1.946 \times 10^{-8}}{4\pi \times 8.85 \times 10^{-12} \times 0.05}$ [1]

$V = 3.5 \times 10^{3}\,\mathrm{V}$ [1]

6 a The electric field is uniform between the plates, with field lines parallel and equally spaced (and perpendicular to the plates). [1]

The electric field strength is constant between the plates. [1]

b $E = \frac{V}{d} = \frac{5000}{0.072}$ [1]

$E = 6.94 \times 10^{4}\,\mathrm{V\,m^{-1}} \approx 6.9 \times 10^{4}\,\mathrm{V\,m^{-1}}$ [1]

c $a = \frac{EQ}{m} = \frac{6.94 \times 10^{4} \times 1.60 \times 10^{-19}}{9.11 \times 10^{-31}} =$

$1.22 \times 10^{16}\,\mathrm{m\,s^{-2}}$ [2]

$v^2 = u^2 + 2as = 0 + 2 \times 1.22 \times 10^{16} \times 0.072$ [1]

$v = 4.2 \times 10^{7}\,\mathrm{m\,s^{-1}}$ [1]

d The final kinetic energy of the electron is Ve. [1]

This is independent of the separation between the plates. [1]

Chapter 23

1 A [1] **2** B [1] **3** B [1]

4 a $f = T^{-1} = 0.020^{-1} = 50\,\mathrm{Hz}$ [1]

b $F = BIL$; $4.0 \times 10^{-3} = 0.060 \times I \times 0.01$ [1]

$I = 6.7$ A [1]

c $F = BIL\sin\theta$ therefore F will decrease as θ is decreased. [1]

The graph will remain sine-shaped, but maximum force will decrease. [1]

5 a $BQv = \frac{mv^2}{r}$ [1]

$mv = p = BQr$ [1]

b i $p = Ber = 0.012 \times 1.60 \times 10^{-19} \times 0.032$ [1]

$p = 6.144 \times 10^{-23}\,\mathrm{kg\,m\,s^{-1}} \approx 6.1 \times 10^{-23}\,\mathrm{kg\,m\,s^{-1}}$ [1]

ii $v = \frac{6.144 \times 10^{-23}}{9.11 \times 10^{-31}} = 6.744 \times 10^{7}\,\mathrm{m\,s^{-1}}$ [1]

$\mathrm{KE} = \frac{1}{2} \times 9.11 \times 10^{-31} \times (6.744 \times 10^{7})^2 =$

$2.07 \times 10^{-15}\,\mathrm{J}$ [1]

KE = 13 keV [1]

6 a When the current is switched on, an increasing magnetic flux links the coil. [1]

The induced e.m.f. is equal to the rate of change of magnetic flux therefore an e.m.f. is induced in the coil for a short period. [1]

When the current is constant, the flux linkage is constant and therefore no e.m.f. is induced in the coil. [1]

b Field pattern with concentric circles and separation between field lines increasing as shown. [2]

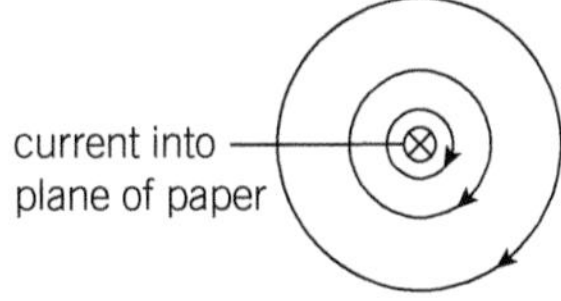

Chapter 24

1 A [1] **2** B [1] **3** A [1] **4** C [1]

5 a R is proportional to $A^{\frac{1}{3}}$, therefore

$\frac{R}{\sqrt[3]{A}}$ = constant. [1]

A minimum of two data used, e.g., $\frac{5.6}{\sqrt[3]{100}} = 1.2$ and $\frac{7.0}{\sqrt[3]{200}} = 1.2$ [1]

b The mass is proportional to A. [1]

The volume is $\frac{4}{3}\pi r^3$ therefore volume is proportional to A. [1]

The density (which is mass divided by volume) is independent of A. [1]

c mass $= 4.00 \times 1.66 \times 10^{-27}$ kg [1]

$r = 1.2 \times 4^{\frac{1}{3}} = 1.9$ fm $= 1.9 \times 10^{-15}$ m

[The 1.2 fm is from part (a)] [1]

density $= \dfrac{4.00 \times 1.66 \times 10^{-27}}{\frac{4}{3}\pi \times (1.9 \times 10^{-15})^3}$ [1]

density $= 2.3 \times 10^{17}$ kg m^{-3} [1]

6 a The repulsive electrostatic force on the alpha-particle is greater at shorter distances, and therefore has greater deviation in its path. [1]

b electrical potential energy = kinetic energy = 8.2×10^{-13} J [1]

$\dfrac{Qq}{4\pi\varepsilon_0 r} = \dfrac{79 \times 2 \times (1.60 \times 10^{-19})^2}{4\pi \times 8.85 \times 10^{-12} \times r} = 8.2 \times 10^{-13}$ [1]

$r = 4.44 \times 10^{-14}$ m $\approx 4.4 \times 10^{-14}$ m [1]

c $F = \dfrac{Qq}{4\pi\varepsilon_0 r^2} = \dfrac{79 \times 2 \times (1.60 \times 10^{-19})^2}{4\pi \times 8.85 \times 10^{-12} \times (4.4 \times 10^{-14})^2}$ [1]

force = 19 N [1]

Chapter 25

1 B [1] **2** B [1] **3** C [1] **4** B [1]

5 a Exponential decay has a constant-ratio property. [1]

With $\Delta t = 5.0$ s, $\frac{34}{40} \approx \frac{28}{34} \approx \frac{24}{28}$ [1]

b $A_0 = 40 \times 10^{14}$ Bq and $A = 24 \times 10^{14}$ Bq when $t = 15$ s [1]

$A = A_0 e^{-\lambda t}$; $24 = 40 e^{-\lambda \times 15}$ [1]

$\lambda = -\dfrac{\ln(0.60)}{15} = 3.41 \times 10^{-2}$ s^{-1} [1]

$T = \dfrac{\ln 2}{3.41 \times 10^{-2}} = 20.3$ s ≈ 20 s [1]

c ${}^{10}_{6}C \rightarrow {}^{10}_{5}B + {}^{0}_{+1}e + \nu_e$ (Allow ${}^{10}_{5}X$ for ${}^{10}_{5}B$) [2]

d power = activity 3.4×10^{-13} [1]

power $= 40 \times 10^{14} \times 3.4 \times 10^{-13} = 1.36 \times 10^3$ W ≈ 1.3 kW [1]

6 a The count rate is a fraction of the total activity of the source. [1]

Therefore, the suggestion is incorrect. [1]

b i $\lambda = \dfrac{\ln 2}{T} = \dfrac{\ln 2}{6.0 \times 3600} = 3.21 \times 10^{-5}$ s^{-1} [1]

$A = \lambda N = 3.21 \times 10^{-5} \times 2.0 \times 10^9$ [1]

$A = 6.4 \times 10^4$ Bq [1]

ii 12 hours = 2 half-lives [1]

$N = \dfrac{2.0 \times 10^9}{4} = 5.0 \times 10^8$ [1]

iii $N = N_0 e^{-\lambda t} = 2.0 \times 10^9 \times e^{-(3.21 \times 10^{-5} \times 21 \times 3600)}$ [1]

$N = 1.8 \times 10^8$ [1]

Chapter 26

1 A [1] **2** D [1] **3** B [1] **4** C [1]

5 a This isotope has the largest BE per nucleon and therefore it cannot decay. [1]

b i The BE per nucleon of the ${}^{2}_{1}H$ nucleus is less than the BE per nucleon of the ${}^{4}_{2}He$ nucleus. [1]

The BE of the ${}^{4}_{2}He$ nucleus is greater than the total BE of the ${}^{2}_{1}H$ nuclei. [1]

The difference in the BE is released as energy. [1]

ii number of ${}^{2}_{1}H$ 'pairs' $= 3.01 \times 10^{23}$ [1]

difference in BE $\approx 4 \times 7.1 - 4 \times 1.0 = 24.4$ MeV [1]

total energy released $= 3.01 \times 10^{23} \times 24.4 \times 10^6 \times 1.60 \times 10^{-19}$ [1]

total energy $= 1.2 \times 10^{12}$ J [1]

6 a The change in energy ΔE is related to the change in mass Δm by the equation $\Delta E = \Delta mc^2$, where c is the speed of light in a vacuum. Therefore, mass and energy are equivalent. [1]

b i The hydrogen nuclei are positively charged and therefore repel each other. [1]

High temperatures are necessary for the nuclei to have high enough KE to get close enough to each other so that the strong nuclear force can fuse together the hydrogen nuclei. [1]

ii The mass of the helium-3 nucleus is less than the total mass of the hydrogen-2 and hydrogen-1 nuclei. [1]

The difference in the mass is equivalent to the energy released ($\Delta E = \Delta mc^2$). [1]

iii energy $= 5.5 \times 10^6 \times 1.60 \times 10^{-19}$ [1]

$\Delta m = \dfrac{5.5 \times 10^6 \times 1.60 \times 10^{-19}}{(3.00 \times 10^8)^2}$ [1]

$\Delta m = 9.8 \times 10^{-30}$ kg [1]

c In fission, a neutron is absorbed by a 'heavy' nucleus (e.g., uranium-235). This causes the nucleus to *split* into smaller nuclei and fast moving neutrons. [1]

Fusion occurs with 'lighter' nuclei (e.g., hydrogen, helium, etc.). These lighter nuclei *join together* to form a larger nucleus. [1]

Chapter 27

1 B [1] **2** C [1] **3** A [1]

4 a A transducer is used to send a pulse of ultrasound into the patient. [1]

The pulse is reflected at the boundaries between tissues. [1]

The reflected pulse is monitored using the transducer. [1]

The time difference t between the pulse received and sent is used to find the depth x of the boundary; $x = \frac{1}{2}ct$ where c is the (average) speed of sound in the tissues. [1]

4 b acoustic impedance = density of material × speed of ultrasound in material [1]

c i $Z_X = 1500 \times 4100 = 6.15 \times 10^6$ [1]

$Z_Y = 1100 \times 1600 = 1.76 \times 10^6$ [1]

unit: $\text{kg m}^{-2}\,\text{s}^{-1}$ [1]

ii $\dfrac{I_r}{I_0} = \dfrac{(Z_2 - Z_1)^2}{(Z_2 + Z_1)^2} = \dfrac{(6.15 - 1.76)^2}{(6.15 + 1.76)^2}$ [1]

$\dfrac{I_r}{I_0} = 0.31$ [1]

percentage transmitted = 69% [1]

5 a X-rays cause ionisation. [1]

It can damage healthy cells in the patient. [1]

b i $I = I_0 e^{-\mu x}$ therefore $\ln I = \ln I_0 - \mu x$ [1]

Compare with the equation for a straight line, $y = mx + c$. The gradient is $-\mu$. [1]

ii gradient $= 6.7\ \text{cm}^{-1}$ (Allow $\pm 0.1\ \text{cm}^{-1}$) [1]

iii $I = I_0 e^{-\mu x} = 2.0 \times 10^9 e^{-6.7 \times 1.8}$ [1]

$I = 1.157... \times 10^4\ \text{W m}^{-2}$ [1]

power $= 1.157... \times 10^4 \times 6.0 \times 10^{-4} = 6.944$ W [1]

energy $= 1.0 \times 60 \times 6.944 = 420$ J [1]

Answers to summary questions

2.1/2.2

1 distance = 210×10^3 m = 2.1×10^5 m [1]

2 speed = 1.2×10^{-3} m s^{-1} [1]

3 time = 12×10^{-9} s = 1.2×10^{-8} s [1]

4 diameter in pm = $\dfrac{2.3 \times 10^{-10}}{10^{-12}}$ = 230 pm [1]

diameter in nm = $\dfrac{2.3 \times 10^{-10}}{10^{-9}}$ = 0.23 nm [1]

5 unit for work done = unit for force × unit for length

unit for work done = kg m s^{-2} × m [1]

unit for work done = kg m^2 s^{-2} [1]

6 time = $\dfrac{\text{distance}}{\text{speed}} = \dfrac{150 \times 10^9}{300 \times 10^6}$ [2]

time = 500 s (8.3 minutes) [1]

7 1 cm = 10^{-2} m therefore 1 cm^2 = $(10^{-2})^2$ = 10^{-4} m^2 [1]

area = 620×10^{-4} = 6.2×10^{-2} m^2 [1]

2.3/2.4

1 Any two correct scalars (e.g., speed and distance). [1]

Any two correct vectors (e.g., velocity and momentum). [1]

2 A scalar (13 kg) is being added to a vector (force), which is incorrect. [1]

3 maximum velocity = 20 + 5.0 = 25 m s^{-1} [1]

minimum velocity = 20 −5.0 = 15 m s^{-1} [1]

4 Correct triangle drawn. [1]

Correct directions of the arrows and correct labelling. (See below.)

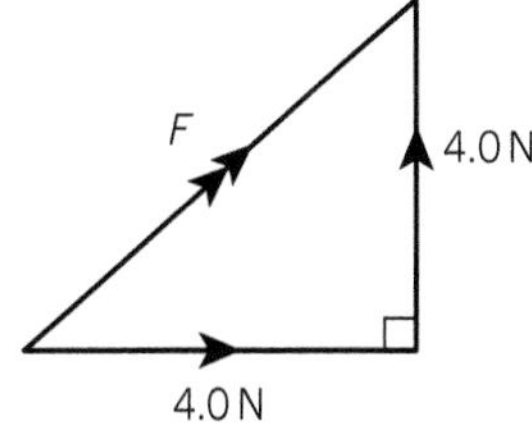

[1]

5 $F^2 = 4.0^2 + 4.0^2$ [1]

F = 5.7 N [1]

6 a Correct triangle drawn. [1]

Correct directions of the arrows and correct labelling. (See below.)

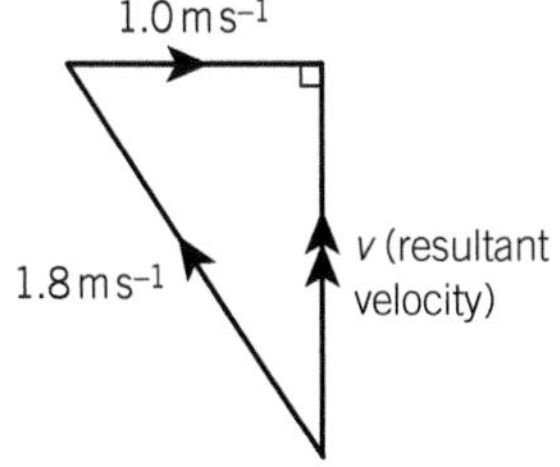

[1]

b $1.8^2 = v^2 + 1.0^2$ [1] v = 1.5 m s^{-1} [1]

2.5/2.6

1 Horizontally: $v = v_0 \cos\theta = 300 \times \cos 65°$
$= 127$ N ≈ 130 N [1]

Vertically: $v = v_0 \sin\theta = 300 \times \sin 65°$
$= 272$ N ≈ 270 N [1]

2 acceleration = 9.81 cos 30° = 8.5 m s^{-2} [1]

3 $100 = F \sin 50°$ [1]

$F = \dfrac{100}{\sin 50°}$ = 130.54 N ≈ 130 N [1]

4 a $F^2 = 2.0^2 + 3.1^2$ [1]

$F = \sqrt{13.61}$ = 3.7 N [1]

b $F^2 = (3.0 - 2.0)^2 + 1.0^2$ [1]

$F = \sqrt{2.0}$ = 1.4 N [1]

5 $F^2 = 100^2 + 200^2 - 2 \times 100 \times 200 \times \cos 70°$ [1]

$F = \sqrt{100^2 + 200^2 - 2 \times 100 \times 200 \times \cos 70°}$ [1]

F = 191 N ≈190 N [1]

6 $F^2 = 100^2 + 200^2 - 2 \times 100 \times 200 \times \cos 110°$ [1]

$F = \sqrt{100^2 + 200^2 - 2 \times 100 \times 200 \times \cos 110°}$ [1]

F = 252 N ≈ 230 N [1]

3.1/3.2

1 $v = \dfrac{1.20}{3.4}$ = 0.35 m s^{-2} [1]

2 distance = 20 × (0.40 × 3600) [1]

distance = 2.88×10^4 m ≈ 2.9×10^4 m [1]

3 A: Stationary. [1]

B: Constant velocity and moving away. [1]

C: Constant velocity and coming back. [1]

4 Correct graph from t = 0 to t = 30 minutes. [1]

Correct graph from t = 30 minutes to t = 60 minutes.

See below:

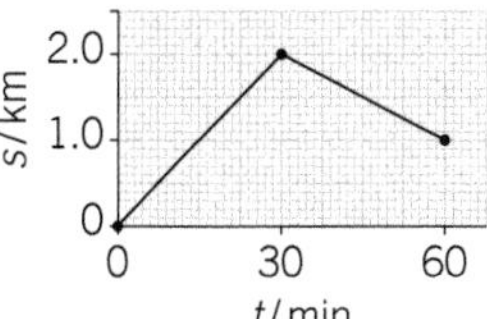

[1]

5 $t = \dfrac{2\pi r}{v} = \dfrac{2\pi \times 0.020}{5.0 \times 10^5}$ [1] $t = 2.5 \times 10^{-7}$ s [1]

6 a The magnitude of the velocities is the same. [1]

However, the ball is travelling in opposite directions. [1]

b The velocity is zero. [1]

The gradient of the graph at t = 1.0 s is zero. [1]

3.3/3.4

1 a Zero acceleration or constant velocity. [1]

b Constant acceleration (starting from a non-zero velocity). [1]

c Constant deceleration (starting from a non-zero velocity). [1]

2 $a = \frac{\Delta v}{\Delta t} = \frac{320 - 100}{20}$ [1]

$a = 11\,\text{m s}^{-2}$ [1]

3 $\Delta v = a\Delta t = 3.0 \times (1.5 \times 60)$ [1]

$\Delta v = 270\,\text{m s}^{-1}$ [1]

4 $\Delta t = \frac{\Delta v}{a} = \frac{10}{400}$ [1]

$\Delta t = 0.025$ s [1]

5 acceleration = gradient [1]

$a = \frac{4.0 - 1.0}{4.0 - 2.0} = 1.5\,\text{m s}^{-2}$ [1]

6 a 0 – 10 s: Straight line of positive gradient starting from the origin. [1]

10 s – 45 s: Horizontal line with $v = 4.0\,\text{m s}^{-1}$ [1]

45 s – 50 s: Straight line of negative gradient from $v = 4.0\,\text{m s}^{-1}$ to $v = 0$

See below

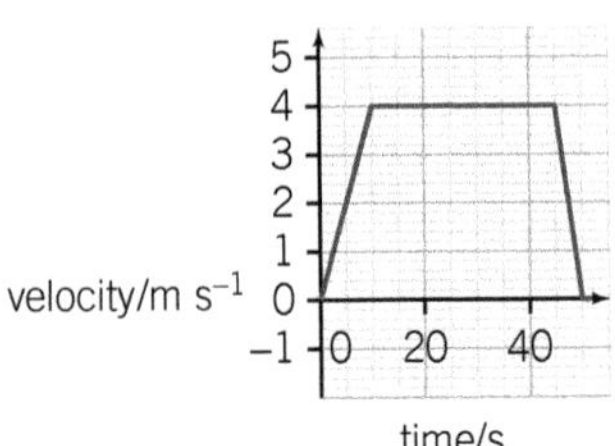

[1]

b distance = area under the graph [1]

distance $= \left(\frac{1}{2} \times 4.0 \times 10\right) + (4.0 \times 35) + \left(\frac{1}{2} \times 4.0 \times 5.0\right)$ [1]

distance = 170 m [1]

c $a = \frac{\Delta v}{\Delta t} = \frac{0 - 4.0}{5.0}$ [1] $a = (-)0.80\,\text{m s}^{-2}$ [1]

3.5/3.6

1 $v = u + at = 15 + (-9.81 \times 1.0)$ [1]

$v = 5.2\,\text{m s}^{-1}$ [1]

2 $s = ut + \frac{1}{2}at^2 = (5.0 \times 7.0) + \frac{1}{2} \times 3.0 \times 7.0^2$ [1]

$s = 108.5\,\text{m} \approx 109\,\text{m}$ [1]

3 thinking distance = 16 × 0.5 = 8.0 m [1]

braking distance $= \frac{1}{2} \times 16 \times 3.0 = 24\,\text{m}$ [1]

stopping distance = 8.0 + 24 = 32 m [1]

4 $s = \frac{1}{2}(u + v)t$ [1]

$80 = \frac{1}{2}(10 + 25)t$ [1]

$t = \frac{160}{35} = 4.6\,\text{s}$ [1]

5 $s = ut + \frac{1}{2}at^2$ with $u = 0$ [1]

$100 = \frac{1}{2} \times 9.81 \times t^2$ [1] $t = 4.5\,\text{s}$ [1]

6 $a = \frac{v - u}{t}$ $v = -14\,\text{m s}^{-1}$ $u = 12\,\text{m s}^{-1}$

$t = 4.0 \times 10^{-3}\,\text{s}$ [1] $a = \frac{-26}{40 \times 10^{-3}}$ [1]

(Note: change in velocity $= v - u = -14 - 12 = -26\,\text{m s}^{-1}$)

Therefore, magnitude of the acceleration $a = 6.5 \times 10^3\,\text{m s}^{-2}$ [1]

3.7

1 All objects have a common acceleration of free fall of $9.81\,\text{m s}^{-2}$ [1]

2 $s = ut + \frac{1}{2}at^2$; $u = 0$ and $g = a = 9.81\,\text{m s}^{-2}$ [1]

$s = \frac{1}{2} \times 9.81 \times 0.30^2 = 0.44\,\text{m}$ [1]

3 $s = ut + \frac{1}{2}at^2$; $u = 0$ and $g = a = 1.6\,\text{m s}^{-2}$ [1]

$s = \frac{1}{2} \times 1.6 \times 0.30^2 = 0.072\,\text{m}$ [1]

4 $s = \frac{1}{2}at^2$ therefore $g = \frac{2s}{t^2}$ [1]

$g = \frac{2 \times 2.5}{0.70^2}$ [1]

$g = 10.2\,\text{m s}^{-2} \approx 10\,\text{m s}^{-2}$ [1]

5 $s = \frac{1}{2}at^2$ therefore $g = \frac{2s}{t^2}$ [1]

$g = \frac{2 \times 0.800}{0.403^2}$ [1]

$g = 9.85\,\text{m s}^{-2}$ [1]

6 gradient $= \frac{g}{2}$ [1]

$g = 2 \times \text{gradient} = 2 \times \frac{4.00 - 1.50}{0.82 - 0.31}$ [1]

$g = 9.8\,\text{m s}^{-2}$ (Allow ± 0.2 m s^{-2}) [1]

7 absolute uncertainty $= \left(\frac{3}{800} + 2 \times \frac{0.002}{0.403}\right) \times 9.85\,\text{m s}^{-2}$ [2]

absolute uncertainty $= 0.13\,\text{m s}^{-2}$ [1]

Therefore, $g = 9.85 \pm 0.13\,\text{m s}^{-2}$ [1]

3.8

1 The horizontal velocity is $5.0\,\text{m s}^{-1}$ [1]

There is no acceleration in the horizontal direction because $g\cos 90° = 0$. [1]

2 displacement = 5.0 × 0.25 = 1.25 m ≈ 1.3 m [1]

3 height $= ut + \frac{1}{2}gt^2$ [1]

height $= 0 + \frac{1}{2} \times 9.81 \times 0.70^2$ (Vertical motion) [1]

height = 2.4 m [1]

4 horizontal displacement = 250 × 0.30 = 75 m [1]

Vertically ⇒ $s = ?$, $u = 0$, $a = 9.81\,\text{m s}^{-2}$, and $t = 0.30\,\text{s}$. [1]

$s = \frac{1}{2} \times 9.81 \times 0.30^2$ [1]

vertical displacement = 0.44 m [1]

5 $s = ?$ $u = 30\sin 45° = 21.2\,\text{m s}^{-1}$

$v = 0$ $a = -9.81\,\text{m s}^{-2}$ [1]

$s = \frac{v^2 - u^2}{2a} = \frac{0 - (30\sin 45)^2}{2 \times -9.81}$ [1]

$s = 23\,\text{m}$ [1]

6 $t = \frac{0 - 30\sin 45}{9.81} = 4.32\,\text{s}$ (t = time of flight) [1]
horizontal displacement = $30\cos 45° \times 4.32$ [1]
horizontal displacement = 92 m [1]

4.1/4.2/4.3

1 $a = \frac{400}{1000} = 0.40\,\text{m s}^{-2}$ [1]

2 $a = \frac{400 - 150}{1000}$ [1]
$a = 0.25\,\text{m s}^{-2}$ [1]

3 The car has zero acceleration. [1]
According to $F = ma$, the resultant force acting on the car must be zero. [1]

4 $F \propto a$ for a given mass therefore $a = \frac{65}{20} \times 40$ [1]
$a = 13\,\text{m s}^{-2}$ [1]

5 resultant force $F = \sqrt{(2.0 \times 10^{-15})^2 + (2.0 \times 10^{-15})^2}$ [1]
$F = 2.83 \times 10^{-15}\,\text{N}$ [1]
$a = \frac{2.83 \times 10^{-15}}{9.1 \times 10^{-31}} = 3.1 \times 10^{15}\,\text{m s}^{-2}$ [1]

6 $s = \frac{1}{2}at^2 = \frac{1}{2} \times 3.1 \times 10^{15} \times (20 \times 10^{-9})^2$ [1]
distance = 0.62 m [1]

4.4

1 The drag force on the truck would be greater because of its large area. [1]

2 The resultant force on the object will be zero. [1]

3 resultant force = weight = $mg = 2.0 \times 9.81 = 20\,\text{N}$
acceleration = $g = 9.81\,\text{m s}^{-2}$ [1]

4 $a = \frac{(70 \times 9.81) - 300}{70}$ [2]
$a = 5.5\,\text{m s}^{-2}$ [1]

5 $D = 0.80v^2$ [1]

6 drag = weight [1]
$0.80v^2 = 0.100 \times 9.81$ [1]
$v = 1.1\,\text{m s}^{-1}$ [1]

4.5/4.6

1 moment = $Fx = 2.0 \times 0.10 = 0.2\,\text{N m}$ [1]

2 force = $\frac{20}{400}$ [1]
force = $5.0 \times 10^{-2}\,\text{m}$ [1]

3 torque = $3.0 \times 0.15 = 0.45\,\text{N m}$ [1]

4 **a** $1.2 \times 20 = 0.80 \times d_2$ [1]
$d_2 = 30\,\text{cm}$ [1]
b force at pivot = 1.20 + 0.80 = 1.2 N [1]

5 Moments about Y:
$Wd_y = S_x(d_x + d_y)$ [1]
$S_x = \frac{Wd_y}{d_x + d_y}$ [1]
Moments about X:
$Wd_x = S_y(d_x + d_y)$ [1]
$S_y = \frac{Wd_x}{d_x + d_y}$ [1]

6 $S_x = \frac{Wd_y}{d_x + d_y} = \frac{1.00 \times 1.20}{2.00} = 60\,\text{N}$ [1]
$S_y = \frac{Wd_x}{d_x + d_y} = \frac{1.00 \times 1.20}{2.00} = 40\,\text{N}$ [1]

4.7

1 Constant velocity implies zero acceleration and therefore the resultant force on the object must be zero. [1]

2 All three forces drawn in the correct direction and form a closed triangle.
All arrows are in the correct directions.
See below.

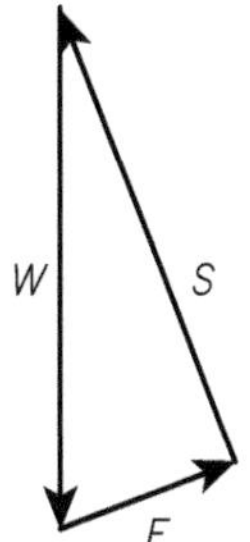

[1]

3 $F^2 = 12^2 + 5.0^2$ [1] $F = 13\,\text{N}$ [1]
$\tan\theta = \frac{12}{5.0}$ [1] $\theta = 67°$ [1]

4 $\cos\theta = \frac{4.8}{9.6}$ [1]
$\theta = 60°$ [1]

5 $\frac{20 \times 9.81}{\sin 140} = \frac{T}{\sin 20}$ [2]
$T = 104\,\text{N} \approx 100\,\text{N}$ [1]

6 $2T\cos 20° = 20 \times 9.81$ [1]
$T = \frac{20 \times 9.81}{2\cos 20°}$ [1]
$T = 104\,\text{N} \approx 100\,\text{N}$ [1]

4.8/4.9

1 $\rho = \frac{m}{V} = \frac{350}{0.50} = 700\,\text{kg m}^{-3}$ [1]

2 $p = \frac{F}{A} = \frac{3.0 \times 10^4 \times 9.81}{1.2}$ [1]
pressure = $2.453 \times 10^5\,\text{Pa} \approx 2.5 \times 10^5\,\text{Pa}$ [1]

3 $F = pA = 1.0 \times 10^5 \times (\pi \times 0.15^2)$ [1]
force = 7100 N [1]

4 total pressure = atmospheric pressure + $h\rho g$ [1]
total pressure = $1.0 \times 10^5 + [3.0 \times 1.0 \times 10^3 \times 9.81]$ [1]
total pressure = $1.3 \times 10^5\,\text{Pa}$ [1]

5 density = $\frac{m}{V} = \frac{6.0 \times 10^{24}}{\frac{4}{3}\pi \times [6.4 \times 10^6]^3}$ [1]
density = $5.5 \times 10^3\,\text{kg m}^{-3}$ [1]

6 $h = \frac{p}{\rho g}$ [1]
height = $\frac{1.0 \times 10^5}{13.6 \times 1.0 \times 10^3 \times 9.81} = 0.75\,\text{m}$ [1]

5.1/5.2

1 $W = Fx = 20 \times 5.0 = 100\,\text{J}$ [1]

2 Energy is conserved, therefore kinetic energy $= 100 - 30 = 70\,\text{J}$ [1]

3 $W = Fx\cos\theta = 100 \times 20 \times \cos 45°$ [1]

work done $= 1.4 \times 10^3\,\text{J}$ [1]

4 The weight of the ball is 90° to the horizontal. [1]

There is no work done by the weight because $Fx\cos 90° = 0$ [1]

5 $\cos\theta = \frac{W}{Fx} = \frac{100}{20 \times 8.0} = 0.625$ [1]

$\theta = \cos^{-1}(0.625) = 51°$ [1]

6 work done by the weight $= 2.0 \times 9.81 \times 10$ [1]

work done by drag force $= [2.0 \times 9.81 \times 10] - 120 = 76.2\,\text{J}$ [1]

$76.2 = F \times 10$ [1] $F = 7.6\,\text{N}$ [1]

5.3

1 $E_p = mgh = 1500 \times 9.81 \times 310 = 4.6 \times 10^6\,\text{J}$ [1]

2 $E_k = \frac{1}{2}mv^2 = \frac{1}{2} \times 1.0 \times 10^{-3} \times (120 \times 10^3)^2$ [1]

$E_k = 7.2 \times 10^6\,\text{J}$ [1]

3 The change in the KE of the stone is the same as the change in its gravitational potential energy (20 J) because energy is conserved. [1]

Assumption: There are no frictional forces and therefore no thermal losses. [1]

4 $v = \sqrt{\frac{2E_k}{m}} = \sqrt{\frac{2 \times 50 \times 10^3}{1000}}$ [1]

$v = 10\,\text{m}\,\text{s}^{-1}$ [1]

5 final KE $= 8.0 \times 10^{-21} + \frac{1}{2} \times 9.1 \times 10^{-31} \times (1.2 \times 10^5)^2$ [1]

final KE $= 1.455 \times 10^{-20}\,\text{J}$ [1]

$v = \sqrt{\frac{2E_k}{m}} = \sqrt{\frac{2 \times 1.455 \times 10^{-20}}{9.1 \times 10^{-31}}}$ [1]

$v = 1.8 \times 10^5\,\text{m}\,\text{s}^{-1}$ [1]

6 a There is loss of GPE and gain in KE from top of swing to the bottom of the swing. [1]

There is gain in GPE and loss in KE from the bottom of the swing to height h. [1]

b $v^2 = 2g(h_0 - h)$ [1]

$v = \sqrt{(2g(h_0 - h)}$ [1]

$v = \sqrt{(2 \times 9.81 \times (3.0 - 0.6)}$ [1]

$v = 6.9\,\text{m}\,\text{s}^{-1}$ [1]

5.4

1 efficiency $= \frac{2.4}{56} \times 100 = 4.3\%$ [1]

2 $W = Pt = 24 \times 3600$ [1]

input energy $= 8.6 \times 10^4\,\text{J}$ [1]

3 $P = Fv$ $F = \frac{P}{v} = \frac{6000}{20}$ [1]

force $= 300\,\text{N}$ [1]

4 input power $= \frac{20}{0.25}$ [1]

input power $= 80\,\text{W}$ [1]

5 a work done = gain in GPE $= mgh$
$= 0.10 \times 9.81 \times 1.20$ [1]

output power $= \frac{0.10 \times 9.81 \times 1.20}{2.4}$ [1]

output power $= 0.49\,\text{W}$ [1]

b input power $= \frac{0.49}{0.12}$ [1]

input power $= 4.1\,\text{W}$ [1]

6 $P = Fv \propto v^2 \times v$; therefore $P \propto v^3$. [1]

As v is doubled, P will increase by a factor of $2^3 = 8$ [2]

6.1/6.2

1 $F = kx$ $x = \frac{F}{k} = \frac{1.2}{25}$ [1]

$x = 4.8 \times 10^{-2}\,\text{m}$ [1]

2 The spring does *not* obey Hooke's law because increasing the force by a factor of 3 (from 10 N to 30 N) does not increase the extension by the same factor. [2]

Extension increases by a factor of $\frac{9.2}{2.5} = 3.8$ [1]

3 $E = \frac{1}{2}kx^2 = \frac{1}{2} \times 120 \times 0.020^2$ [1]

$E = 2.4 \times 10^{-2}\,\text{J}$ [1]

4 $5.4 = \frac{1}{2}kx^2 = \frac{1}{2} \times 120 \times x^2$ [1]

$x = \sqrt{2 \times \frac{5.4}{120}}$ [1]

$x = 0.30\,\text{m}$ [1]

5 $E = \frac{1}{2}kx^2 \propto x^2$ [1]

The stored energy will increase by a factor of $5^2 = 25$ [1]

6 kinetic energy of ball = elastic potential energy [1]

$\frac{1}{2}mv^2 = \frac{1}{2}kx^2$ [1]

$v = \sqrt{\frac{150 \times 0.09^2}{0.020}}$ [1]

$v = 7.8\,\text{m}\,\text{s}^{-1}$ [1]

6.3/6.4

1 stress → Pa; strain → none; Young modulus → Pa [1]

2 strain $= \frac{x}{L} = \frac{0.05}{0.100}$ [1]

strain = 0.50 or 50% [1]

3 stress $= \frac{F}{A} = \frac{12}{\pi \times (1.0 \times 10^{-3})^2}$ [1]

stress $= 3.82 \times 10^6\,\text{Pa} \approx 3.8 \times 10^6\,\text{Pa}$ [1]

4 strain $= \frac{\text{stress}}{E}$ [1]

strain $= \frac{3.82 \times 10^6}{1.8 \times 10^{11}}$ [1]

strain $= 2.1 \times 10^{-5}$ [1]

5 stress $= \frac{F}{A} = \frac{5.0 \times 9.81}{\pi \times (0.7 \times 10^{-3})^2} = 3.19 \times 10^7\,\text{Pa}$ [1]

strain $= \frac{x}{L} = \frac{3.0 \times 10^{-3}}{3.2} = 9.38 \times 10^{-4}$ [1]

$E = \frac{\text{stress}}{\text{strain}} = \frac{3.19 \times 10^7}{9.38 \times 10^{-4}}$ [1]

$E = 3.4 \times 10^{10}\,\text{Pa}$ [1]

6 stored energy $= \frac{1}{2}Fx$

$= \frac{1}{2} \times 5.0 \times 9.81 \times 3.0 \times 10^{-3}$

$= 7.36 \times 10^{-2}\,\text{J}$ [1]

energy stored per unit volume $= \frac{7.36 \times 10^{-2}}{3.2 \times \pi \times (0.7 \times 10^{-3})^2}$ [1]

energy stored per unit volume $= 1.5 \times 10^4\,\text{J}\,\text{m}^{-3}$ [1]

7.1/7.2

1 According to Newton's third law, the force on the centre of the Earth is opposite to the weight and equal to 700 N. [1]

2 $p = mv = 900 \times 30 = 2.7 \times 10^4\,\text{kg}\,\text{m}\,\text{s}^{-1}$ [1]

3 final velocity $v = at = 9.81 \times 3.0$ [1]

$v = 29.43\,\text{m}\,\text{s}^{-1}$ [1]

$p = mv = 1.2 \times 29.43 = 35\,\text{kg}\,\text{m}\,\text{s}^{-1}$ [1]

4 momentum of shell = momentum of cannon [1]

$20 \times 240 = 1200 \times v$ [1]

$v = 4.0\,\text{m}\,\text{s}^{-1}$ [1]

5 change in momentum $= -mv - mu$ [1]

change in momentum $= -m(v + u)$ [1]

6 final momentum $= (1100 \times 30) - 1.3 \times 10^4$ [1]

$1100v = 2.0 \times 10^4$ [1] $v = 18\,\text{m}\,\text{s}^{-1}$ [1]

7 a total initial momentum = total final momentum [1]

$0.040 \times 80 = 0.340 \times v$ [1]

$v = 9.4\,\text{m}\,\text{s}^{-1}$ [1]

b loss in KE $= \frac{1}{2} \times 0.040 \times 80^2 - \frac{1}{2} \times 0.340 \times 9.4^2$ [1]

loss in KE = 110 J [1]

7.3/7.4/7.5

1 impulse = change in momentum $= 0.12\,\text{kg}\,\text{m}\,\text{s}^{-1}$ [1]

2 $\Delta p = 900 \times (10 - 30) = -1.8 \times 10^4\,\text{kg}\,\text{m}\,\text{s}^{-1}$ [1]

force $= \frac{\Delta p}{\Delta t} = \frac{1.8 \times 10^4}{5.0}$ [1]

force $= 3.6 \times 10^3\,\text{N}$ [1]

3 impulse = area under the graph $= \frac{1}{2} \times 3000 \times 4.0$ [1]

impulse = 6000 N s [1]

4 change in momentum = impulse $= 6000\,\text{kg}\,\text{m}\,\text{s}^{-1}$

$1000 \times \Delta v = 6000$ [1]

change in velocity $= 6.0\,\text{m}\,\text{s}^{-1}$ [1]

5 $\Delta p = -0.040 \times (30 + 30) = 2.4\,\text{kg}\,\text{m}\,\text{s}^{-1}$ [1]

force $= \frac{\Delta p}{\Delta t} = \frac{2.4}{2.0 \times 10^{-3}}$ [1]

force $= 1.2 \times 10^3\,\text{N}$ [1]

6 The final momentums p_1 and p_2 add vectorially to give a resultant momentum equal to the initial momentum p in the x-direction. [1]

Correct diagram showing p_1, p_2, and p, with correct angles shown. [2]

See below.

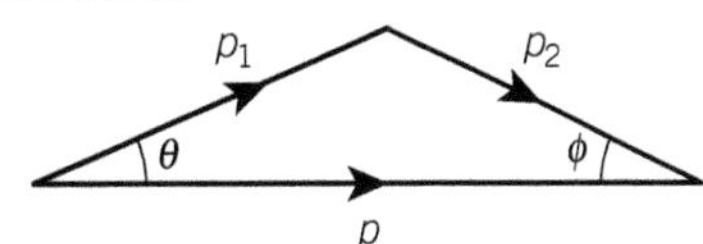

8.1/8.2

1 Electrons in a metal and ions in electrolytes. [1]

2 a This charge is half the charge on an electron, which is not possible. The student is incorrect. [1]

b $I = \frac{\Delta Q}{\Delta t} = \frac{12}{60}$ [1]

current = 0.20 A [1]

3 charge $= 8 \times 1.60 \times 10^{-19}$ [1]

charge $= 1.28 \times 10^{-18}\,\text{C}$ [1]

4 charge $= I \times \Delta t = 0.020 \times (4.0 \times 3600)$ [1]

charge $= 288\,\text{C} \approx 290\,\text{C}$ [1]

5 number of electrons $= \frac{288}{1.60 \times 10^{-19}}$ [1]

number of electrons $= 1.8 \times 10^{21}$ [1]

6 $I = \frac{\Delta Q}{\Delta t} = \frac{10^6 \times 1.60 \times 10^{-19}}{60}$ [1]

current $= 2.7 \times 10^{-15}\,\text{A}$ [1]

8.3/8.4

1 The same number of 2 billion electrons would leave the point because of Kirchhoff's first law and the conservation of charge. [1]

2 a 1 A b 2 A c 7 A [3]

3 $v \propto I$ therefore as the current decreases by a factor of 4, the mean drift velocity will also decrease by the same factor. [1]

mean drift velocity $= 0.5\,\text{mm}\,\text{s}^{-1}$ [1]

4 $I \propto \frac{1}{A}$ [2]

As electrons travel from B to A, the cross-sectional area of the conductor increases therefore the mean drift velocity of the electrons decreases. [1]

5 number of electrons = N × volume

$= 8.5 \times 10^{28} \times [1.0 \times 10^{-2} \times \pi \times (0.5 \times 10^{-3})^2]$ [1]

number of electrons $= 6.7 \times 10^{20}$ [1]

6 $v = \frac{I}{Aen} = \frac{3.0 \times 10^{-6}}{[1.0 \times 10^{-3}]^2 \times 1.6 \times 10^{-19} \times 5.0 \times 10^{28}}$ [2]

$v = 3.8 \times 10^{-10}\,\text{m}\,\text{s}^{-1}$ [1]

9.1/9.2/9.3

1 Any two sources of e.m.f. (e.g., chemical cell, solar cell, thermocouple, etc.) [1]

2 $V = \frac{W}{Q} = \frac{20}{4.0}$ [1]

$V = 5.0\,V$ [1]

3 $W = \mathcal{E}Q = 100 \times 60$ [1]

$W = 6000\,J$ [1]

4 $W = VQ = 1000 \times 1.6 \times 10^{-19}$ [1]

$W = 1.6 \times 10^{-16}\,J$ [1]

5 charge $= I\Delta t = 0.25 \times 30 = 7.5\,C$ [1]

$W = VQ = 6.0 \times 7.5 = 45\,J$ [1]

rate of energy transfer $= \frac{45}{30} = 1.5\,J\,s^{-1}$ [1]

6 $eV = \frac{1}{2}mv^2$ [1]

$v = \sqrt{\frac{2 \times 10 \times 10^3 \times 1.6 \times 10^{-19}}{1.67 \times 10^{-27}}}$ [1]

$v = 1.4 \times 10^6\,m\,s^{-1}$ [1]

9.4/9.5/9.6

1 Any one from: diode, LED, and filament lamp. [1]

2 $V = IR = 120 \times 0.030 = 3.6\,V$ [1]

3 The p.d. is also doubled, but resistance, which is $\frac{\text{p.d.}}{\text{current}}$, remains the same. [1]

4 $R = \frac{V}{I}$ so when $I = 0$ the resistance R will be a very large value (infinite). [1]

5 final current $= \frac{2.1}{20} = 0.1050\,A$ [1]

initial current $= \frac{1.9}{200} = 0.0095\,A$ [1]

change in current $= 0.1050 - 0.0095$
$= 0.096\,A$ (2 s.f.) [1]

6 $I = Anev$ therefore number density $n \propto$ current I. [1]

percentage change $= \frac{0.1050 - 0.0095}{0.0095} \times 100$
$= 1000\%$ [1]

9.7/9.8/9.9

1 $R \propto L$ therefore the resistance increases by a factor of 4. [1]

resistance $= 2.3 \times 4 = 9.2\,\Omega$ [1]

2 $A = 1.0\,m^2$ and $L = 1.0\,m$ [1]

$R = \frac{\rho L}{A} = \frac{2.7 \times 10^{-8} \times 1.0}{1.0} = 2.7 \times 10^{-8}\,\Omega$ [1]

3 $R = \frac{\rho L}{A} = \frac{1.6 \times 10^{-8} \times 1.00}{1.2 \times 10^{-6}}$ [1]

$R = 0.013\,\Omega$ [1]

4 $R = \frac{\rho L}{A} = \frac{2.7 \times 10^{-8} \times 1000}{\pi \times 0.010^2}$ [2]

$R = 0.086\,\Omega$ [1]

5 0°C to 20°C

% decrease $= \frac{3.5 - 1.8}{3.5} \times 100 \approx 50\%$ [1]

20°C to 40 °C

% decrease $= \frac{1.8 - 1.0}{1.8} \times 100 \approx 40\%$ [1]

There is a smaller % change in the resistance as temperature increases. [1]

6 $R \propto \frac{L}{A}$ [1]

The volume of the wire remains constant, so when L is 10 times longer, the area A is 10 times smaller. [1]

$R = 12 \times \frac{10}{0.1} = 1200\,\Omega$ [1]

9.10/9.11

1 $J\,s^{-1}$ [1]

2 $I = \frac{P}{V} = \frac{2000}{240}$ [1]

$I = 8.3\,A$ [1]

3 $P = \frac{V^2}{R} = \frac{240^2}{100}$ [1]

$P = 580\,W$ [1]

4 cost $= 3.0 \times 4.0 \times 8.1$ [1]

cost $= 97$p [1]

5 $W = 60 \times (10 \times 3600)$ [1]

$W = 2.16 \times 10^6\,J \approx 2.2 \times 10^6\,J$ [1]

6 $R = \frac{V^2}{P} = \frac{12^2}{36}$ [1]

$R = 4.0\,\Omega$ [1]

7 energy = p.d. × charge, and current $= \frac{\text{charge}}{\text{time}}$ [1]

power $= \frac{\text{energy}}{\text{time}} = \frac{\text{charge} \times \text{p.d.}}{\text{time}}$ [1]

power $= \frac{\text{charge}}{\text{time}} \times$ p.d = current × p.d.,

therefore $P = VI$ [1]

8 The volume of the wire is the same, so as the length is doubled the cross-sectional area is halved. [1]

$R \propto \frac{L}{A}$ therefore the resistance of the wire is quadrupled. [1]

$P = \frac{V^2}{R}$. The p.d. provided by the power supply is the same. The power dissipated decreases by a factor of 4 because $P \propto \frac{1}{R}$ [1]

The power dissipated is $\frac{10}{4} = 2.5\,W$ [1]

10.1/10.2/10.3

1 series: $R = 120 \times 2 = 240\,\Omega$ [1]

parallel: $R = (120^{-1} + 120^{-1})^{-1} = 60\,\Omega$ [2]

2 p.d. $= \frac{6.0}{10} = 0.60\,V$ [1]

3 $R = 2.0 + 1.0 + 7.0 = 10.0\,\Omega$ [1]

total e.m.f. $= 3.0 + 2.0 = 5.0\,V$ [1]

4 current $= \frac{5.0}{10} = 0.50\,A$ [1]

$V = IR = 0.50 \times 7.0$ [1]

$V = 3.5\,V$ [1]

5 $\frac{1}{R} + \frac{1}{100} = \frac{1}{70}$ [1]

$\frac{1}{R} = \frac{1}{70} - \frac{1}{100} = \frac{30}{7000}$ [1]

$R = \frac{7000}{30} = 233\,\Omega \approx 230\,\Omega$ [1]

6 As temperature decreases, the resistance of the thermistor increases. [1]

Therefore, the current I in the circuit decreases. [1]

$V = IR$; since R is constant at $100\,\Omega$, the p.d. across this resistor will decrease. [1]

10.4/10.5/10.6

1 The 'lost volts' is the p.d. across the internal resistance. [1]

2 There is a p.d. across the internal resistor because of the current in the circuit. [1]

There is a p.d. of 0.4 V across the internal resistance therefore the terminal p.d. cannot be equal to the e.m.f. [1]

3 $I = \frac{1.0}{2.0} = 0.050\,\text{A}$ [1]

$r = \frac{0.4}{0.05} = 8.0\,\Omega$ [1]

4 $V = \frac{R_2}{R_1 + R_2} \times V_{in} = \frac{68}{100 + 68} \times 1.5$ [1]

$V = 0.61\,\text{V}$ [1]

5 $V_{out} = \frac{R_2}{R_1 + R_2} \times V_{in} = \frac{4R}{R + 4R} V_{in}$ [1]

$V_{out} = 0.80 V_{in}$ [1]

6 The total resistance of the voltmeter and the $68\,\Omega$ thermistor in parallel would be less than $68\,\Omega$. [1]

The total resistance of the circuit would be less and the current in the circuit would be greater. [1]

There is a greater p.d. across the variable resistor and therefore the p.d. across the thermistor would be less (than 0.61 V). [1]

11.1/11.2

1 A frequency of 10 Hz means 10 wavelengths passing through a point per second or 10 oscillations of a medium particle per second. [1]

2 The distance travelled is 2.0 cm in one period. [1]

3 $v = f\lambda = 30 \times 0.04 = 1.2\,\text{m}\,\text{s}^{-1}$ [1]

4 $\lambda = \frac{v}{f} = \frac{340}{150} \times 10^3$ [1]

$\lambda = 2.3 \times 10^{-3}\,\text{m}$ [1]

5 **a** separation = 0.5λ [1]

Therefore, phase difference = 180° (or π rad) [1]

b separation = 0.75λ [1]

Therefore, phase difference = 270° (or $\frac{3\pi}{2}$ rad) [1]

6 $\frac{x}{20} \times 360 = 300$ [1]

$x = 17\,\text{cm}$ [1]

11.3/11.4

1 Speed, wavelength, and frequency all remain the same. [1]

2 Sound is a longitudinal wave, and not a traverse wave, so it cannot be polarised. [1]

3 **a** The wavelength of 500 nm is much smaller than 3.0 cm, so very little diffraction. [1]

b The wavelength of 3.0 cm is similar to 3.0 cm, so there is diffraction. [1]

4 Speed or wavelength. [1]

5 Correct diagram showing the curving around of wavefronts at the edge of the obstacle, see below.

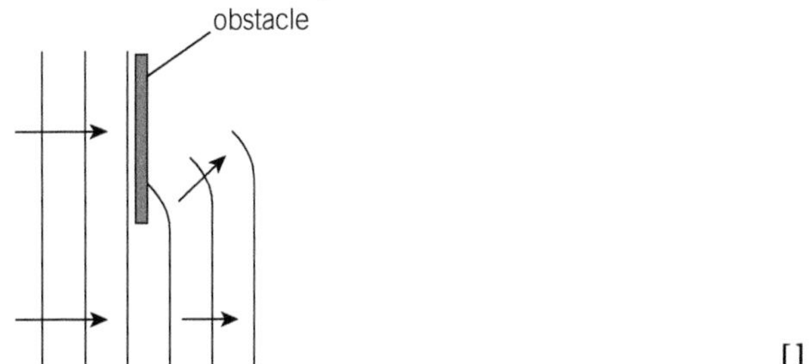

[1]

6 The wavelength is longer in water. [1]

The speed $v = f\lambda$ and frequency f remain constant. [1]

Therefore, this implies that the speed of light in water is greater than that in glass. [1]

11.5/11.6/11.7

1 power → [W] and intensity → [W m^{-2}] [1]

2 They can be polarised because they are all transverse waves. [1]

3 intensity $\propto$ amplitude^2 [1]

Therefore, the intensity decreases by a factor of 4 when the amplitude is halved. [1]

4 $\lambda = \frac{c}{f} = \frac{3.0 \times 10^8}{1.5 \times 10^{17}} = 2.0 \times 10^{-9}\,\text{m}$ [2]

The electromagnetic waves must be X-rays (according to Figure 1). [1]

5 $P = 4\pi r^2 \times I$ [1]

$P = 4\pi \times (1.5 \times 10^{11})^2 \times 1050$ [1]

$P = 3.0 \times 10^{26}\,\text{W}$ [1]

6 incident power = $1050 \times 1.2 = 1.26 \times 10^3\,\text{W}$ [1]

output power = $0.20 \times 1.26 \times 10^3 = 252\,\text{W}$ [1]

number of panels = $\frac{1000}{252} = 4.0$ [1]

11.8/11.9

1 The refractive index of vacuum is 1. [1]

2 $n = \frac{c}{0.5c} = 2.0$ [1]

3 The refractive index of air is less than the refractive index of glass (1.5). [1]

The statement is incorrect because light has to travel from glass in the direction of air for TIR. [1]

4 $v = \frac{3.0 \times 10^8}{1.50}$ [1]

$v = 2.0 \times 10^8\,\text{m s}^{-1}$ [1]

5 $n_1 \sin\theta_1 = n_2 \sin\theta_2$; $1.00 \times \sin 70° = N \times \sin 35°$ [1]

$n = 1.64$ [1]

6 $n_1 \sin\theta_1 = n_2 \sin\theta_2$

$2.42 \times \sin C = 1.00 \times \sin 90°$ [1]

$\sin C = \frac{1.00}{2.42} = 0.413$ [1]

$C = 24°$ [1]

12.1/12.2/12.3

1 Sources that emits waves with a constant phase difference. [1]

2 The path difference must be a whole number of wavelengths or the phase difference must be zero. [2]

3 **a** path difference = 11.0 − 7.0 = 4.0 cm [1]

This is equal to two whole wavelengths therefore constructive interference. [1]

b path difference = 7.2 − 6.2 = 1.0 cm [1]

This is equal to half a wavelength therefore destructive interference. [1]

4 $\lambda = \frac{ax}{D} = \frac{0.10 \times 10^{-3} \times 3.1 \times 10^{-2}}{6.20}$ [1]

$\lambda = 5.0 \times 10^{-7}\,\text{m}$ [1]

5 $\lambda \propto x$ therefore $\lambda = \frac{4.1}{5.2} \times 650$ [1]

$\lambda = 510\,\text{nm}$ [1]

6 uncertainty in $a = \pm 0.01$ mm, uncertainty in $x = \pm 0.1$ cm, and uncertainty in $D = \pm 0.01$ m [1]

fractional uncertainty in λ

$= \frac{0.01}{0.10} + \frac{0.1}{3.1} + \frac{0.1}{6.20} = 0.134$ [1]

absolute uncertainty in $\lambda = (5.0 \times 10^{-7}) \times 0.134$

$= 0.7 \times 10^{-7}\,\text{m}$ [1]

12.4/12.5/12.6

1 The phase difference is zero. [1]

2 The separation between node and antinode is $\frac{\lambda}{4}$ [1]

wavelength = 4 × 10 = 40 cm [1]

3 The microwaves are reflected off the metal plates. [1]

The superposition of these reflected microwaves and those coming directly from the transmitter produce the stationary wave. [1]

4 $\frac{\lambda}{2} = 28$ [1]

$\lambda = 2 \times 28 = 56\,\text{cm}$ [1]

$v = f\lambda = 60 \times 0.56 = 34\,\text{m s}^{-1}$ [1]

5 length of tube $= \frac{\lambda}{4} = 30$ [1]

$\lambda = 4 \times 30 = 120\,\text{cm}$ [1]

$f = \frac{v}{\lambda} = \frac{340}{1.20} = 280\,\text{Hz}$ [1]

6 $\lambda = \frac{v}{f} = \frac{340}{500} = 0.68\,\text{m}$ [1]

length of tube $= 3\frac{\lambda}{2} = 0.68$ [1]

$\lambda = 0.45\,\text{m}$ [1]

13.1/13.2

1 joule (J) [1]

2 $100\,\text{eV} = 100 \times 1.60 \times 10^{-19} = 1.60 \times 10^{-17}\,\text{J}$ [1]

3 1000 eV [1]

4 The energy of a photon of visible light must be less than the work function of the metal. [1]

5 $\lambda = \frac{hc}{E} = \frac{6.63 \times 10^{-34} \times 3.00 \times 10^8}{8.0 \times 10^{-19}}$ [2]

$\lambda = 2.5 \times 10^{-7}\,\text{m}$ [1]

6 $E = \frac{hc}{\lambda} = \frac{6.63 \times 10^{-34} \times 3.00 \times 10^8}{560 \times 10^{-9}} = 3.552 \times 10^{-19}\,\text{J}$ [2]

number of photons per second $= \frac{0.040}{3.552} \times 10^{-19}$ [1]

number of photons per second $= 1.1 \times 10^{17}\,\text{s}^{-1}$ [1]

7 energy of photon:

$E = \frac{hc}{\lambda} = \frac{6.63 \times 10^{-34} \times 3.00 \times 10^8}{400 \times 10^{-9}} = 4.97 \times 10^{-19}\,\text{J}$ [2]

work function: $\phi = 4.30 \times 1.60 \times 10^{-19}$

$= 6.88 \times 10^{-19}\,\text{J}$ [1]

The photon energy is less than the work function, therefore no emission of photoelectrons. [1]

13.3/13.4

1 Energy. [1]

2 $KE_{max} = (7.2 - 6.9) \times 10^{-19}$ [1]

$KE_{max} = 3.0 \times 10^{-20}\,\text{J}$ [1]

3 $\frac{1}{2}mv^2 = 3.0 \times 10^{-20}$

$\frac{1}{2} \times 9.11 \times 10^{-31} \times v^2 = 3.0 \times 10^{-20}$ [1]

$v = 2.6 \times 10^5\,\text{m s}^{-1}$ [1]

4 work function = energy of photon $= \frac{hc}{\lambda}$ [1]

work function $= 6.63 \times 10^{-34} \times \frac{3.00 \times 10^8}{2.1 \times 10^{-7}}$ [1]

work function $= 9.5 \times 10^{-19}\,\text{J}$ [1]

5 energy of photon:

$E = \frac{hc}{\lambda} = \frac{6.63 \times 10^{-34} \times 3.00 \times 10^8}{320 \times 10^{-9}} = 6.22 \times 10^{-19}\,\text{J}$ [2]

work function:

$\phi = 2.3 \times 1.60 \times 10^{-19} = 3.68 \times 10^{-19}\,\text{J}$ [1]

$KE_{max} = (6.22 - 3.68) \times 10^{-19}\,\text{J} \approx 2.5 \times 10^{-19}\,\text{J}$ [1]

6 $\frac{1}{2} \times 9.11 \times 10^{-31} \times v^2 = 1.5 \times 1.60 \times 10^{-19}$ [1]
$v = 7.26 \times 10^5\,\mathrm{m\,s^{-1}}$ [1]
$\lambda = \frac{h}{mv} = \frac{6.63 \times 10^{-34}}{9.11 \times 10^{-31} \times 7.26 \times 10^5}$ [1]
$\lambda = 1.0 \times 10^{-9}\,\mathrm{m}$ [1]

14.1/14.2

1 Any sensible comment, e.g., temperature is measured in °C or K and heat energy is measured in J. [1]

2 Insert the thermometer into pure *melting* ice. The level of mercury in the glass will be the 0 °C mark. [1]

3 **a** 3 K [1]
b 221 K [1]
c 293 K [1]
d 373 K [1]
e 2273 K [1]

4 **a** −270 °C [1]
b 0 °C [1]
c 107 °C [1]
d 227 °C [1]
e 5227 °C [1]

5 The potential energy of the molecules is greater in the water phase because of the increased separation between the molecules. [2]
The kinetic energy of the molecules is also greater in the water phase because of the increase in the temperature. [2]

6 The mean kinetic energy of the molecules and the smoke particles is the same at a specific temperature. [1]
KE $= \frac{1}{2}mv^2$ therefore $v \propto \frac{1}{\sqrt{m}}$ [1]
The mass m of the molecules << mass of the smoke particles therefore mean speed of molecules >> mean speed of the smoke particles. [1]

14.3/14.4/14.5

1 Absolute zero is temperature of 0 K. The internal energy of a substance at absolute zero is a minimum. [1]

2 $E = mL_f = 0.010 \times 3.3 \times 10^4$ [1]
$E = 330\,\mathrm{J}$ [1]

3 $E = mc\Delta\theta = 0.300 \times 4200 \times (100 - 20) = 1.01 \times 10^5\,\mathrm{J}$ [1]
time $= \frac{\text{energy}}{\text{power}} = \frac{1.01 \times 10^5}{200}$ [1]
time = 504 s (8.4 minutes) [1]

4 $E = mL_v = 0.300 \times 2.3 \times 10^6 = 6.9 \times 10^5\,\mathrm{J}$ [1]
time $= \frac{\text{energy}}{\text{power}} = \frac{6.9 \times 10^5}{200}$ [1]
time $= 3.45 \times 10^3$ s (58 minutes) [1]

5 energy lost by water $= E = mc\Delta\theta =$
$0.120 \times 4200 \times (100 - 87) = 6552\,\mathrm{J}$ [1]
$6552 = 0.200 \times c \times (87 - 20)$ [1]
$c = \frac{6552}{13.4}$ [1]
$c = 490\,\mathrm{J\,kg^{-1}\,K^{-1}}$ [1]

6 energy needed to change phase $= 0.500 \times 3.3 \times 10^4 =$
$1.65 \times 10^4\,\mathrm{J}$ [1]
energy needed to warm water $= 0.500 \times 4200 \times 20 =$
$4.2 \times 10^4\,\mathrm{J}$ [1]
time $= \frac{\text{energy}}{\text{power}} = \frac{1.65 \times 10^4 + 4.2 \times 10^4}{120}$ [1]
time = 490 s (8.1 minutes) [1]

15.1/15.2

1 number of molecules $= 3.0 \times 6.02 \times 10^{23} = 1.8 \times 10^{24}$ [1]

2 mass of molecule $= \frac{0.032}{6.02} \times 10^{23}$ [1]
mass of molecule $= 5.3 \times 10^{-26}\,\mathrm{kg}$ [1]

3 $p \propto T$ [1]
T increases by a factor of $\frac{473}{373} = 1.27 \approx 1.3$ [1]
Therefore, the pressure p also increases by a factor of 1.3. [1]

4 Assuming the temperature of the water is constant,
volume $\propto \frac{1}{\text{pressure}}$ [1]
As the bubble rises, the pressure exerted on it by the water decreases ($\rho = h\rho g$). [1]
Therefore, the volume of the bubble increases as it rises. [1]

5 $pV = nRT$ [1]
$V = \frac{nRT}{p} = \frac{1 \times 8.31 \times [273 - 50]}{1.0 \times 10^5}$ [1]
$V = 1.853 \times 10^{-2}\,\mathrm{m^3}$ [1]
density $= \frac{0.029}{1.853} \times 10^{-2} = 1.6\,\mathrm{kg\,m^{-3}}$ [1]

6 Any reasonable estimates: $V = 25\,\mathrm{m^3}$, $T = 293\,\mathrm{K}$ (20 °C), and $p = 1.0 \times 10^5\,\mathrm{Pa}$. [1]
$n = \frac{pV}{RT} = \frac{1.0 \times 10^5 \times 25}{8.31 \times 293} = 1.027 \times 10^3\,\mathrm{mol}$ [2]
mass $= 0.029 \times 1.027 \times 10^3 \approx 30\,\mathrm{kg}$ [1]

15.3/15.4

1 mean velocity $= \frac{-100 - 200 + 150 + 200 + 300}{5}$
$= +70\,\mathrm{m\,s^{-1}}$ [1]
mean speed $= \frac{100 + 200 + 150 + 200 + 300}{5}$
$= 190\,\mathrm{m\,s^{-1}}$ [1]

2 mean square speed
$= \frac{100^2 + 200^2 + 150^2 + 200^2 + 300^2}{5} = 40\,500\,\mathrm{m^2\,s^{-2}}$ [1]
r.m.s. speed $= \sqrt{40500} = 200\,\mathrm{m\,s^{-1}}$ [1]

3 mean KE of molecules

$= \frac{3}{2}kT = \frac{3}{2} \times 1.38 \times 10^{-23} \times (273 + 200)$ [1]

mean KE of molecules = 9.8×10^{-21} J [1]

4 The smoke particles and the air molecules have the same mean kinetic energy at a specific temperature. [1]

As the mass of the smoke particle >> mass of the air molecule, according to $\overline{c^2} = \frac{3kT}{m}$, the r.m.s. speed of the 'heavier' smoke particles << r.m.s. of the air molecules. [1]

5 $\frac{1}{2}m\overline{c^2} = \frac{3}{2}kT = 9.8 \times 10^{-21}$ [1]

$\frac{1}{2} \times 4.8 \times 10^{-26} \times \overline{c^2} = 9.8 \times 10^{-21}$ [1]

r.m.s. speed = $640\,\text{m s}^{-1}$ [1]

6 internal energy = $2.0 \times 6.02 \times 10^{23} \times \frac{3}{2}kT$ [1]

internal energy

$= 2.0 \times 6.02 \times 10^{23} \times \frac{3}{2} \times 1.38 \times 10^{-23} \times 273$ [1]

internal energy = 6.8×10^{3} J [1]

16.1/16.2

1 **a** 0.79 rad [1] **b** 0.087 rad [1] **c** 7.3 rad [1]

2 $v = \omega r = 15 \times 0.50$ [1]

$v = 7.5\,\text{m s}^{-1}$ [1]

3 $45° = 0.785$ rad and $\omega = \frac{\theta}{t}$ [1]

$\omega = \frac{0.785}{2.5} = 0.31\ \text{rad s}^{-1}$ [1]

4 $v = \omega r$; $150 = \omega \times 12 \times 10^{3}$ [1]

$\omega = 0.0125\ \text{rad s}^{-1} \approx 0.013\ \text{rad s}^{-1}$ [1]

$a = \frac{v^2}{r} = \frac{150^2}{12 \times 10^3}$ [1]

$a = 1.9\,\text{m s}^{-2}$ [1]

5 $a = g = \frac{v^2}{r}$; $9.81 = \frac{20^2}{r}$ [1]

$r = 41$ m [1]

6 The centripetal force is at right angles to the velocity. [1]

The centripetal force has no component in the direction of the velocity ($F\cos 90 = 0$) – no work is done on the object and therefore the speed does not change. [1]

16.3

1 Gravitational force between the planet and Sun. [1]

2 $F = \frac{mv^2}{r} = \frac{0.20 \times 3.2^2}{0.50}$ [1]

$F = 4.1$ N [1]

3 $4300 = \frac{mv^2}{r} = \frac{900 \times v^2}{30}$ [1]

$v = 12\,\text{m s}^{-1}$ [1]

4 $\frac{mv^2}{r} = Mg$ or $v^2 = \frac{gr}{m} \times M$ [1]

The equation for a straight line through the origin is $y = mx$ therefore the gradient is equal to $\frac{gr}{m}$. [1]

5 $v = \frac{2\pi r}{T} = \frac{2\pi \times 3.8 \times 10^8}{27 \times 24 \times 3600} = 1024\,\text{m s}^{-1}$ [1]

$F = \frac{mv^2}{r} = \frac{7.3 \times 10^{22} \times 1024^2}{3.8 \times 10^8}$ [1]

$F = 2.0 \times 10^{20}$ N [1]

6 $L\sin\theta = \frac{mv^2}{r}$ and $L\cos\theta = W = mg$ [1]

Divide these two equations to give $\frac{\sin\theta}{\cos\theta} = \frac{v^2}{rg}$ [1]

$\tan\theta = \frac{\sin\theta}{\cos\theta}$ therefore $\tan\theta = \frac{v^2}{rg}$ [1]

17.1

1 $\omega = \frac{2\pi}{T} = \frac{2\pi}{2.0} = 3.14\,\text{rad s}^{-1}$ [1]

2 At the equilibrium position ($x = 0$), the acceleration is zero and it increases to a maximum when it is maximum displacement. [1]

The direction of the acceleration is always towards the equilibrium position. [1]

3 $\omega = 2\pi f = 2\pi \times 1500$ [1]

$\omega = 9400\ \text{rad s}^{-1}$ [1]

$a_{max} = \omega^2 A = (2\pi \times 1500)^2 \times 0.60 \times 10^{-3}$ [1]

$a_{max} = 5.3 \times 10^{4}\,\text{m s}^{-2}$ [1]

4 $\phi = 2\pi \times \left(\frac{\Delta t}{T}\right) = 2\pi \times \frac{1}{8}$ [1]

$\phi = 0.79$ rad

5 $\omega^2 = 400$ [1]

$\frac{2\pi}{T} = \sqrt{400}$ [1]

$T = 0.31$ s [1]

6 $\omega^2 = \frac{g}{L}$ [1]

$a_{max} = \frac{g}{L} \times A$ [1]

$F_{max} = \frac{mgA}{L} = \frac{1.2 \times 9.81 \times 1.6}{5.0}$ [1]

$F_{max} = 3.8$ N [1]

17.2/17.3

1 The maximum speed is directly proportional to the amplitude ($v_{max} = \omega A$). [1]

2 The acceleration against time graph is an 'inverted' displacement–time graph. [1]

Therefore, acceleration ∝ –displacement, illustrating SHM. [1]

3 $v_{max} = \omega A = \frac{2\pi}{0.040} \times 3.0 \times 10^{-3}$ [1]

$v_{max} = 0.47\,\text{m s}^{-1}$ [1]

4 $v = \omega\sqrt{A^2 - x^2} = \frac{2p}{0.040} \times (0.003^2 - 0.001^2)^{\frac{1}{2}}$ [1]

$v = 0.44\,\text{m s}^{-1}$ [1]

5 $E_k = \frac{1}{2} m\omega^2 (A^2 - x^2)$ therefore $E_p = \frac{1}{2}m\omega^2x^2$ [1]

$E_p = \frac{1}{2}\times 0.180\times\left(\frac{2p}{0.80}\right)^2\times 0.10^2$ [1]

$E_p = 0.056\,\text{J}$ [1]

6 $x = A\cos\omega t$; $0.075 = 0.200\times\cos\frac{2p}{0.80\times t}$ [1]

$7.86t = \cos^{-1}(0.375)$

$t = 0.15\,\text{s}$ [1]

17.4/17.5

1 An oscillator is in resonance when it reaches maximum amplitude when being forced to oscillate. [1]

2 The amplitude of the oscillations decreases with time (because of frictional forces). [1]

3 The sharpness of the resonance decreases with increased damping. [1]

4 The resonant frequency becomes less than the natural frequency as the amount of damping is increased. [1]

5 $\frac{1}{8} = \frac{1}{2^3}$ therefore number of oscillations = 10×3 [1]

number of oscillations = 30 [1]

6 total energy = $\frac{1}{2} m\omega^2A^2 \propto A^2$ [1]

total energy = $10^2\times 0.0126$ [1]

total energy = $1.26\,\text{J} \approx 1.3\,\text{J}$ [1]

18.1/18.2/18.3

1 $g = \frac{F}{m} = \frac{190}{50} = 3.8\,\text{N kg}^{-1}$ [1]

2 $F = mg = 600\times 3.0 = 1.8\times 10^3\,\text{N}$ [1]

3 $F = \frac{GMm}{r^2} = \frac{6.67\times 10^{-11}\times 2.0\times 10^{30}\times 3.3\times 10^{23}}{(5.8\times 10^{10})^2}$ [1]

$F = 1.3\times 10^{22}\,\text{N}$ [1]

4 $F \propto \frac{1}{r^2}$ [1]

The distance r has increased by a factor of **4**. [1]

force = $\frac{F}{4^2} = \frac{F}{16}$ [1]

5 $g = \frac{GM}{r^2}$; $r = \sqrt{\frac{GM}{g}}$ [1]

$r = \sqrt{\frac{6.67\times 10^{-11}\times 8.7\times 10^{25}}{10}}$ [1]

radius = $2.4\times 10^7\,\text{m}$ [1]

6 $g \propto \frac{M}{r^2}$ [1]

$g = 9.81\times\frac{320}{11^2}$ [1]

$g = 26\,\text{N kg}^{-1}$ [1]

18.4/18.5

1 The square of the orbital period of any planet is directly proportional to the cube of the mean distance from the Sun. [1]

2 The orbital period is 1 day. [1]

3 $\frac{GMm}{r^2} = \frac{mv^2}{r}$ [1]

$v^2 = \frac{GM}{r} = \frac{6.67\times 10^{-11}\times 6.0\times 10^{24}}{(6400\times 10^3 + 3000\times 10^3)}$ [2]

$v = 6.5\times 10^3\,\text{m s}^{-1}$ [1]

4 period = $\frac{\text{circumference}}{\text{speed}}$ [1]

period = $\frac{2\pi\times(6400 + 3000)\times 10^3}{6.5\times 10^3}$ [1]

period = 9100 s (2.5 hours) [1]

5 Kepler's third law: $T^2 = kr^3$ [1]

Take logs of both sides: $\lg T^2 = \lg k + \lg r^3$

$2\lg T = \lg k + 3\lg r$

$\lg T = \frac{\lg k}{2} + 1.5\lg r$ [1]

The equation for a straight line is $y = mx + c$, therefore a graph of $\lg T$ against $\lg r$ will be a straight line of gradient 1.5. [1]

6 $T^2 = \left(\frac{4\pi^2}{GM}\right)r^3$ [1]

$(24\times 3600)^2 = \left(\frac{4\pi^2}{6.67\times 10^{-11}\times 6.0\times 10^{24}}\right)\times r^3$

$r = 4.23\times 10^7\,\text{m}$ [1]

$r = \frac{4.23\times 10^7}{6400\times 10^3} = 6.6.$

The radius is 6.6 times the Earth radius. [1]

18.6/18.7

1 Zero. [1]

2 54 MJ [1]

3 $V_g = -\frac{GM}{r}$; $54\times 10^6 = \frac{6.67\times 10^{-11}\times M}{6.1\times 10^6}$ [1]

$M = 4.9\times 10^{24}\,\text{kg}$ [1]

4 $V_g = -\frac{GM}{r}$; $V_g = -\frac{6.67\times 10^{-11}\times 6.0\times 10^{24}}{6400\times 10^3}$ [1]

$V_g = -6.3\times 10^7\,\text{J kg}^{-1}$ [1]

5 $\Delta V_g =$

$-6.67\times 10^{-11}\times 6.0\times 10^{24}\left(\frac{1}{9400\times 10^3} - \frac{1}{6400\times 10^3}\right)$ [2]

$\Delta V_g = 2.0\times 10^7\,\text{J kg}^{-1}$ [1]

6 change in GPE = $2.0\times 10^7\times 1200$ [1]

change in GPE = $2.4\times 10^{10}\,\text{J}$ [1]

19.1/19.2/19.3

1 1.44 × solar mass [1]

2 $10\times 2.0\times 10^{30} = 2.0\times 10^{31}\,\text{kg}$ [1]

3 The mass of the core. (The remnant is a neutron star if the mass of the core $< 3M_\odot$ and a black hole if $> 3M_\odot$.) [1]

4 Red giants have lower temperature and larger luminosity than white dwarfs. [2]

5 $\text{density} = \frac{\text{mass}}{\text{volume}} = \frac{4.0 \times 10^{30}}{\frac{4}{3}\pi \times 11000^3} = 7.2 \times 10^{17}\,\text{kg m}^{-3}$ [1]

mass = density × volume = $7.2\times10^{17}\times10^{-9}$ [1]

mass = 7.2×10^{8} kg [1]

6 $v = \sqrt{\frac{2GM}{r}}$, $v = c$ [1]

$r = \frac{2GM}{c^2} = \frac{2 \times 6.67 \times 10^{-11} \times 5 \times 2.0 \times 10^{30}}{(3.0 \times 10^8)^2}$ [1]

radius = 15 km [1]

19.4/19.5

1 An energy level is one of the discrete set of energies of an electron inside an atom. [1]

2 energy of photon = 6.0 − 2.0 = 4.0 eV [1]

3 The energy of the electron increases therefore a photon is absorbed by the electron. [1]

4 energy of photon = $4.0\times1.6\times10^{-19} = 6.4\times10^{-19}$ J [1]

$\frac{hc}{\lambda} = \Delta E;\ \lambda = \frac{6.63 \times 10^{-34} \times 3.0 \times 10^8}{6.4 \times 10^{-19}}$ [1]

$\lambda = 3.1\times10^{-7}$ m [1]

5 There are 6 possible spectral lines. [1]

See energy level diagram below.

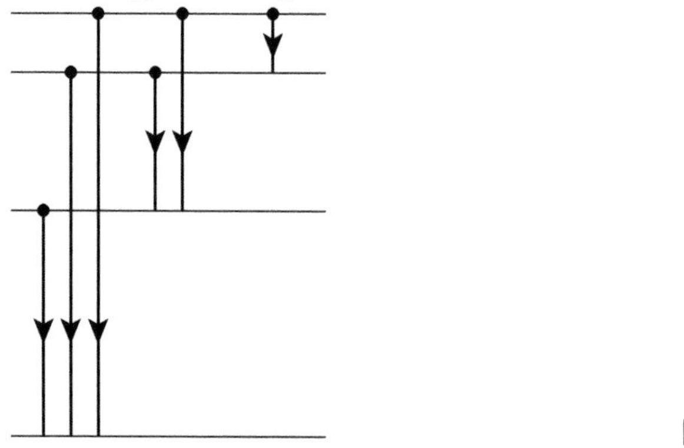

[1]

6 $\frac{hc}{\lambda} = \Delta E;\ \Delta E = \frac{6.63 \times 10^{-34} \times 3.0 \times 10^8}{400 \times 10^{-9}} = 4.97 \times 10^{-19}\,\text{J}$ [2]

$\Delta E = \frac{4.97 \times 10^{-19}}{1.60 \times 10^{-19}} = 3.11\,\text{eV}$ [1]

energy level = −12.2 + 3.11

energy level = −9.1 eV [1]

19.6/19.7

1 $d = \frac{10^{-3}}{80} = 1.25\times10^{-5}\,\text{m}$ [1]

2 luminosity ∝ (thermodynamic temperature)^4 [1]

3 $d = \frac{10^{-3}}{800} = 1.25\times10^{-6}\,\text{m}$ [1]

$d\sin\theta = n\lambda;\ \sin\theta = \frac{1 \times 6.4 \times 10^{-7}}{1.25 \times 10^{-6}}$ [1]

$\theta = 30.8° \approx 31°$ [1]

4 $\lambda_{max}T$ = constant; $500\times5800 = \lambda_{max}\times12\,000$ [1]

$\lambda_{max} = 240\,\text{nm}$ [1]

5 $L = 4\pi r^2\sigma T^4 \propto r^2$ [1]

Red giants have greater radius (and therefore surface area). Therefore, the total power emitted is much greater than the smaller main sequence stars. [1]

6 For the Sun: $L_\odot = 4\pi R_\odot^2\sigma\times5800^4$ [1]

For Rigel: $L = 4\pi(79\,R_\odot)^2\sigma\times12\,000^4$ [1]

$L = \frac{79^2 \times 12000^4}{5800^4}L_\odot$ [1]

$L = 1.14\times10^5 L_\odot$ [1]

20.1/20.2

1 a distance = $5.2\times1.5\times10^{11} = 7.8\times10^{11}$ m [1]

b distance = $1.6\times9.5\times10^{15} = 1.52\times10^{16}\,\text{m} \approx 1.5\times10^{16}$ m [1]

c distance = $2600\times9.5\times10^{15} = 2.47\times10^{19}\,\text{m} \approx 2.5\times10^{19}$ m [1]

d distance = $95\times3.1\times10^{16} = 2.945\times10^{18}\,\text{m} \approx 2.9\times10^{18}$ m [1]

2 angle = $\frac{0.78}{3600} = 2.2\times10^{-4}°$ [1]

3 distance = $8.0\times10^{3}\times3.1\times10^{16} = 2.48\times10^{20}$ m [1]

time = $\frac{2.48 \times 10^{20}}{3.0 \times 10^8}$ [1]

time = 8.3×10^{11} s ($2.6\times10^4 y$) [1]

4 $\frac{\Delta\lambda}{\lambda} \approx \frac{v}{c} = \frac{7.6 \times 10^3}{3.0 \times 10^8} = 2.5 \times 10^{-5}$ [1]

% change = 2.5×10^{-3} % [1]

5 distance in pc = $8.6\times\frac{9.5 \times 10^{15}}{3.1 \times 10^{16}}$ [1]

distance = 2.64 pc [1]

$p = \frac{1}{d} = \frac{1}{2.64} = 0.38$ arcseconds [1]

6 $\frac{\Delta\lambda}{\lambda} \approx \frac{v}{c} = \frac{5.3 \times 10^6}{3.0 \times 10^8} = 1.767 \times 10^{-2}$ [1]

$\Delta\lambda = 1.767\times10^{-2}\times119.5 = 2.1$ nm [1]

Observed wavelength = 119.5 + 2.1 = 121.6 nm [1]

20.3/20.4/20.5

1 2.7 K [1]

2 When viewed on a large enough scale, the Universe is homogeneous and isotropic, and the laws of physics are universal. [1]

3 $v = H_0 d = 70\times200$ [1]

$v = 14\,000\,\text{km s}^{-1}$ [1]

4 $5200 = 70\times d$ [1]

$d = 74$ Mpc [1]

5 $70\,\text{km}\,\text{s}^{-1}\,\text{Mpc}^{-1} = \dfrac{70 \times 10^3}{10^6 \times 3.1 \times 10^{16}}$ [2]

$70\,\text{km}\,\text{s}^{-1}\,\text{Mpc}^{-1} = 2.26 \times 10^{-18}\,\text{s}^{-1}$ [1]

6 $\lambda_{max}T$ = constant; T = 5800 K and $\lambda_{max} = 5.0 \times 10^{-7}$ m for the Sun [1]

$\lambda_{max} = \dfrac{5.0 \times 10^{-7} \times 5800}{2.7} = 1.1 \times 10^{-3}\,\text{m}$ [1]

This wavelength is in the microwave region of the EM spectrum. [1]

21.1/21.2/21.3

1 $Q = VC = 9.0 \times 500 \times 10^{-6}$ [1]

$Q = 4.5 \times 10^{-3}\,\text{C}$ [1]

2 energy $= \dfrac{1}{2}V^2C = \dfrac{1}{2} \times 9.0^2 \times 500 \times 10^{-6}$ [1]

energy $= 2.0 \times 10^{-2}\,\text{J}$ [1]

3 maximum: parallel combination with C = 100 + 100 + 100 = 300 μF [2]

minimum: series combination with $C = \dfrac{100}{3}$ = 33 μF [2]

4 parallel branch: C = 300 + 100 = 400 μF [1]

complete circuit: $C = (400^{-1} + 500^{-1})^{-1}$ = 220 μF ≈ 220 μF [2]

5 charge on the 500 μF capacitor:

$Q = 6.0 \times 222 \times 10^{-6} = 1.33 \times 10^{-3}\,\text{C}$ [1]

p.d. $= \dfrac{Q}{C} = \dfrac{1.33 \times 10^{-3}}{500} \times 10^{-6}$ [1]

p.d. = 2.7 V [1]

6 charge $= 10 \times 1000 \times 10^{-6} = 1.0 \times 10^{-2}\,\text{C}$ [1]

total capacitance = C = 1000 + 500 = 1500 μF [1]

The p.d. across each capacitor is the same. [1]

$V = \dfrac{Q}{C} = \dfrac{1.0 \times 10^{-2}}{1500 \times 10^{-6}} = 6.7\,\text{V}$ [1]

21.4/21.5/21.6

1 time constant = $CR = \tau$

$\tau = 100 \times 10^{-6} \times 150 \times 10^3 = 15\,\text{s}$ [1]

2 There is a current in the circuit and charge is removed from the capacitor. [1]

Since charge ∝ p.d., the p.d. across the capacitor decreases. [1]

3 $CR = 100 \times 10^{-6} \times 200 \times 10^3 = 20\,\text{s}$ [1]

$V = V_0 e^{-\frac{t}{CR}} = 10 \times e^{-\frac{38}{20}}$ [1]

V = 1.5 V [1]

4 $t = 5CR$ [1]

$Q = Q_0 e^{-\frac{t}{CR}} = Q_0 \times e^{-5}$ [1]

$Q = 0.0067\,Q_0$ therefore charge left is 0.67% < 1%. [1]

5 $CR = 500 \times 10^{-6} \times 100 \times 10^6 = 50\,\text{s}$ [1]

$V_C = V_0(1 - e^{-\frac{t}{CR}}) = 10 \times (1 - e^{-\frac{80}{50}}) = 8.0\text{V}$ [1]

$V_R = 10.0 - 8.0$ [1]

$V_R = 2.0\,\text{V}$ [1]

6 $V = V_0 e^{-\frac{t}{CR}}$; CR = 20 s and $V = 0.5\,V_0$ [1]

$0.5 = e^{-\frac{t}{20}}$ [1]

$\ln 0.5 = -\dfrac{t}{20}$ [1]

t = 14 s [1]

22.1/22.2

1 $E = \dfrac{F}{Q} = \dfrac{8.0 \times 10^{-14}}{1.60 \times 10^{-19}}$ [1]

$E = 5.0 \times 10^5\,\text{N}\,\text{C}^{-1}$ [1]

2 $F = EQ = 6.0 \times 10^4 \times 1.60 \times 10^{-19}$ [1]

$F = 9.6 \times 10^{-15}\,\text{N}$ [1]

3 $F = \dfrac{Qq}{4\pi\varepsilon_0 r^2} = \dfrac{(1.60 \times 10^{-19})^2}{4\pi \times 8.85 \times 10^{-12} \times (2.0 \times 10^{-10})^2}$ [2]

$F = 5.76 \times 10^{-9}\,\text{N} \approx 5.8 \times 10^{-9}\,\text{N}$ [1]

4 $F \propto \dfrac{1}{r^2}$ therefore the force will increase by a factor of 4. [1]

force $= 4 \times 5.76 \times 10^{-9} = 2.3 \times 10^{-8}\,\text{N}$ [1]

5 $E = \dfrac{Q}{4\pi\varepsilon_0 r^2}$

$r = \sqrt{\dfrac{Q}{4\pi\varepsilon_0 E}} = \sqrt{\dfrac{3.8 \times 10^{-9}}{4\pi \times 8.85 \times 10^{-12} \times 5.0 \times 10^4}}$ [2]

r = 0.026 m (2.6 cm) [1]

6 $E = \dfrac{Q}{4\pi\varepsilon_0 r^2} = \left(\dfrac{Q}{4\pi r^2}\right) \times \dfrac{1}{\varepsilon_0}$

(surface area of sphere = $4\pi r^2$) [1]

Therefore, $E = \dfrac{\sigma}{\varepsilon_0}$ [1]

$\sigma = \varepsilon_0 E$ [1]

22.3/22.4

1 $E = \dfrac{V}{d} = \dfrac{2000}{0.01}$ [1]

$E = 2.0 \times 10^5\,\text{V}\,\text{m}^{-1}$ [1]

2 $Q = VC = 2000 \times 10 \times 10^{-12} = 2.0 \times 10^{-8}\,\text{C}$ [1]

3 $C = \dfrac{\varepsilon_0 A}{d}$; $10 \times 10^{-12} = \dfrac{8.85 \times 10^{-12} \times A}{0.01}$ [1]

$A = 1.13 \times 10^{-2}\,\text{m}^2 \approx 1.1 \times 10^{-2}\,\text{m}^2$ [1]

4 $C = \dfrac{\varepsilon_r \varepsilon_0 A}{d} = \dfrac{4.0 \times 8.85 \times 10^{-12} \times 6.24 \times 10^{-2}}{0.070 \times 10^{-3}}$

= 31.6 nF [1]

$C = 3.16 \times 10^{-8}\,\text{F}$ [1]

C = 32 nF [1]

5 $F = EQ = \dfrac{500}{0.01} \times 1.60 \times 10^{-19} = 8.0 \times 10^{-15}\,\text{N}$ [1]

$a = \dfrac{8.0 \times 10^{-15}}{9.11 \times 10^{-31}} = 8.782 \times 10^{15}\,\text{m}\,\text{s}^{-2}$ [1]

$s = \dfrac{1}{2}at^2$; $0.01 = \dfrac{1}{2} \times 8.782 \times 10^{15} \times t^2$ [1]

$t = 1.5 \times 10^{-9}\,\text{s}$ [1]

6 $E = \frac{1}{2}\frac{Q^2}{C}$ and $C = \varepsilon_0 \frac{A}{d}$ [1]

$E \propto d$ (Q and A are constants) [1]

As the separation is doubled, the final energy stored is $2E_0$. [1]

22.5

1 energy = 100 J [1]

2 $V = \frac{Q}{4\pi\varepsilon_0 r} \propto \frac{1}{r}$ [1]

The distance from the centre is doubled therefore $V = \frac{2000}{2} = 1000\,\text{V}$ [1]

3 $V = \frac{Q}{4\pi\varepsilon_0 r} = \frac{1.60 \times 10^{-19}}{4\pi \times 8.85 \times 10^{-12} \times 1.2 \times 10^{-10}}$ [1]

$V = 12\,\text{V}$ [1]

4 $V = \frac{Q}{4\pi\varepsilon_0 r}$; $Q = V \times 4\pi\varepsilon_0 r$ [1]

$Q = 100 \times 4\pi \times 8.85 \times 10^{-12} \times 0.050$ [1]

$Q = 5.6 \times 10^{-10}\,\text{C}$ [1]

5 $C = 4\pi\varepsilon_0 R = 4\pi \times 8.85 \times 10^{-12} \times 0.025$ [1]

$C = 2.78 \times 10^{-12}\,\text{F}$ [1]

$Q = VC = 5000 \times 2.78 \times 10^{-12}$ [1]

$Q = 1.4 \times 10^{-8}\,\text{C}$ [1]

6 $Q = V \times 4\pi\varepsilon_0 r$ [1]

$Q = 3000 \times 4\pi \times 8.85 \times 10^{-12} \times 0.015 = 5.00 \times 10^{-9}\,\text{C}$ [1]

$F = \frac{Qq}{4\pi\varepsilon_0 r^2} = \frac{(5.00 \times 10^{-9})^2}{4\pi \times 8.85 \times 10^{-12} \times 0.050^2}$ [1]

$F = 9.0 \times 10^{-5}\,\text{N}$ [1]

23.1/23.2

1 The magnetic flux density is constant. [1]

2 The magnetic field is similar to that of a bar magnetic. [1]

However, within the core, the magnetic field is uniform (parallel and equally spaced magnetic field lines). [1]

3 $F = BIL = 0.020 \times 5.0 \times 0.040$ [1]

$F = 4.0 \times 10^{-3}\,\text{N}$ [1]

4 $F = BIL \propto BI$ [1]

Therefore, the force on the wire will be $\frac{3}{2}F$. [1]

5 Each wire lies in the magnetic field of the other wire. [1]

Therefore, each wire will experience a magnetic force.

6 $\sin\theta = \frac{F}{BIL} = \frac{1.5 \times 10^{-3}}{0.02 \times 5.0 \times 0.040}$ [1]

$\theta = 22°$ [1]

23.3

1 The force is perpendicular to the velocity of the electron. [1]

Therefore, no work is done on the electron – its speed remains constant. [1]

2 $F = BQv = 0.080 \times 1.60 \times 10^{-19} \times 4.0 \times 10^6$ [1]

$F = 5.12 \times 10^{-14}\,\text{N} \approx 5.1 \times 10^{-14}\,\text{N}$ [1]

3 $a = \frac{F}{m} = \frac{5.12 \times 10^{-14}}{9.11 \times 10^{-31}}$ [1]

$a = 5.62 \times 10^{16}\,\text{m s}^{-2} \approx 5.6 \times 10^{16}\,\text{m s}^{-2}$ [1]

4 $a = \frac{v^2}{r}$; $5.62 \times 10^{16} = \frac{(4.0 \times 10^6)^2}{r}$ [1]

$r = 2.8 \times 10^{-4}\,\text{m}$ [1]

5 $r = \frac{mv}{BQ} = \frac{6.7 \times 10^{-27} \times 6.0 \times 10^5}{0.720 \times 2 \times 1.60 \times 10^{-19}}$ [2]

$r = 0.017\,\text{m}$ [1]

6 $T = \frac{2\pi m}{Be}$ (see worked example) [1]

$T = \frac{2\pi \times 1.673 \times 10^{-27}}{0.900 \times 1.60 \times 10^{-19}}$ [1]

$T = 7.3 \times 10^{-8}\,\text{s}$ [1]

23.4/23.5/23.6

1 $1\,\text{Wb} = 1\,\text{T m}^2$ [1]

2 The secondary (output) coil has more turns than the primary (input) coil. [1]

3 The magnetic flux linking the coil is constant. [1]

There is no induced e.m.f. because the rate of change of magnetic flux linkage is zero. [1]

4 $\frac{n_s}{n_p} = \frac{V_s}{V_p} = \frac{5}{230} = \frac{1}{46}$ [1]

The primary coil has 46 times more turns than the secondary coil. [1]

5 The rate of change of magnetic flux increases. [1]

The maximum e.m.f. would increase. [1]

The frequency of the output signal would also increase. [1]

6 $\varepsilon = \frac{\Delta(N\phi)}{\Delta t} = \frac{\Delta(BAN)}{\Delta t}$ (magnitude only) [1]

$0.010 = \frac{B \times 2.0 \times 10^{-4} \times 500}{0.20}$ [1]

$B = 0.020\,\text{T}$ [1]

24.1/24.2

1 There are 8 protons and 7 neutrons. [2]

2 The nucleus is 10^5 times smaller than the atom. [1]

The chance of an alpha particle getting close to the nucleus, and therefore be deflected, is very small. [1]

3 Helium nucleus: $R = r_0 A^{\frac{1}{3}} = 1.2 \times 4^{\frac{1}{3}} = 1.9\,\text{fm}$ [1]

Uranium nucleus: $R = r_0 A^{\frac{1}{3}} = 1.2 \times 235^{\frac{1}{3}} = 7.4\,\text{fm}$ [1]

4 The radius of an atom is about $10^{-10}\,\text{m}$ or diameter of about $2 \times 10^{-10}\,\text{m}$. [1]

If connected end-to-end:

number of atoms $= \frac{0.010}{2 \times 10^{-10}} = 5 \times 10^7$ atoms [1]

5 At the minimum separation, the alpha particle stops momentarily therefore initial kinetic energy = final electric potential energy. [1]

kinetic energy = 7.7 MeV = $7.7\times10^{6}\times1.60\times10^{-19}$ = 1.232×10^{-12} J [1]

$E = \frac{Qq}{4\pi\varepsilon_0 r}$ or $r = \frac{Qq}{4\pi\varepsilon_0 E}$ [1]

$r = \frac{13\times2\times(1.60\times10^{-19})^2}{4\pi\times8.85\times10^{-12}\times1.232\times10^{-12}} = 4.86\times10^{-15}\,\text{m} \approx$ 4.9×10^{-15} m [1]

6 $F = \frac{Qq}{4\pi\varepsilon_0 r^2} = \frac{13\times2\times(1.60\times10^{-19})^2}{4\pi\times8.85\times10^{-12}\times(4.86\times10^{-15})^2}$ [2]

$F = 250$ N [1]

24.3/24.4/24.5

1 A fundamental particle has no internal structure and cannot be sub-divided into smaller particles. [1]

Any two correct examples, e.g., quark and electron. [1]

2 Quarks experience the strong nuclear force. [1]

3 ${}^{1}_{0}\text{n} \rightarrow {}^{1}_{1}\text{p} + {}^{0}_{-1}\text{e} + \overline{\nu_e}$ [2]

Quantities conserved: nucleon number and proton number (or charge). [1]

4 The strong nuclear force experienced by the protons is short-ranged (~10^{-15} m) and attractive. [1]

The gravitational force on the protons has an infinite range and is attractive. [1]

5 neutron → u d d [1]

charge = $\left(\frac{2}{3}-\frac{1}{3}-\frac{1}{3}\right)e = 0$ [1]

6 u u d → u d d + ${}^{0}_{+1}\text{e} + \nu_e$ [2]

7 Antimatter coming into contact with matter will annihilate each other. [1]

Since we observe matter around us and in the Universe, there must be more matter than antimatter. [1]

8 $F_G = \frac{GMm}{r^2}$ and $F_E = \frac{Qq}{4\pi\varepsilon_0 r^2}$ [1]

ratio = $\frac{4\pi\varepsilon_0 GMm}{Qq}$ [1]

ratio = $\frac{4\pi\times8.85\times10^{-12}\times6.67\times10^{-11}\times(1.7\times10^{-27})^2}{(1.6\times10^{-19})^2}$ [1]

ratio ≈ 8×10^{-37} [1]

25.1/25.2

1 ${}^{0}_{-1}\text{e}$ is an electron. [1]

$\overline{\nu_e}$ is an electron antineutrino. [1]

2 a The two numbers conserved are proton number and nucleon number. [1]

b number of neutrons in magnesium-28 nucleus = 28 − 12 = 16 [1]

number of neutrons in aluminium-28 nucleus = 28 − 13 = 15 [1]

3 a ${}^{204}_{82}\text{Pb} \rightarrow {}^{200}_{80}\text{Hg} + {}^{4}_{2}\text{He}$ [1]

b ${}^{249}_{98}\text{Cf} \rightarrow {}^{245}_{96}\text{Cm} + {}^{4}_{2}\text{He}$ [1]

4 a ${}^{19}_{8}\text{O} \rightarrow {}^{19}_{9}\text{F} + {}^{0}_{-1}\text{e} + \overline{\nu_e}$ [1]

b ${}^{21}_{11}\text{Na} \rightarrow {}^{21}_{10}\text{Ne} + {}^{0}_{+1}\text{e} + \nu_e$ [2]

5 The number of gamma photons emitted per second from the source is spread equally over a sphere of radius 30 cm and the GM tube detects a fraction of this count rate. [1]

number of photons per second = $\frac{4\pi\times0.30^2}{2.0\times10^{-4}}\times120$ [1]

number of photons per second = $6.8\times10^{5}\,\text{s}^{-1}$ [1]

6 10 eV = $10\times1.6\times10^{-19}$ J = 1.6×10^{-18} J [1]

kinetic energy of alpha particle = $1.6\times10^{-18}\times10^{4}\times30 = 4.8\times10^{-13}$ J [1]

$\frac{1}{2}\times6.6\times10^{-27}\times v^2 = 4.8\times10^{-13}$ [1]

$v = 1.2\times10^{7}\,\text{m s}^{-1}$ [1]

25.3/25.4

1 The decay constant is inversely proportional to half-life. [1]

2 number of alpha particles = 120 × 2 = 240 [1]

Assumption: The activity remains constant over the period of 2.0 s. [1]

3 a 4.0 mins is 2 half-lives. [1]

number of nuclei left = $\frac{8.0\times10^{15}}{2^2} = 2.0\times10^{15}$ [1]

b 6.0 mins is 8 half-lives. [1]

number of nuclei left = $\frac{8.0\times10^{15}}{2^3} = 1.0\times10^{15}$ [1]

number of nuclei decayed = $(8.0-1.0)\times10^{15}$ = 7.0×10^{15} [1]

4 1.5 MeV = $1.5\times10^{6}\times1.60\times10^{-19}$ J [1]

power = activity × energy of each α-particle [1]

power = $3.4\times10^{10}\times1.5\times10^{6}\times1.60\times10^{-19}$ = 8.16×10^{-3} W ≈ 8.2 mW [1]

5 number of nuclei = $\frac{1.5\times10^{-6}}{0.234}\times6.02\times10^{23} = 3.86\times10^{18}$ [1]

$\lambda = \frac{\ln(2)}{6.7\times3600} = 2.87\times10^{-5}\,\text{s}^{-1}$ [1]

$A = \lambda N = 2.87\times10^{-5}\times3.86\times10^{18}$ [1]

$A = 1.1\times10^{14}$ Bq [1]

6 $\lambda = \frac{\ln(2)}{5730} = 1.21\times10^{-4}\,\text{y}^{-1}$ [1]

$A = A_0 e^{-\lambda t}$; $0.72 = e^{-1.21\times10^{-4}t}$ [1]

$\ln(0.72) = -1.21\times10^{-4}t$ [1]

$t = \frac{\ln(0.72)}{-1.21\times10^{-4}} = 2720\,\text{y}$ [1]

25.5/25.6

1 The activity of the source is not constant over such a long period of time. [1]

2 **a** $\lambda = \frac{\ln(2)}{10} = 0.0693\,s^{-1} \approx 0.069\,s^{-1}$ [1]

b number of nuclei decaying = $\Delta N = (\lambda \Delta t)\,N$ [1]

$\Delta N = 0.0693 \times 0.10 \times 1000 = 6.9$ (allow 7) [1]

3 The activity has decreased by a factor of 4 – this is equivalent to 2 half-lives. [1]

age = $5730 \times 2 \approx 11500\,y$ [1]

4 $\lambda = \frac{\ln(2)}{5730} = 1.21 \times 10^{-4}\,y^{-1}$ [1]

$A = \lambda N;\ N = \frac{1.7}{1.21 \times 10^{-4}}$ [1]

$N = 1.4 \times 10^4$ [1]

5 $\lambda = \frac{\ln(2)}{5730} = 1.21 \times 10^{-4}\,y^{-1}$ [1]

$A = A_0 e^{-\lambda t}$; $0.32 = 1.6\,e^{-1.21 \times 10^{-4} t}$ [1]

$t = -\frac{\ln(0.20)}{1.21 \times 10^{-4}}$ [1]

$t = 13300\,y \approx 13000\,y$ [1]

6 $\lambda = \frac{\ln(2)}{49 \times 10^9} = 1.41 \times 10^{-11}\,y^{-1}$ [1]

$\frac{N}{N_0} = e^{-\lambda t}$; $0.94 = e^{-1.41 \times 10^{-11} t}$ [1]

$t = -\frac{\ln(0.94)}{1.41 \times 10^{-11}}$ [1]

$t = 4.4 \times 10^9\,y$ (4.4 billion years) [1]

26.1/26.2

1 $\Delta E = \Delta mc^2 = 1.0 \times 10^{-6} \times (3.00 \times 10^8)^2$ [1]

energy = $9.0 \times 10^{10}\,J$ [1]

2 $\Delta E = \Delta mc^2 = 9.11 \times 10^{-31} \times (3.0 \times 10^8)^2$ [1]

energy = $8.20 \times 10^{-14}\,J$ [1]

3 **a** The (thermal) energy of the lump of iron decreases. [1]

Therefore, its mass will *decrease* because $\Delta E \propto \Delta m$. [1]

b The (kinetic) energy of the electron decreases. [1]

Therefore, its mass will *decrease* because $\Delta E \propto \Delta m$. [1]

c The (kinetic) energy of the proton increases. [1]

Therefore, its mass will *increase* because $\Delta E \propto \Delta m$. [1]

4 **a** BE per nucleon = 1.0 MeV [1]

BE = $1.0 \times 2 = 2.0$ MeV [1]

b BE per nucleon = 7.1 MeV [1]

BE = $7.1 \times 4 = 28.4$ MeV [1]

c BE per nucleon = 7.5 MeV [1]

BE = $7.5 \times 238 \approx 1800$ MeV [1]

5 mass of 8 protons = $8 \times 1.673 \times 10^{-27}$ kg [1]

and mass of 8 neutrons = $8 \times 1.675 \times 10^{-27}$ kg [1]

BE = $[8 \times 1.673 \times 10^{-27} + 8 \times 1.675 \times 10^{-27} - 2.656 \times 10^{-26}] \times (3.00 \times 10^8)^2$

BE = 2.016×10^{-11} J [1]

BE = $\frac{2.016 \times 10^{-11}}{1.60 \times 10^{-19}} = 126$ MeV

BE per nucleon = $\frac{126}{16} = 7.9$ MeV [1]

6 $\Delta m = 5.8 \times 10^{-3} \times 1.661 \times 10^{-27}$ kg [1]

energy released = $\Delta mc^2 = 5.8 \times 10^{-3} \times 1.661 \times 10^{-27} \times (3.00 \times 10^8)^2$ [1]

energy released = 8.67×10^{-13} J [1]

total energy released = $8.67 \times 10^{-13} \times 6.02 \times 10^{23} \approx 5.2 \times 10^{11}$ J [1]

26.3/26.4

1 Energy is released in both fission and fusion reactions. [1]

2 **a** Z: left-hand side = 1 + 1 = 2 and right-hand side = 2 [1]

A: left-hand side = 2 + 1 = 3 and right-hand side = 3 [1]

b Z: left-hand side = 92 + 0 = 92 and right-hand side = 56 + 36 + 0 = 92 [1]

A: left-hand side = 1 + 235 = 236 and right-hand side = 141 + 92 + 3 = 236 [1]

3 The nuclei are positive and therefore repel each other. [1]

The (mean) kinetic energy of the nuclei is greater at higher temperatures and this means that nuclei can get close enough for the strong nuclear force to bind the nuclei together. [1]

4 energy = number of nuclei $\times 2.0 \times 10^{-11}$ [1]

energy = $\frac{1.0}{0.235} \times 6.02 \times 10^{23} \times 2.0 \times 10^{-11}$ [1]

energy = 5.1×10^{13} J [1]

5 The protons, positron, and neutrino are lone particles and therefore have no BE. [1]

energy released = BE of ${}^{2}_{1}H$ = $2 \times 1.0 = 2.0$ MeV [1]

energy released = $2.0 \times 10^6 \times 1.60 \times 10^{-19} = 3.2 \times 10^{-13}$ J [1]

6 1.0 kg of hydrogen-1 has $\frac{1}{2} \times \frac{1.0}{0.001} \times 6.02 \times 10^{23} = 3.01 \times 10^{26}$ 'pairs' [1]

energy = $3.01 \times 10^{26} \times 3.2 \times 10^{-13}$ [1]

energy = 9.6×10^{13} J [1]

27.1/27.2

1 Simple scatter and Compton scattering. [1]

2 Photoelectric effect. [1]

3 $\frac{I}{I_0} = e^{-\mu x}$ [1]

$\frac{I}{I_0} = e^{-0.21\times3.0} = 0.53$ [1]

53% of the original intensity is transmitted through 3.0 cm of muscle. [1]

4 The maximum energy of the electron is 120 keV. [1]

maximum energy of photon energy = 120 keV = $120\times10^3\times1.60\times10^{-19}$ [1]

maximum energy of photon energy = 1.9×10^{-14} J [1]

5 $eV = \frac{hc}{\lambda}$ [1]

$\lambda = \frac{hc}{eV} = \frac{6.63\times10^{-34}\times3.00\times10^{8}}{1.60\times10^{-19}\times200\times10^{3}}$ [1]

$\lambda = 6.2\times10^{-12}$ m [1]

6 For the photoelectric effect mechanism, $\mu \propto Z^3$. [1]

Therefore the attenuation of barium is $\left(\frac{56}{7}\right)^3 \approx 510$ times greater than that of soft tissues. [1]

This makes it a much better absorber of X-rays than soft tissues. [1]

27.3/27.4/27.5

1 A three-dimensional image of the patient can be produced. [1]

2 X-rays are ionising or X-rays can damage cells. [1]

3 Lead absorbs gamma rays. [1]

Having narrow and long tubes allows only gamma photons from a specific direction to form the image (and hence produces a clearer image). [1]

4 PET scanning can be used. [1]

The FDG (or fluorine-18) will accumulate where rate of respiration is high and hence the **function** of the brain can be monitored. [1]

5 The time difference between the arrival times is $\frac{0.005}{3.0\times10^{8}} \approx 2\times10^{-11}$ s. [1]

This means that the computer software has to process the information (or digital signals) very quickly. [1]

6 Tc-99m produces gamma photons that can pass through the patient. [1]

The gamma photons do not interact much with the atoms /cells of the patient. [1]

The short half-life of 6.0 hours means that the activity reduces to safe levels in a short period of time (e.g., the activity drops to about 6% of the initial activity after 1 day). [1]

27.6/27.7/27.8

1 A high-frequency alternating p.d. is applied to a piezoelectric material which makes it vibrate. [1]

The vibration of the material in air produces the ultrasound. [1]

2 In the time t the total distance travelled by the ultrasound is twice the thickness. [1]

3 $\Delta f \propto \cos\theta$; the change in frequency is zero when $\cos 90° = 0$. [1]

4 $\frac{I_r}{I_0} = \frac{(Z_2 - Z_1)^2}{(Z_2 + Z_1)^2}$ and $Z_2 = 2Z_1$ (or $Z_1 = 2Z_1$) [1]

$\frac{I_r}{I_0} = \left(\frac{1}{3}\right)^2$ [1]

$\frac{I_r}{I_0} = \frac{1}{9} \approx 0.11$ [1]

5 $\frac{I_r}{I_0} = \frac{(Z_{air} - Z_{skin})^2}{(Z_{air} + Z_{skin})^2}$ and $Z_{air} << Z_{skin}$ [1]

Therefore, the ratio $\frac{I_r}{I_0}$ is almost 1.0, with most of the ultrasound reflected at the air–skin boundary. [1]

6 $\Delta f = \frac{2fv\cos\theta}{c}$ [1]

$3200 = \frac{2\times18\times10^{6}\times v\times\cos 60°}{1600}$ [1]

$v = 0.28\,\text{m s}^{-1} = 28\,\text{cm s}^{-1}$ [1]